AF389390

Future Powertrain Technologies

Future Powertrain Technologies

Editors

Stephan Rinderknecht
Philippe Jardin
Arved Esser

MDPI • Basel • Beijing • Wuhan • Barcelona • Belgrade • Manchester • Tokyo • Cluj • Tianjin

Editors
Stephan Rinderknecht
Technische Universität Darmstadt
Germany

Philippe Jardin
Technische Universität Darmstadt
Germany

Arved Esser
Technische Universität Darmstadt
Germany

Editorial Office
MDPI
St. Alban-Anlage 66
4052 Basel, Switzerland

This is a reprint of articles from the Special Issue published online in the open access journal *Vehicles* (ISSN 2624-8921) (available at: https://www.mdpi.com/journal/vehicles/special_issues/Powertrain).

For citation purposes, cite each article independently as indicated on the article page online and as indicated below:

LastName, A.A.; LastName, B.B.; LastName, C.C. Article Title. *Journal Name* **Year**, *Volume Number*, Page Range.

ISBN 978-3-03943-753-5 (Hbk)
ISBN 978-3-03943-754-2 (PDF)

Cover image courtesy of Institute for Mechatronic Systems in Mechanical Engineering, TU Darmstadt.

Contents

About the Editors

Stephan Rinderknecht has served as Full Professor of Mechatronic Systems in Mechanical Engineering at the Technical University of Darmstadt since 2009. His research is focused on mechatronic vehicle systems, energy systems, general vibration systems and, recently, also on robotic systems. In some of his research fields, he works in very close cooperation with the industry, such as for example innovative drivetrains and propulsion systems for e-mobility as well as highly integrated designs for kinetic energy storage systems. There, recent key topics are sectoral integration and AI methods. For other research fields like active piezoelectric systems to reduce bending vibrations of flexible rotors, the projects approach more fundamental problems. Before becoming university professor, Stephan worked in the automotive transmission industry for more than 13 years and left the position of Senior Vice President of Research and Development. He graduated and received his doctoral degree at University of Stuttgart in the faculty of Aerospace Engineering.

Philippe Jardin has served as Research Associate of Mechatronic Systems in Mechanical Engineering at the Technical University of Darmstadt since 2017. His research is focused on mechatronic vehicle systems and driving comfort, where he applies machine learning methods jointly together with conventional development methods. Philippe graduated in Mechanical Engineering at TU Darmstadt and studied at University of California, Berkeley. During his studies, he focused on the control of mechatronic systems in practical applications.

Arved Esser has served as Research Associate for Mechatronic Systems in Mechanical Engineering at the Technical University of Darmstadt since 2016. The main focus of his research is on comparative evaluation of the ecological potential of vehicle powertrain concepts based on real fleet driving data using a big data approach. Arved studied Mechanical Engineering at TU Darmstadt and Institut National Polytechnique in Grenoble. During his studies, he focused on structural dynamics and vibration measurement technologies with applications in automotive engineering.

 *vehicles*

Editorial

Special Issue on Future Powertrain Technologies

Philippe Jardin *, Arved Esser * and Stephan Rinderknecht *

Department of Mechanical Engineering, Institute for Mechatronic Systems in Mechanical Engineering, Technical University of Darmstadt, 64287 Darmstadt, Germany
* Correspondence: jardin@ims.tu-darmstadt.de (P.J.); esser@ims.tu-darmstadt.de (A.E.); rinderknecht@ims.tu-darmstadt.de (S.R.)

Received: 25 September 2020; Accepted: 28 September 2020; Published: 30 September 2020

Beside others, climate change and digitalization are trends of huge public interest, which highly influence the development process of future powertrain technologies. To handle these trends, new disruptive technologies are integrated into the development process. They open up space for diverse research, which is distributed over the entire vehicle design process. Recent research on this topic incorporates results for selecting and designing the powertrain topology [1–7] with their vehicle operating strategy [8,9], as well as results for handling the reliability of new powertrain components [10–12].

In [7], the optimal passenger car vehicle fleet transformation is developed based on lifecycle assessment. The results differ for short-range and long-range requirements, where battery electric vehicles are best for short-range. For longer distances, and at least until 2040, plug-in hybrid electric vehicles (PHEV) offer the greatest potential. In accordance with these results, ref. [6] optimizes an aftermarket hybridization kit, which can convert a combustion engine vehicle into a PHEV. Such kits may accelerate the passenger car vehicle fleet transition. With [1], a method for measuring non-volatile particle emissions is given. This method is highly relevant in the context of upcoming vehicle regulations within the transformation process.

In case of battery electric vehicles (BEV), ref. [5] investigates the influence of increasing the speed of the electric machine. Whilst this increases the system complexity through a higher gear ratio, a better power density of the overall powertrain system may be achieved. Different powertrain complexities in case of vehicles with dedicated hybrid transmissions are compared in [4]. The results show that a high complexity in the transmission may lead to a higher potential in reducing the vehicle's CO_2 emissions. However, an increased powertrain complexity also increases the cost. In [2], a low-end multipurpose vehicle is proposed. Here, the combustion engine is designed only to meet the constant driving power requirements. On short distances, this vehicle drives purely electric. For longer distances, the constant driving power is delivered by the combustion engine.

In [3], the environmental and ecological benefits of a dual-battery BEV are considered. The authors propose to decrease the size of the Lithium-Ion battery and use a Zinc-Air battery pack as range extender. The results show that the vehicle cost can be reduced significantly by introducing a second, less costly battery.

PHEV offer new possibilities for intelligent operating strategies. A user-specific operating strategy that adapts during operating is proposed in [9]. The strategy uses supervised machine learning methods, and achieves lower consumption compared to a reference. Furthermore, new vehicle concepts, such as fuel cell vehicles, may introduce additional requirements regarding their operating strategy. In [8], an operating strategy for a fuel cell bus with a supercapacitor is developed based on real life data from London, UK. Here, it has to be assured that the capacitor is never over- or undercharged, and that the power requirements can be met.

Future powertrain technologies must be equipped with new and reliable system components. For example, in [12], condition monitoring and remaining useful lifetime estimation methods for switching devices in the electric powertrain are developed. The authors validate their results with a 100 kW traction inverter. The overall system reliability also depends on the mechanical components. With [10], a systematic study of machine learning based reliability analysis methods is given. The authors integrate ensemble learning strategies into mechanical reliability estimations and compare the results that show the high potential of these methods. A crucial powertrain component is given by the clutch system. In [11], a novel method for clutch sensor fault diagnosis is developed and tested.

Author Contributions: All authors contributed equally to this work. All authors have read and agreed to the published version of the manuscript.

Funding: This research received no external funding.

Conflicts of Interest: The authors declare no conflict of interest.

References

1. Giechaskiel, B.; Melas, A.D.; Lähde, T.; Martini, G. Non-Volatile Particle Number Emission Measurements with Catalytic Strippers: A Review. *Vehicles* **2020**, *2*, 342–364. [CrossRef]
2. Zhen, Y.; Bao, Y.; Zhong, Z.; Rinderknecht, S.; Zhou, S. Development of a PHEV Hybrid Transmission for Low-End MPVs Based on AMT. *Vehicles* **2020**, *2*, 236–248. [CrossRef]
3. Tran, M.-K.; Sherman, S.; Samadani, E.; Vrolyk, R. Environmental and Economic Benefits of a Battery Electric Vehicle Powertrain with a Zinc–Air Range Extender in the Transition to Electric Vehicles. *Vehicles* **2020**, *2*, 398–412. [CrossRef]
4. Sieg, C.; Küçükay, F. Benchmarking of Dedicated Hybrid Transmissions. *Vehicles* **2020**, *2*, 100–125. [CrossRef]
5. Schweigert, D.; Gerlach, M.E.; Hoffman, A.; Morhard, B. On the Impact of Maximum Speed on the Power Density of Electromechanical Powertrains. *Vehicles* **2020**, *2*, 365–397. [CrossRef]
6. Eckert, J.J.; Santiciolli, F.M.; Silva, L.C.d.A.e.; Corrêa, F.C.; Dedini, F.G. Design of an Aftermarket Hybridization Kit: Reducing Costs and Emissions Considering a Local Driving Cycle. *Vehicles* **2020**, *2*, 210–235. [CrossRef]
7. Belmonte, B.B.; Esser, A.; Weyand, S.; Franke, G.; Schebek, L.; Rinderknecht, S. Identification of the Optimal Passenger Car Vehicle Fleet Transition for Mitigating the Cumulative Life-Cycle Greenhouse Gas Emissions until 2050. *Vehicles* **2020**, *2*, 75–99. [CrossRef]
8. Partridge, J.S.; Wu, W.; Bucknall, R.W. Investigation on the Impact of Degree of Hybridisation for a Fuel Cell Supercapacitor Hybrid Bus with a Fuel Cell Variation Strategy. *Vehicles* **2020**, *2*, 1–17. [CrossRef]
9. Harold, C.K.D.; Prakash, S.; Hofman, T. Powertrain Control for Hybrid-Electric Vehicles Using Supervised Machine Learning. *Vehicles* **2020**, *2*, 267–286. [CrossRef]
10. You, W.; Saidi, A.; Zine, A.-m.; Ichchou, M. Mechanical Reliability Assessment by Ensemble Learning. *Vehicles* **2020**, *2*, 126–141. [CrossRef]
11. Lv, Z.; Wu, G. A Novel Method for Clutch Pressure Sensor Fault Diagnosis. *Vehicles* **2020**, *2*, 191–209. [CrossRef]
12. Kundu, A.; Balamurali, A.; Korta, P.; Iyer, K.L.V.; Kar, N.C. An Approach for Estimating the Reliability of IGBT Power Modules in Electrified Vehicle Traction Inverters. *Vehicles* **2020**, *2*, 413–423. [CrossRef]

 *vehicles*

Article

Identification of the Optimal Passenger Car Vehicle Fleet Transition for Mitigating the Cumulative Life-Cycle Greenhouse Gas Emissions until 2050

Benjamin Blat Belmonte [1], Arved Esser [1], Steffi Weyand [2], Georg Franke [1], Liselotte Schebek [2] and Stephan Rinderknecht [1,*]

[1] Department of Mechanical Engineering, Institute for Mechatronic Systems in Mechanical Engineering, Technische Universität Darmstadt, 64287 Darmstadt, Germany; blatbelmonte@ims.tu-darmstadt.de (B.B.B.); esser@ims.tu-darmstadt.de (A.E.); franke@ims.tu-darmstadt.de (G.F.)

[2] Department of Civil and Environmental Engineering Sciences, Institute IWAR, Chair of Material Flow Management and Resource Economy, Technische Universität Darmstadt, 64287 Darmstadt, Germany; s.weyand@iwar.tu-darmstadt.de (S.W.); l.schebek@iwar.tu-darmstadt.de (L.S.)

* Correspondence: rinderknecht@ims.tu-darmstadt.de; Tel.: +49-6151-16-23-250

Received: 19 December 2019; Accepted: 20 January 2020; Published: 24 January 2020

Abstract: We present an optimization model for the passenger car vehicle fleet transition—the time-dependent fleet composition—in Germany until 2050. The goal was to minimize the cumulative greenhouse gas (GHG) emissions of the vehicle fleet taking into account life-cycle assessment (LCA) data. LCAs provide information on the global warming potential (GWP) of different powertrain concepts. Meta-analyses of batteries, of different fuel types, and of the German energy sector are conducted to support the model. Furthermore, a sensitivity-analysis is performed on four key influence parameters: the battery production emissions trend, the German energy sector trend, the hydrogen production path trend, and the mobility sector trend. Overall, we draw the conclusion that—in any scenario—future vehicles should have a plug-in option, allowing their usage as fully or partly electrical vehicles. For short distance trips, battery electric vehicles (BEVs) with a small battery size are the most reasonable choice throughout the transition. Plug-in hybrid electric vehicles (PHEVs) powered by compressed natural gas (CNG) emerge as promising long-range capable solution. Starting in 2040, long-range capable BEVs and fuel cell plug-in hybrid electric vehicles (FCPHEVs) have similar life-cycle emissions as PHEV-CNG.

Keywords: fleet transition; optimization; life-cycle assessment; greenhouse gas; global warming potential; vehicle powertrain concepts

---

1. Introduction

The world community concurred to the goal of limiting global temperature rise to ideally 1.5 °C compared with the pre-industrial age during the United Nations climate conference in Paris in 2015. According to this, the German government set the goals of reducing greenhouse gas (GHG) emissions by 40% in 2020 (compared to 1990). The climate protection report of 2018 states that these climate targets will be missed. In contrast to the two main contributors—the energy sector and the industry sector—that will achieve almost a 40% GHG emission reduction by 2020, the GHG emissions of the mobility sector have only been reduced by 5% [1]. Consequently, in this sector there still is a lot of unlocked potential for climate protection.

This study addresses this issue focusing on the vehicle fleet transition and minimizing its GHG emissions. In contrast to this study, previous studies have made prognostics of the German vehicle fleet transition following different approaches without minimizing the GHG emissions. Supported by the Transport Emission Model (TREMOD), a project of the German Federal Environment Agency (UBA), Knörr et al. predict 7.5 and 44 million vehicles with plug-in option for the years 2020 and 2035 respectively [2]. In 2035, this corresponds to roughly 90% of the fleet according to their calculations. This model has been developed over 20 years ago and since then been improved and expanded several times. Its main goal was to calculate the emissions of the German transport sector. The German Aerospace Center (DLR) has developed the simulation model VECTOR21 [3]. They predict that by 2040 only 35% of the fleet will have a plug-in option. The simulations are based on the customer perspective and reflect the limited acceptance these powertrain classes currently have. The model includes a total cost of ownership approach, which calculates the lifelong costs which have to be covered by the respective owners. VECTOR21 minimizes this cost and this way determines the future fleet composition. Interestingly, Harrison et al. conclude that even assuming high user acceptance, by 2050 only 50% of the fleet will have a plug-in option [4]. Their model was specifically developed to gain insights in influences on the adoption rates of new technologies that can be applied in a policy context. Therefore, they modeled several agents of the automobile sector including their interactions. Plötz et al. determine the vehicle fleet transition with special focus on the resulting electricity demand. Similar to VECTOR21, their model relies on a total cost of ownership approach. They base the distribution of annual vehicle kilometers travelled over different vehicle classes on a large dataset of driving profiles from Germany. They predict that in 2050 approx. 60%—or 25 million vehicles—will have a plug-in option [5]. Overall, the prognostics have very differing outcomes depending on who performed them and what their modelling approach was. When comparing different studies and prognostics regarding the German vehicle fleet, one finds that most of them predict a transition to more hybrid electric vehicles (HEVs) without plug-in option, plug-in hybrid electric vehicles (PHEVs), battery electric vehicles (BEVs), and fuel cell electric vehicles (FCEVs). However, as the outcomes of most studies predict, even in the next 20 to 30 years, these alternative powertrain classes will not replace conventional internal combustion engines vehicles (ICEVs) completely. In some studies, the purchasing behavior plays an important role. According to the studies, only a limited customer segment feels comfortable switching to PHEVs, BEVs, and FCEVs.

In the present paper, we focus on the minimum achievable GHG emission potential—hereinafter referred to as ecological potential—of passenger car vehicles in the vehicle fleet. For this purpose, we research available life-cycle assessment (LCA) data. An essential difficulty in LCAs of passenger car vehicles lies in the many varying factors that influence its GHG emissions. Especially for PHEVs the differing charging behavior of different users can lead to very different outcomes when looking at the GHG emissions. This uncertainty adds to the fact that the GHG emissions caused by the electrical consumption differ depending on the energy sector composition in every region.

A series of studies have performed LCAs on series production vehicles, which is a common approach to assess the life-cycle GHG emissions of current state-of-the-art powertrain technologies [6–8]. However, when explicitly comparing the ecological potential of different powertrain concepts, the meaningfulness of these studies has to be questioned. This is because series production vehicles were not dedicatedly designed to minimize their GHG emissions and, therefore, will not represent the true ecological potential of their powertrain concept. Further, a powertrain concept can have different parametrizations, thus, the LCA of one specific series production vehicle does not allow generalized conclusions on the powertrain concept. In [9] an optimization framework was introduced, that allows to design the powertrain concepts so that the ecological potential is achieved. Using this approach, we ensure that no powertrain concepts are excluded from the fleet transitions due to non-optimal powertrain parametrizations.

Our team conducts comprehensive analyses of the production phase, the operational phase—taking into account the energy mix for electricity production—and the end-of-life (EoL) phase. We model the German fleet transition until 2050 and calculate the cumulative GHG emissions, including all life-cycle phases of all vehicles. The cumulative GHG emissions serve as objective value for an optimization problem, which enables us to determine the GHG-optimal fleet transition. This way, we are able to make a precise time-based assessment of the GHG emissions of different powertrain concepts taking into account the dynamic interaction of the vehicle fleet with the energy sector. Furthermore, the results serve as guideline as to which powertrain concepts should be introduced at which point in time in order to reduce the GHG emissions of the German vehicle fleet as a whole. Unlike other prognostics, this study is not intended to show the most likely transition of the German vehicle fleet. Instead, the GHG-optimal transition paths for different scenarios are identified and analyzed to gain comprehensive knowledge about similarities in different scenarios that can be used as a guideline for decision makers.

In Section 2.1, we discuss relevant technology parameters, such as the specific energy and the GHG emissions of battery production. In addition, different fuel types are analyzed regarding their GWP_{100}—the GWP over a time span of 100 years—and the expected energy sector transition to more renewable energy sources in Germany will be addressed. Section 2.2 covers the different future scenarios that will be optimized and 2.3 describes the vehicle and its life-cycle GHG emissions. In Section 3, we identify the optimal fleet transitions for various scenarios and in Section 4 we draw our conclusions.

2. Optimization Model of the Vehicle Fleet Transition

As mentioned earlier, the present paper aims at identifying the GHG-optimal fleet transition in Germany. On one hand, our model shows some similarities with the models of previous studies mentioned in the introduction. In fact, we are talking about a replacement model in all cases, because existing vehicles are replaced by new vehicles once they achieve their specific lifetime mileage. On the other hand, our model differs significantly in other aspects. In particular, user acceptance has a very limited effect in our model. The only relevant aspect is that most users expect their vehicles to have a minimum range. Other than that, we make an accurate assessment of the GHG emissions stemming from electricity production considering the interaction with the electricity demand of the vehicle fleet. This way, we can precisely assess the GHG emissions of the operational phase of electrified powertrain classes. A detailed discussion on the German energy sector can be found in Section 2.1.3.

The powertrain concepts that we consider for the fleet transition are listed in Table 1. Under the powertrain class of conventional ICEV, four concepts are taken into consideration. The ICEV-E10 and ICEV-B7, which are powered by E10 and B7 respectively. E10 is a blend of petrol with 10% bioethanol and B7 is a blend of diesel and 7% biodiesel. Then we consider the ICEV-CNG and ICEV-LPG, which are powered by compressed natural gas (CNG) and liquefied petroleum gas (LPG) respectively. We only consider one hybrid powertrain concept without plug-in option: the mild hybrid electric vehicle powered by E10 (mHEV-E10). Further, we account for PHEV variants that are powered by E10, CNG, and hydrogen. For E10 and CNG there are three variants of the powertrain class with varying battery size (PHEV10 possesses a 10 kWh battery). Finally, we have the BEV with two different battery size variants and two fuel cell electric vehicles (FCEVs), one of them having a plug-in option and a battery size of 20 kWh (FCPHEV20). Regarding the different concepts, the BEV20 is the only concept that is not considered as long-range capable, which will become relevant to fulfill the user acceptance constraints during the fleet optimization. Furthermore, as explained in the introduction, each powertrain concept is parametrized to achieve minimal GHG emissions using the approach presented in [9]. This way, we ensure that no powertrain concepts are excluded due to non-optimal powertrain parametrizations.

Table 1. Powertrain concepts considered for the identification of greenhouse gas (GHG)-optimal vehicle fleet transitions.

Powertrain Class	Internal Combustion Engines Vehicles (ICEV)	Hybrid Electric Vehicles (HEV)	Plug-In Hybrid Electric Vehicles (PHEV)	Battery Electric Vehicles (BEV)	Full Cell Electric Vehicles (FCEV)
Powertrain concepts	ICEV-E10 ICEV-B7 ICEV-CNG ICEV-LPG	mHEV-E10	PHEV10-E10 PHEV15-E10 PHEV20-E10 PHEV10-CNG PHEV15-CNG PHEV20-CNG	BEV20 BEV100	FCHEV FCPHEV20

2.1. Meta-Analysis of Life-Cycle GHG Emissions

The following subsections give an overview of meta-analysis results regarding the battery production GHG emissions and the fuel-related GHG emissions. In addition, the emission factors of the German energy sector and their contribution to the mean specific GHG emissions factor are described.

2.1.1. Meta-Analysis of the Life-Cycle GHG Emissions of Batteries

In general, for powertrain classes with a high voltage battery—which includes all vehicles with plug-in option in this study—the GHG emissions of the battery production represent a very important factor. Relevant battery characteristics for the considered use case—a mobile energy storage unit—are the following: specific energy, specific power, time to fully charge, self-discharge, memory-effect, and lifetime expectancy. Since lithium-ion (li-ion) batteries have proven by far as the most apt technology in the market regarding these requisites, we focus on them [10]. The two most important factors relevant for the present study are the production emissions and the specific energy. In the Supplementary Data, Table S1 [11], we have performed a meta-analysis of different studies regarding characteristics of li-ion batteries used in vehicles as traction batteries. These data points allow for a confident estimation of the GWP$_{100}$ and the specific energy in present and future scenarios. The left side of Figure 1 shows the distribution of researched values regarding the GHG emissions of current and future battery production in 2025. The median value is approx. 16 kgCO$_2$-equivalents (CO$_2$-eq.) per kg battery. On the right side, we find the distribution of researched values regarding the specific energy of li-ion battery systems that can be reached nowadays. Further, prognostics for 2030 and 2040 predict specific energies of 500 and up to 700 Wh/kg battery. These prognostics include lithium-sulfur batteries—a different, lithium-based battery type. The median values marked in red are used in Section 2.2.1 as reference values for the first years of the fleet transition. The consideration of future technological improvements is also presented in Section 2.2.1.

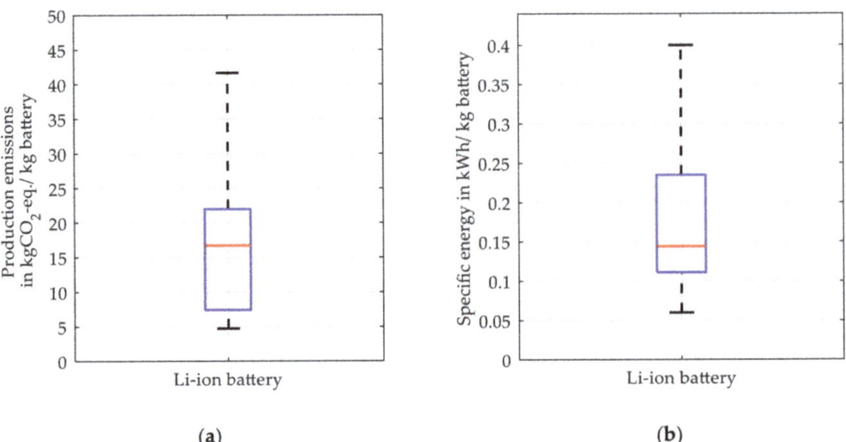

Figure 1. (**a**) Boxplot of the GWP$_{100}$ values of battery production considered relevant for this work; (**b**) boxplot of the specific energy of li-ion batteries. The boxes include all values between the 25% quantile and the 75% quantile of the data. Minima and maxima are displayed with a short horizontal dash. Refer to Supplementary Data, Table S1 for more details [11].

2.1.2. Meta-Analysis of the Life-Cycle GHG Emissions of Fuels

For the optimization process in Section 3, we need to gain insights regarding the operational phase of the vehicles. Therefore, we are comparing the characteristics of conventional fuels, i.e., petrol and diesel, with CNG, LPG, and hydrogen resulting from a meta-analysis of available research. Furthermore, we are considering biofuels and conventional fuels blended with biofuels. Supplementary

Data, Table S2 contains all researched data points regarding the fuels [12]. Figure 2 shows the lower heating value for all fuel types considered. Only in the case of CNG, there are significant variations in the literature. This is due to the fact that the chemical composition of CNG slightly varies depending on its region of origin.

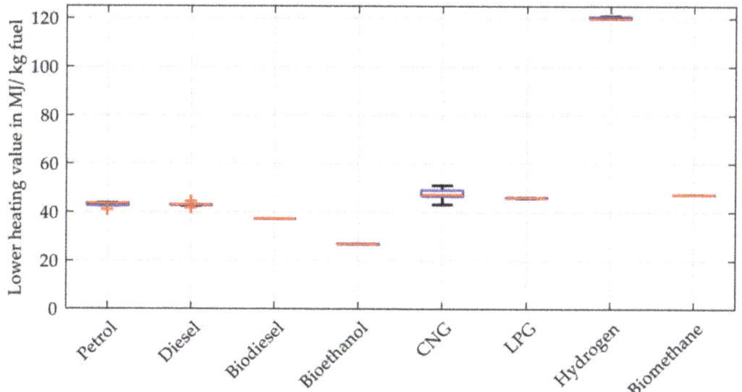

Figure 2. Boxplots of the data points regarding the lower heating values of all fuels considered. Refer to Supplementary Data, Table S2 for more details [12].

The densities of liquefied fuels are listed in Table 2.

Table 2. Densities of liquefied fuels [2,13–15].

Fuel Type	Petrol	Diesel	Bioethanol	Biodiesel	Liquefied Petroleum Gas (LPG)
Density	0.745 kg/L	0.837 kg/L	0.786 kg/L	0.879 kg/L	0.590 kg/L

In Figure 3 we compare the GHG emissions of the state-of-the-art fuel supply for all fuels. Our research lead to relatively accurate results for most fuels, except for hydrogen, which has the highest and most scattering specific emission values during fuel supply in the literature. The values vary strongly, because they highly depend on the chosen production path.

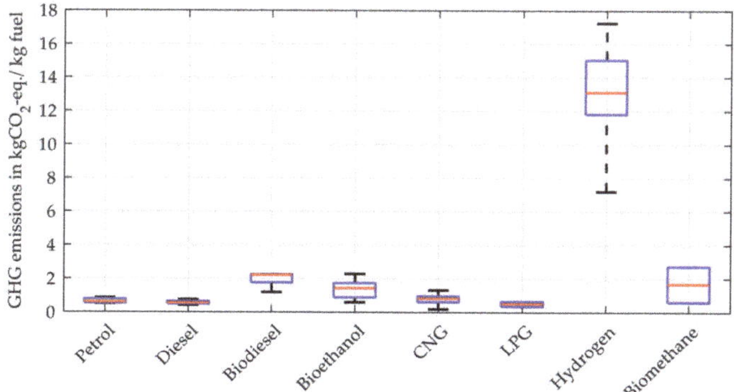

Figure 3. Boxplots of the data points regarding the fuel supply emissions. Refer to Supplementary Data, Table S2 for more details [12].

Hydrogen production nowadays has a very high GWP_{100} when compared to other liquid fossil fuels, due to the currently dominant production paths. The three main production paths are steam gas reformation, coal gasification, and electrolysis. All of these processes are very energy intensive and add to the energy expenses for subsequent compression. In theory, electrolysis powered with renewable energy would lead to the lowest production emissions of all fuels. Unfortunately, the German energy sector is not yet carbon neutral and there is currently not enough capacity for electrolysis on a large scale within Germany. In 2006, electrolysis represented only 4% of global hydrogen production [16]. There are several studies predicting an efficiency increase of electrolysis from 65% in 2017 to 68% in 2025 [17,18]. Optimistic studies even predict a 95% efficiency in 2050. One main advantage of hydrogen is that there are no direct emissions during the operational phase of a FCEV.

Very optimistic low hydrogen supply emission values, i.e., electrolysis with carbon neutral electricity, are not displayed in Figure 3, because this production path is not representative of current hydrogen production. Its GHG emissions would be reduced to transport and compression—approx. 2.4 kgCO$_2$-eq./kg hydrogen. This represents a great opportunity for future reduction of its GWP$_{100}$ during fuel supply.

Figure 4 compares the direct emissions of the different fuels. Because of their chemical composition, gaseous fuels like LPG and CNG have lower GHG emissions per energy unit than liquid fossil fuels. CNG, whose main component is methane, has the lowest direct emissions—apart from hydrogen.

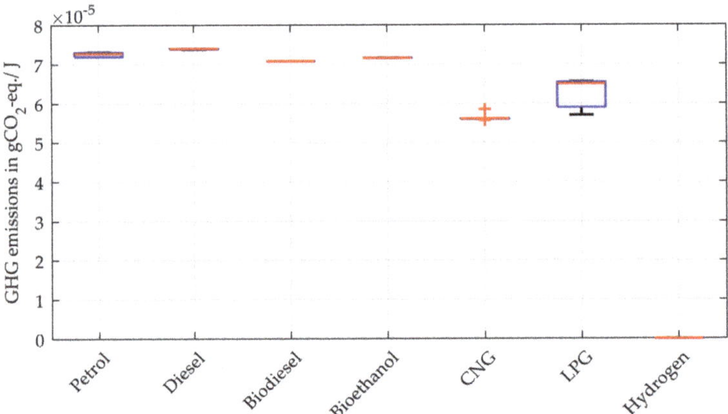

Figure 4. Boxplots of the data points regarding direct fuel emissions. Refer to Supplementary Data, Table S2 for more details [12].

In the context of the fleet transition modeled within this work, GHG emissions from fuel supply and the direct fuel emissions remain constant over the years with the exception of hydrogen. Because of its high variability during fuel supply depending on the chosen production path and further increases in electrolysis efficiency, hydrogen related GHG emissions are modelled separately. Detailed insights can be found in Section 2.2.3.

The chemical composition of synthetic natural gas (SNG)—either biomethane or methane synthesized via electricity-powered methanation (e-methane)—is almost identical to the one of CNG. This means that it can theoretically replace CNG in fuel tanks and gas pipes by 100% [19]. Similar to hydrogen, if the production of SNG is powered by carbon neutral electricity, there is ecological potential to be exploited. Since in 2012 biomethane was only present in niche markets representing 0.1% of total fuel consumption in Germany and there is not enough e-methane production capacity in Germany, it is not explicitly modelled in Section 2.2 [20].

2.1.3. Emissions Factors of the German Energy Sector from Today till 2050

The modelling of the German energy sector is of central relevance for conducting a reliable analysis and optimization. Figure 5 displays the energy mix in terms of installed capacity. Dominant trends are the gradual reduction of installed capacity related to coal, the out-phasing of nuclear energy, and the further expansion of wind power [21–23].

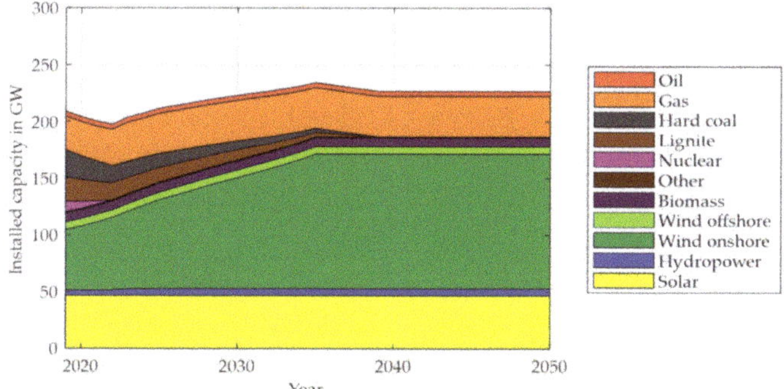

Figure 5. Development of the installed capacity in the German energy sector divided by all energy sources [21–23]. Other includes energy sources like waste incineration, which only account for a negligible part of electricity production.

The actual energy output of a power plant over a given period of time divided by the theoretically maximum possible energy output over that same period of time is defined as load factor. Figure 6 displays the load factors of German power plants for 2018 divided by energy source. The electricity production of renewable energies precedes other energy sources by law and, therefore, is mostly independent of economic criteria. Thus, the load factors of renewable energies are mainly influenced by climatological phenomena and not by economic considerations. We assume them to be constant over the time horizon considered in this study. The remaining load factors depend on several other factors. Important to mention are the flexibility of different power plants in reacting to demand variations and the cost of operating different power plants. A closer look at the actual electricity production in 2018 reveals that hard coal, gas, and oil show greater variations than other non-renewable energy sources [24]. This leads to the conclusion that they are more likely to be used for satisfying demand variations. Consequently, in our model, most load factors remain constant—as in 2018—while the load factors of hard coal, gas, and oil will be adapted on a yearly basis in order to match the actual electricity demand.

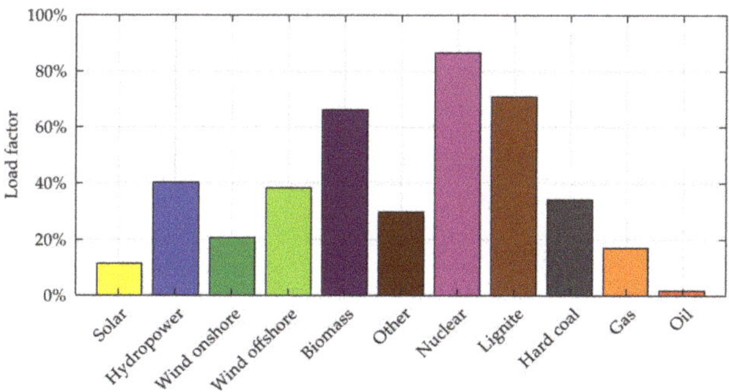

Figure 6. Load factors of German power plants in 2018 divided by energy source [25].

The UBA expects a slight decrease of electricity demand in Germany from 512 TWh in 2018 to 465.8 TWh in 2050 [26]. We assume a linearly decreasing trend until 2050. In Figure 7, we calculate the actual electricity production that matches the expected demand. Note that this does not include additional electricity demand from the vehicle fleet, which will be considered in the results in Section 3. For all energy sources—except hard coal, gas, and oil—the available installed capacity is multiplied by the load factor, which gives us the electricity production on a yearly basis. This electricity production partly covers the annual electricity demand. In order to satisfy the remaining electricity demand, the load factors of hard coal, gas, and oil are slightly adapted in every year, based on the reference load factors of 2018. As can be seen in Figure 7, this leads to the electricity production exactly matching the demand. Figure 7 also reveals the yearly mean specific GHG emissions of the energy sector resulting from the energy mix. As we will see in Section 3, considering the additional energy demand of the vehicle fleet leads to an increase of mean specific GHG emissions of the energy sector.

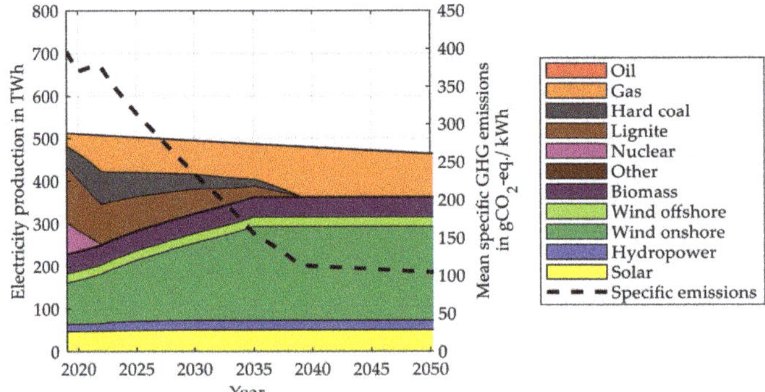

Figure 7. Electricity production in Germany matching electricity demand projected by the German Federal Environment Agency (UBA) divided by energy source and the resulting mean specific GHG emissions of the energy sector.

The out-phasing of hard coal and lignite allows for a drastic reduction of the mean specific GHG emissions of the energy sector. The underlying emission factors of all energy sources are listed in Table 3.

Table 3. Emission factors of power plants in Germany divided by energy source [27–29].

Energy Source	Emission Factor in 2018 in gCO$_2$-eq./kWh
Lignite	944.23
Hard coal	805.29
Oil	651.94
Other [1]	520.00
Gas	386.75
Solar	93.19
Hydropower	38.00
Biomass	32.49
Nuclear	22.37
Wind onshore	9.62
Wind offshore	5.10

[1] Includes energy sources like waste incineration, which only account for a negligible part of electricity production.

2.2. Scenarios for the Sensitivity-Analysis of Key Influence Parameters

Given the fact that some parameters of the vehicle fleet transition are subject to great uncertainty, we introduce different scenarios—pessimistic, neutral, and optimistic—for four parameters with key influence on the outcome of the optimization process. In the following four subsections, we take a closer look at the three possible scenarios for each key influence parameter.

2.2.1. Battery Production Scenarios

The GHG emissions during battery production per battery capacity constitute the first key influence parameter. Based on our research data in Supplementary Data, Table S1 [11], we generate the neutral scenario which can be seen in Figure 8. In case of a more pessimistic or optimistic scenario we adapt the emission reduction rate accordingly to be slower or faster.

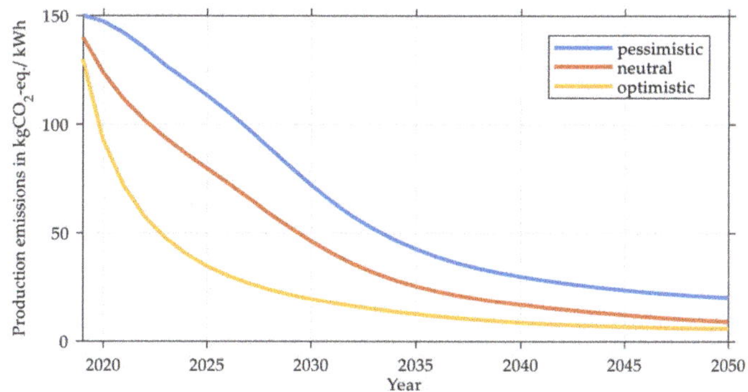

Figure 8. Scenarios of production emissions per capacity of li-ion batteries.

2.2.2. Energy Sector Scenarios

The second key influence parameter is the energy sector transition in Germany. Our neutral scenario assumes that political projects and promises like the out-phasing of nuclear and coal power take place as planned, refer to Figure 9. Power plants that are currently in their planning or construction phase are taken into account. A market research on the possible expansion of wind power conducted by the Federal Ministry for Economic Affairs and Energy (BMWi) plays an important role in our scenarios [21]. In pessimistic and optimistic scenarios, we slightly adapt the out-phasing and construction dates accordingly, as well as the magnitude of expansion or reduction.

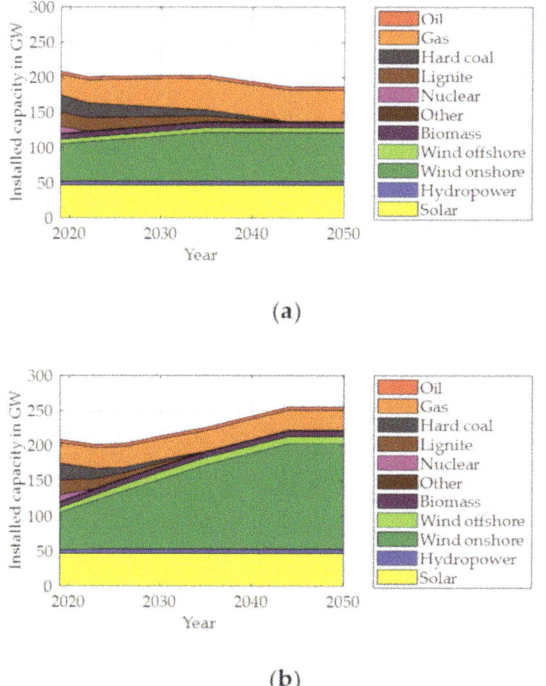

(**a**)

(**b**)

Figure 9. Development of installed capacity in the German energy sector (**a**) for the pessimistic scenario; (**b**) for the optimistic scenario. The neutral scenario is shown in Figure 5.

Our model adapts the yearly electricity production to the demand from the vehicle fleet added to the projected demand presented in Figure 7. The mean specific GHG emissions of the electricity production is thereby changing every year. The biggest influence stems from the German energy transition, which results in a drastic lowering of the GHG emissions until 2050. The additional electricity demand of the vehicle fleet counters this trend, since it is satisfied using conventional energy sources as explained in Section 2.1.3. With this model, we are able to assess the GWP_{100} values for different powertrain concepts in every year until 2050. The GWP_{100} of the electrical consumption of vehicles is based on the mean specific GHG emissions of the energy sector. This contrasts with other assessment methods that exclusively allocate the GHG emissions of the marginal energy source covering the additional electricity demand to the consumer causing it—in this case the vehicle fleet. The resulting transition of the electricity production in the energy sector is presented for each of the investigated scenarios in Section 3.

2.2.3. Hydrogen Production Path Scenarios

The third key influence parameter is the production of hydrogen, because it has a significant influence on the GHG emissions of FCEV. We assume that in future scenarios a bigger share of hydrogen is produced via electrolysis, see Figure 10a. Again, we generate three scenarios. On one hand, this leads to less GHG emissions from fossil fuels during hydrogen production. On the other hand, this further increases electricity demand when hydrogen is produced, although the latter is attenuated by an increase in electrolysis efficiency, displayed in Figure 10b.

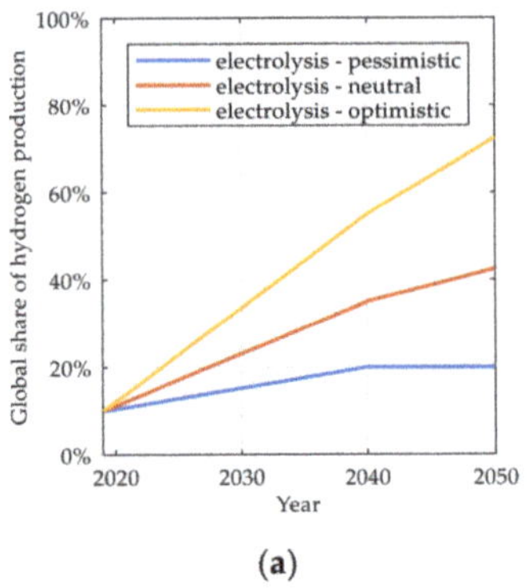

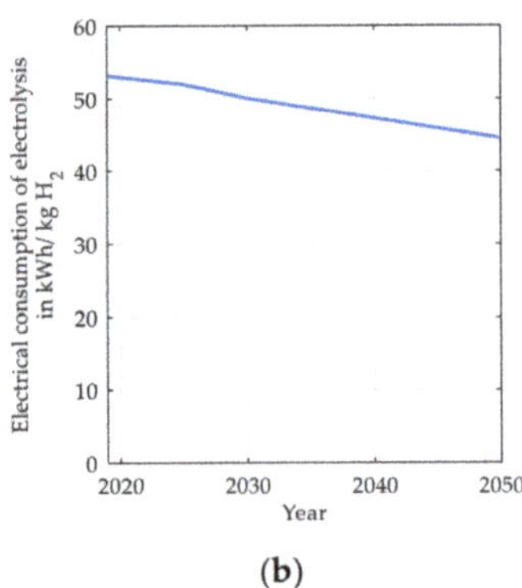

Figure 10. (a) Share of electrolysis in hydrogen production; (b) projected electrical consumption of electrolysis subject to efficiency increases.

As for the additional electricity demand of the vehicle fleet, the electricity demand of electrolysis is covered by the energy sector and, thereby, worsens its GWP_{100}.

2.2.4. Mobility Trend Scenarios

The German vehicle fleet comprises approx. 47 million vehicles in 2019 [30]. For the optimization, we need to determine the trend in absolute vehicle numbers until 2050. Assuming that in Germany there is a yearly passenger-distance to be satisfied, one can split it according to the means of transportation that serve it. The German Ministry of Transport and Digital Infrastructure (BMVI) provides short and mid-term prognostics for the main means of transportation—individual motorized mobility, railway, flights, and public transport [31]. For our work, we focus on the individual motorized mobility, which mainly refers to passenger car vehicles. In 2019, the cumulative passenger-distance of all passenger car vehicles is approx. 950 billion passenger-kilometers. The effect of people switching to other means of transportation and the further expansion of carsharing in Germany—on average more people utilizing the same vehicle—is subject to high uncertainty. This is especially true for prognostics over the next

30 years. Consequently, we generate three scenarios of future passenger-distance demand for passenger car vehicles, shown in Figure 11. This is our fourth key influence parameter.

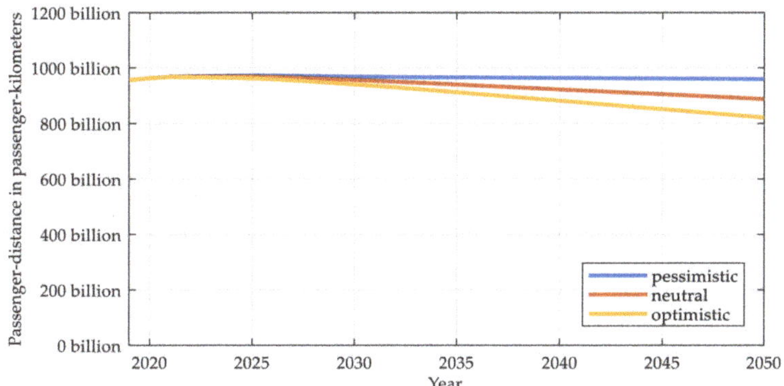

Figure 11. Three scenarios of passenger-distance carried out by passenger car vehicles.

In addition, the penetration of carsharing in the German market leads to less vehicles serving the same amount of passenger-distance, because these vehicles have a higher rate of utilization during their lifetime.

The initial situation of having approx. 30 million petrol and 15 million diesel-powered vehicles in the German vehicle fleet in 2019 has a great influence on the future transition. Having 47 million vehicles in 2019 and an average lifetime expectancy of 12.5 years, approx. 3.76 million vehicles would have to be replaced every year, if the number of vehicles were to stay the same. The initially present vehicles leave the fleet according to their age and their expected lifetime.

2.3. Modelling the Vehicle Behavior and Its Life-Cycle GHG Emissions

At this point, we will take a closer look at the different life-cycle phases of vehicles and the GHG emissions that have to be allocated to each one of them.

The GHG emissions of the production of vehicle components is based on the optimization framework for the comparative evaluation of the eco-impact of powertrain concepts [9]. This framework parametrizes the vehicle components in a way that leads to the minimum GHG emissions over all life-cycle phases. Main achievement of this method is that all vehicle parameters are adapted in such a way that their combined GWP_{100}—production, operational, and EoL phase—is minimized. This approach enables a proper comparison of the ecological potential of different powertrain concepts on a common evaluation basis. We further use this idea within this work to identify the GHG-optimal transition paths of the German vehicle fleet. Table 4 contains the production emissions associated to the different powertrain concepts. Since the battery production emissions are subject to more detailed modelling, the production emissions of Table 4 exclude them.

The lifetime expectancy of the vehicles is assumed to be equal for all different powertrain concepts. The probability of reaching its EoL is described by a Weibull curve. Plötz et al. estimated this distribution based on data from the Federal Motor Transport Authority (KBA) [32]. With the shape parameter of the Weibull function being 2.4 and the scale parameter 14.1, the average lifetime expectancy becomes 12.5 years, as shown in Figure 12.

Table 4. Production emissions of vehicle components excluding the battery.

Powertrain Concept	Production Emissions (Without Battery) in kg CO$_2$-eq.	Powertrain Concept	Production Emissions (Without Battery) in kg CO$_2$-eq.
ICEV-E10	8130	PHEV10-CNG	9349
ICEV-B7	8171	PHEV15-CNG	9349
ICEV-CNG	8130	PHEV20-CNG	9349
ICEV-LPG	8130	BEV20	8945 (8802) [1]
mHEV E10	9410	BEV100	9247 (8941) [1]
PHEV10-E10	9381	FCHEV	8957
PHEV15-E10	9381	FCPHEV20	9195
PHEV20-E10	9381	-	-

[1] Values in parenthesis represent projected values in 2050.

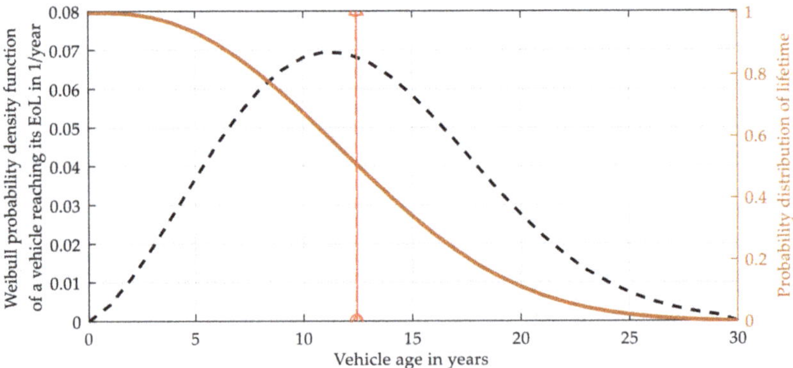

Figure 12. Probability density function of a vehicle reaching its end-of-life (EoL) and the resulting vehicle lifetime.

With an average lifetime expectancy of 12.5 years and knowing that the average annual mileage is 13,922 km, we conclude that the average lifetime mileage is approx. 173,906 km [33]. Additional data from the KBA regarding the annual mileage vehicles perform at different stages of their lifetime is also considered. As can be seen in Figure 13, once a vehicle enters the market, it will perform a high share of its lifetime mileage in the first three years [34]. Thereafter, mileage slowly decreases year by year. For optimization purposes we smoothened the data.

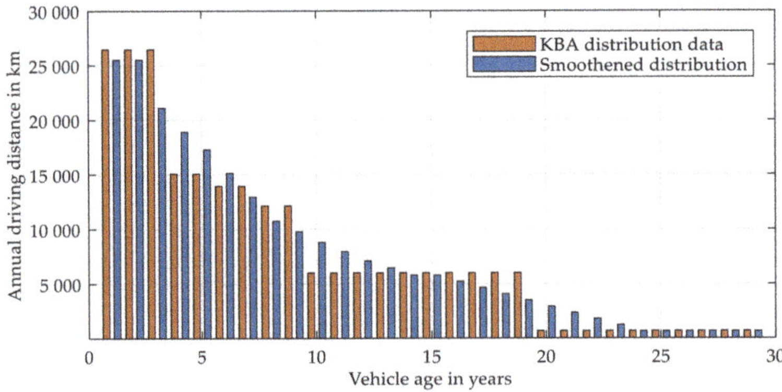

Figure 13. Annual driving distance of vehicles depending on their age.

During the operational phase of the vehicles, we have to consider the fuel consumption, which leads to direct GHG emissions preceded by the emissions during fuel supply. Electrical consumption does not lead to direct GHG emissions. However, the electrical consumption leads to an increase in electricity demand, which ultimately causes additional GHG emissions in the energy sector. These emissions will be allocated to the vehicle, according to the mileage it performs in every year, using the mean specific GHG emissions of the energy sector in the corresponding years, as explained in Section 2.2.2.

The consumption values are also determined with the optimization framework of Esser et al. mentioned earlier [9]. For this work, we use driving cycles resulting from the ARTEMIS project, in which real world data was used to generate representative driving cycles [35]. The driving data is segmented into different driving cycles, depending on the driving situation. For the purpose of this work, we differentiated a short-range driving profile and an all-range driving profile, weighting the driving cycles as in Table 5.

Table 5. Distance travelled on each driving cycle and weighting of the driving cycles for each driving profile.

ARTEMIS-Driving Cycles	Distance Travelled	All-Range Profile	Short-Range Profile
City cycle	0–60 km	29.3%	39.6%
Rural cycle	100 km	44.7%	60.4%
Highway cycle (150 km/h version)	200 km	26%	-

Each powertrain concept has to perform both driving profiles. This way, we obtain four different consumption values for each powertrain concept—two electrical and two fuel consumption values. Obviously, for ICEVs the electrical consumption is zero and for BEVs the fuel consumption is zero. In the case of PHEVs, the fuel and electrical consumption vary relatively strong depending on the driving profile. For example, the PHEVs can drive with a higher electrical percentage on the short-range profile, compared to the all-range profile, which leads to a reduced electrical demand and increased fuel consumption on the all-range profile. Important to notice is that, for determining the consumption values of PHEVs, we assume them to be fully charged before starting out. Table 6 presents all consumption values for all powertrain concepts.

Table 6. Fuel and electrical consumption values for all powertrain concepts in present day.

Powertrain Concept	Short Range Driving Profile		All Range Driving Profile	
	Fuel Consumptionin in kg/100 km	Electrical Consumptionin in kWh/100 km	Fuel Consumptionin in kg/100 km	Electrical Consumptionin in kWh/100 km
ICEV-E10	4.81	0	4.59	0
ICEV-B7	4.86	0	4.46	0
ICEV-CNG	4.24	0	4.03	0
ICEV-LPG	4.36	0	4.15	0
mHEV E10	3	0	3.35	0
PHEV10-E10	1.1	10.8	1.91	8.27
PHEV15-E10	0.2	16	1.19	12.4
PHEV20-E10	0.0745	17.4	0.932	14.1
PHEV10-CNG	1.02	10.6	1.69	8.27
PHEV15-CNG	0.21	15.9	1.05	12.4
PHEV20-CNG	0.013	17.4	0.815	14.2
BEV20	0	17.4 (16.6) [1]	-	-
BEV100	0	23.4 (20.9) [1]	0	23.1 (20.4) [1]
FCHEV	0.874	0	0.935	0
FCPHEV20	0	20.4	0.227	16.2

[1] Values in parenthesis represent projected consumption values in 2050.

The consumption values are constant over the years with the exception of BEVs. It is considered that innovative technologies, i.e., BEVs with two speed transmissions, lead to more efficiency. Less power is lost during energy transmission, the operation points of electric machines can be better adapted to the load behavior, and, consequently, other vehicle components can also be adapted to be more efficient. For the BEV20 we assume that, on fleet-average, a consumption value of 16.6 kWh/100 km will be achieved by 2050. Regarding the BEV100, in 2050 we evaluate a short-range consumption value of 20.9 kWh/100 km and an all-range consumption value of 20.4 kWh/100 km using the optimization framework. Between 2019 and 2050, the average consumption values of BEVs in the fleet are assumed to follow a linear trend.

Considering that in Germany 75% of vehicle-distance is performed over distances under 100 km and only 25% over larger distances, we allocate the short-range consumption values presented in Table 6 to the number of vehicles responsible for the former 75% of the vehicle-distance per year [36]. Consequently, the remaining vehicles have the all-range consumption level. This distribution is equal for all powertrain concepts except the BEV20, which can only drive short distances.

Finally, the EoL phase of the vehicle has comparatively low GHG emissions, which is modelled as 10% of the production emissions in Table 4. This value represents the GHG emissions that are released after the operational phase of the vehicle when it is disassembled. Again, this refers to all vehicle components except batteries. The EoL emissions of batteries are modelled with 1.2 $kgCO_2$-eq./kWh.

3. Identification of the Optimal Vehicle Fleet Transitions

The optimization process is implemented in MatLab. We make use of the nonlinear optimization function fmincon. The objective value of the optimization is the cumulative amount of GHG emissions until 2050 stemming from all life-cycle phases of the German vehicle fleet. Further details on the optimization process are to be found in the MathWorks description [37].

In the first place, we ran our model assuming that there would not be any transition in the vehicle fleet. In this business-as-usual scenario, considering everything remains as is until 2050, there would be approx. 15 million diesel and 30 million petrol powered vehicles in any given year. Alternative powertrain concepts would only represent a negligible part. We assume the mobility trend evolves according to the neutral scenario. The remaining key influence parameters do not affect the cumulative GHG emissions of this fleet transition. In this case, taking into account only vehicles introduced into the market since the beginning of 2019, the cumulative life-cycle GHG emissions of all vehicles until 2050 amount to 4299 million tons of CO_2-eq.—operational and EoL phases beyond 2050 are also considered.

Secondly, we ran our model for three specific scenarios. In our base scenario, all key influence parameters evolve according to their neutral scenario. In addition, we analyzed both cases in which all key influence parameters evolve according to the pessimistic or optimistic scenarios respectively. We refer to them worst- and best-case scenario.

Thirdly, we ran our model for all 81 possible scenario cross-combinations of the four key influence parameters. The results of this exhaustive sensitivity-analysis are presented in aggregated manner in Section 3.4.

3.1. Optimal Vehicle Fleet Transition for the Base Scenario

In the following, we take a closer look at the base scenario, in order to better understand the interconnected behavior of the vehicle fleet and the energy sector. As can be seen in the optimization results shown in Figure 14, apart from the vehicles of the initial situation in 2019, no more conventional ICEVs are introduced into the fleet.

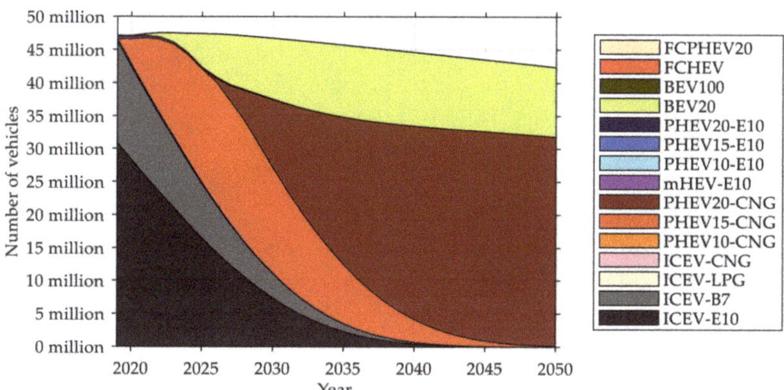

Figure 14. Optimal vehicle fleet transition for the base scenario.

PHEVs play a very important role during the fleet transition. Two types of PHEVs powered by electricity and CNG are the most relevant concepts. At the start, the concept with the 15 kWh battery is introduced, and, beginning in 2026 onwards, the concept with a slightly higher battery capacity—20 kWh—takes over. Shortly after 2019, BEV20 are gaining their share in the fleet until they reach their maximum share of 25% implemented in the mobility trend scenario. This constraint represents the limited user acceptance for short-ranged vehicles in the base scenario.

Our model permits analyzing the yearly life-cycle emissions for all powertrain concepts for every given fleet transition scenario. This way, regardless of being introduced into the fleet, we can estimate the life-cycle emissions of any given powertrain concept. In fact, for this base scenario, we identify that BEV100 and FCPHEV20 are causing a similar amount of GHG emissions as PHEV20-CNG when looking between 2040 and 2050. Even though the use of PHEV-CNG concepts is optimal for this scenario, the introduction of BEV100 or FCPHEV20 starting in 2040 would lead to similar cumulative emissions. Figure 15 reveals that the production emissions during this time interval barely differ between the powertrain concepts. Further, the long-range capable vehicles PHEV20-CNG, BEV100, and FCPHEV20 are mainly driven electrically during their operational phase. Given the similarity of their electrical consumption, they all interact in a similar way with the energy sector, only slightly changing its mean specific GHG emissions. This renders the vehicles interchangeable, with little impact on the objective value. In summary, the ecological potential of the BEV100 and FCPHEV20 concept should not be underestimated—especially considering the uncertainty that accompanies the prediction of technical boundary parameters over long time periods. Figure 15 shows the life-cycle emissions of the PHEV15-CNG (a), the PHEV20-CNG (b), the BEV20 (c), the BEV100 (d), the FCPHEV20 (e), and the ICEV-E10 (f) in every year for the base scenario transition.

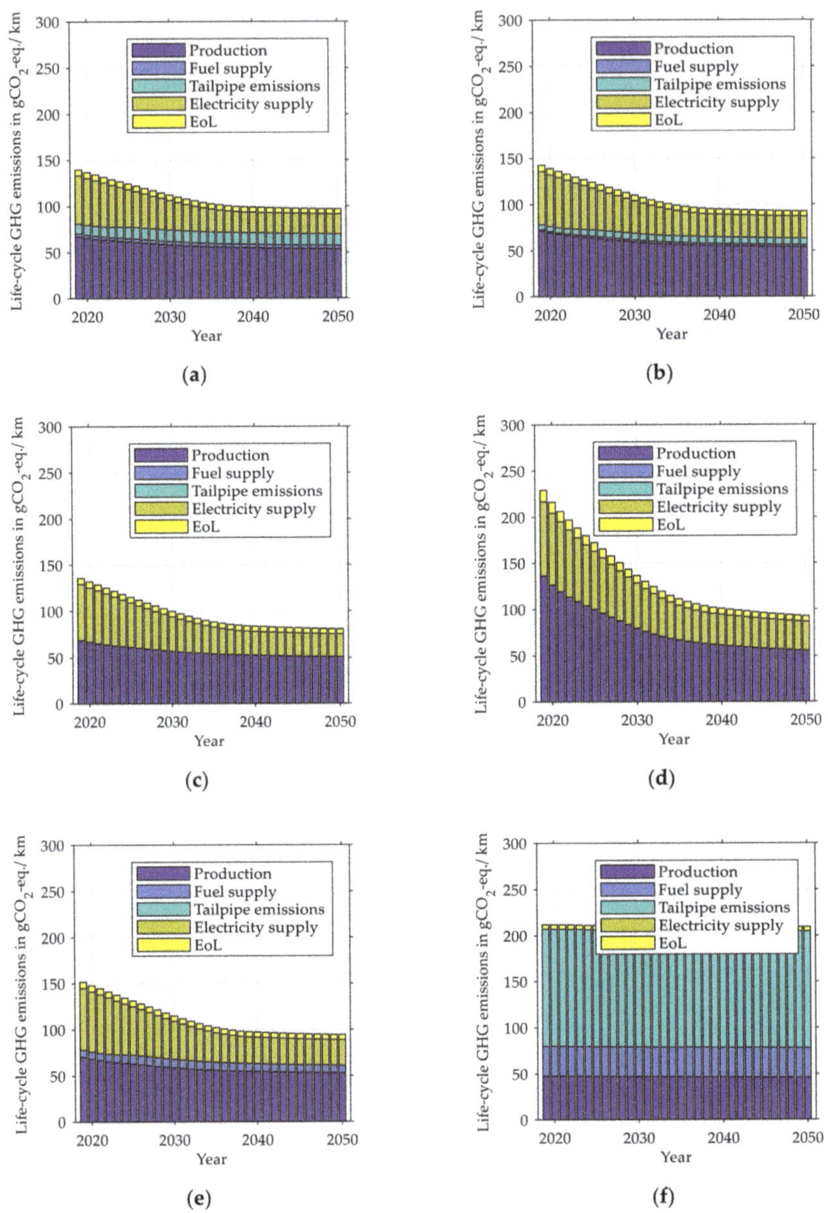

Figure 15. Life-cycle emissions of (**a**) the PHEV15-CNG; (**b**) the PHEV20-CNG; (**c**) the BEV20; (**d**) the BEV100; (**e**) the FCPHEV20; and (**f**) the ICEV-E10 for the base scenario fleet transition.

Table 7 contains the life-cycle emissions of the PHEV15-CNG, the PHEV20-CNG, the BEV20, the BEV100, and the FCPHEV20 in the year 2050. The small magnitude of the differences between the long-range capable powertrain concepts remarks the fact that BEV100 and FCPHEV20 could also be considered at the end of the time-horizon in the context of a near-optimal fleet transition instead of PHEV20-CNG.

Table 7. Life-cycle GHG emissions of the PHEV15-CNG, the PHEV20-CNG, the BEV20, the BEV100, and the FCPHEV20 in 2050 for the base scenario.

Powertrain Concept	PHEV15-CNG	PHEV20-CNG	BEV20	BEV100	FCPHEV20
Life-cycle GHG emissions in kg CO_2-eq./km	96.83	92.49	80.60	92.71	94.23

We already saw the electricity demand in Germany corresponding to expected scenarios of the UBA in Figure 7, Section 2.1.3. Figure 16 shows us the combined electricity demand of the projection and the additional demand from the vehicle fleet. The additional energy demand, caused by the vehicles with a plug-in option, is mainly satisfied by an increased usage of gas power plants. As can be seen, the mean specific GHG emissions of electricity production are higher than in Figure 7, but still follow the same trend of declining sharply during the out-phasing of coal-based power plants.

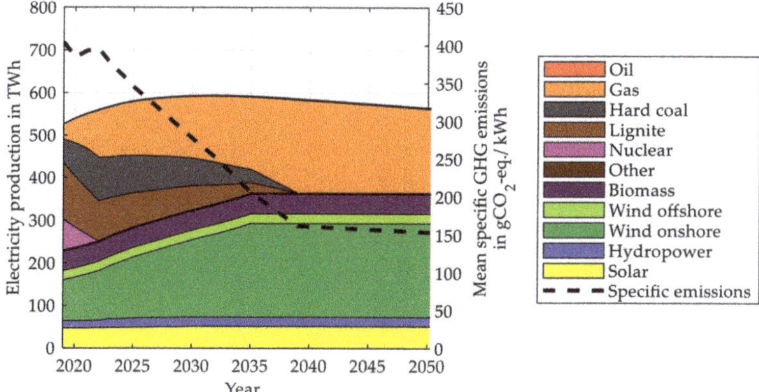

Figure 16. Electricity production for the base scenario including the additional electricity demand of the vehicle fleet.

Figure 17 shows the cumulative GHG emissions caused by the entire vehicle fleet starting in 2019. The cumulative GHG emissions of ICEVs soon reach a plateau, because they are all replaced.

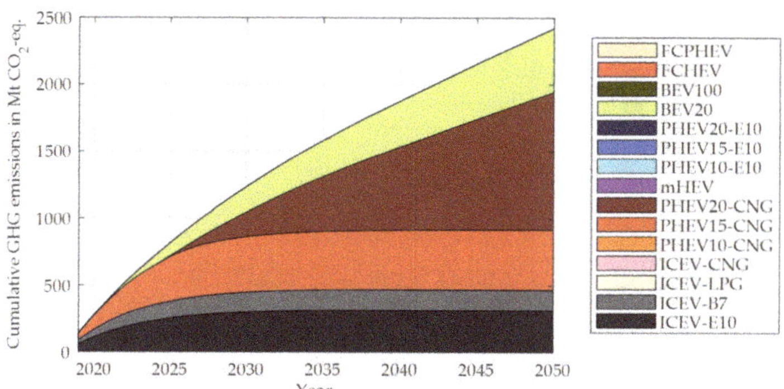

Figure 17. Cumulative greenhouse gas emissions for the base scenario.

For this transition, the cumulative amount of GHG emissions of all life-cycle phases of the fleet until 2050 amounts to approx. 2046 million tons CO_2-eq. This is less than 50% of the cumulative emissions compared to the business-as-usual scenario.

3.2. Optimal Vehicle Fleet Transition for the Worst-Case Scenario

In the worst-case scenario—all key influence parameters evolving according to their pessimistic scenario—the optimal fleet transition identified by the developed model looks similar to the base scenario, see Figure 18. Regarding the BEV20, we notice two main effects compared to the base scenario. They are introduced later and their maximum share in the fleet is limited. The former effect is due to a slower energy transition, which leads to higher GHG emissions resulting from electrical consumption, and the fact that battery production emissions cannot be reduced as fast as in the base scenario. The second effect is due to limited user acceptance of short-ranged vehicles implemented by a 10% cap for these concepts.

As in the base scenario, the rest of the fleet is constituted by PHEV-CNG. However, we notice a shift in the introduction of PHEV20-CNG when comparing to the base scenario. This shift results for the same reasons that lead to the later introduction of BEV20. Overall, the optimal fleet transition in the worst-case scenario leads to approx. 2,463 million tons of CO_2-eq.—which is a reduction of about 40% compared to the business-as-usual scenario.

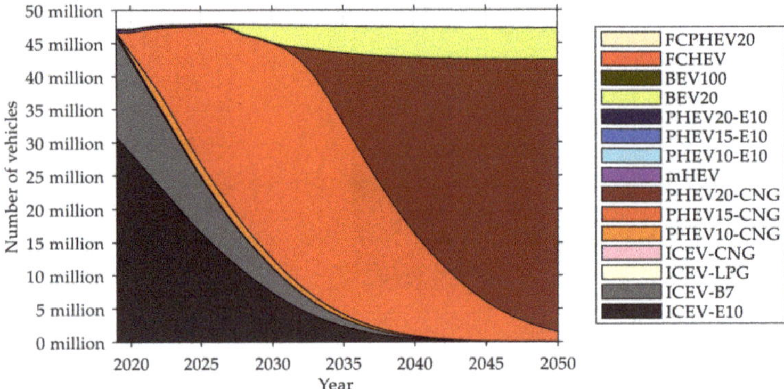

Figure 18. Optimal vehicle fleet transition for the worst-case scenario.

3.3. Optimal Vehicle Fleet Transition for the Best-Case Scenario

Now we have a look at the best-case scenario, in which all key influence parameters evolve according to their optimistic scenario. The mobility trend in Germany has a major influence on the transition. Firstly, people switch to other means of transport and thereby lower the demand for vehicles. Secondly, carsharing penetrates the market and thereby raises the average lifetime mileage of the vehicles putting a higher weight on the operational phase of the vehicles. Thirdly, a higher user acceptance for short-ranged vehicles increases their maximum fleet-share to 50%. As can be seen in Figure 19, the optimal transition mainly consists of BEVs. In fact, BEV20s gain a big share of the fleet very quickly. The fast developments of battery technologies and the effective energy transition to more renewables reduce the role of PHEVs in favor of BEVs. Nevertheless, the former are still relevant for the fleet transition. The PHEV15-CNG now takes a minor role and beginning in 2025 the PHEV20-CNG replaces the former achieving a significant fleet-share of around 50%. Furthermore, production emissions of batteries can be reduced in such a way and the expansion of wind power is such, that, starting in 2039, BEV100 are part of the optimal fleet transition.

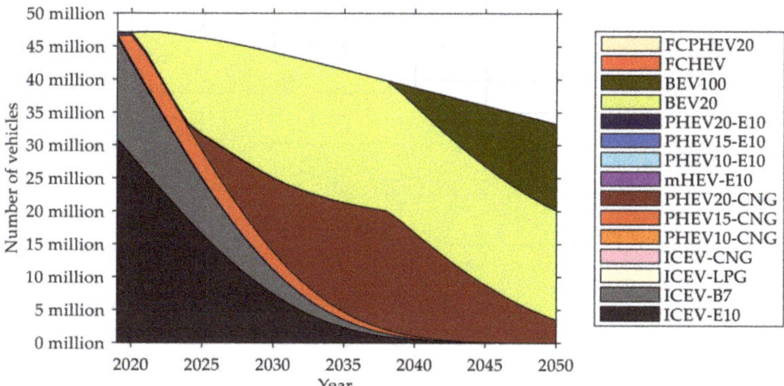

Figure 19. Optimal vehicle fleet transition for the best-case scenario.

Overall, the best-case scenario would lead to approx. 1680 million tons of CO_2-eq. as cumulative emissions until 2050—which is a reduction of about 60% compared to the business-as-usual scenario.

3.4. Sensitivity-Analysis of Key Influence Parameters

The results suggest that future powertrain concepts need to be equipped with a plug-in option. ICEV concepts and even HEV concepts are not part of the presented optimal transition scenarios. PHEVs are present in all three transition scenarios previously considered. They offer a great compromise between electrical and fuel consumption. PHEV-CNG appear more advantageous than PHEV-E10, because their direct emissions are lower. Concerning the short-range vehicles, BEV20 are introduced for all considered scenarios. BEV100 and FCPHEV20 powertrain concepts both show potential to reduce GHG emissions in the optimal transition for the best-case scenario with optimistic boundary conditions and would lead to almost similar results in the base scenario if introduced after 2040.

At this point, we conduct a sensitivity-analysis of all possible future scenario-combinations of the four key influence parameters. Each parameter having three possible scenarios leads to a total of 81 possible combinations for future scenarios. We performed the analysis for all 81 combinations and found 81 optimal fleet transitions. Sequentially, we analyze the four key influence parameters by grouping the results according to their scenarios in every instance. For example, when looking at the mobility trend, there are 27 scenarios in which the trend evolves in a pessimistic way, 27 for the neutral scenario, and 27 for the optimistic scenario. In the end, we have four different ways of grouping the results into three groups of 27 each.

First, we analyze the objective value of the optimization function: the cumulative GHG emissions of the vehicle fleet until the year 2050. Generally, as can be seen in Figure 20, the groups containing the results for the pessimistic scenario of the key influence parameters correlate with higher cumulative emissions. After comparing the median values and their variation within one key influence parameter, we conclude that the mobility trend, followed by the trend in the energy sector are the primal determinants to reduce GHG. This is no surprise since less vehicles result in lower cumulative GHG emissions. In addition, more renewable energy for electricity production has a positive impact on PHEVs, BEVs, and FCPHEV20s, which have reduced GHG emissions during their operational phase. Battery production plays an important role as well. The sooner production emissions per kWh can be reduced, the sooner vehicles with a higher battery capacity can be introduced into the fleet. The electrical range from PHEVs can be increased, thereby reducing the distance travelled with fossil fuels. The key influence parameter regarding the hydrogen production path has no impact on the fleet transition, because no FCEVs are introduced in any scenario. This is due to the significant emissions during the supply of hydrogen as presented in Figure 3, that are not compensated by the positive developments presented in Section 2.2.3.

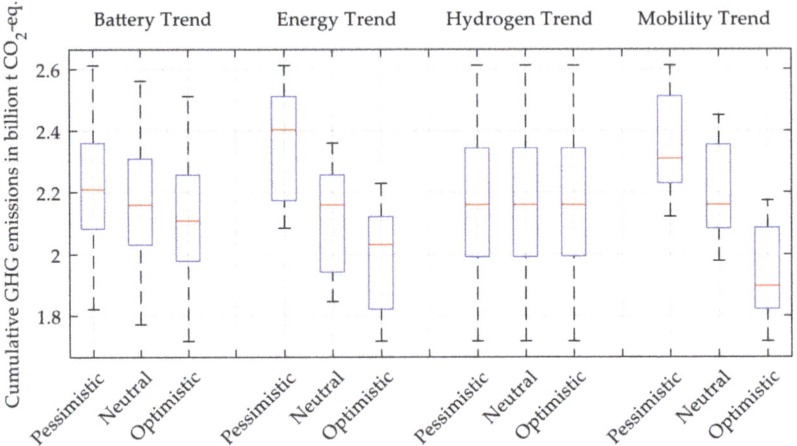

Figure 20. Sensitivity-analysis of the cumulative GHG emission values of 81 optimization processes. The boxes include all values between the 25% quantile and the 75% quantile of the data. Minima and maxima are displayed with a short horizontal dash.

Next, in Figure 21, we present the composition of the fleet transitions. Since analyzing all 81 fleet transitions one by one would exceed the scope of this work, we merged 27 scenarios at a time following the same grouping procedure as above. Prior to merging, we normed the absolute vehicle number in all years to one, because vehicle numbers for the same year differ between different scenarios. The key influence parameter regarding the hydrogen production path is no longer part of our sensitivity-analysis, since there is no influence in our study, as shown by the former results.

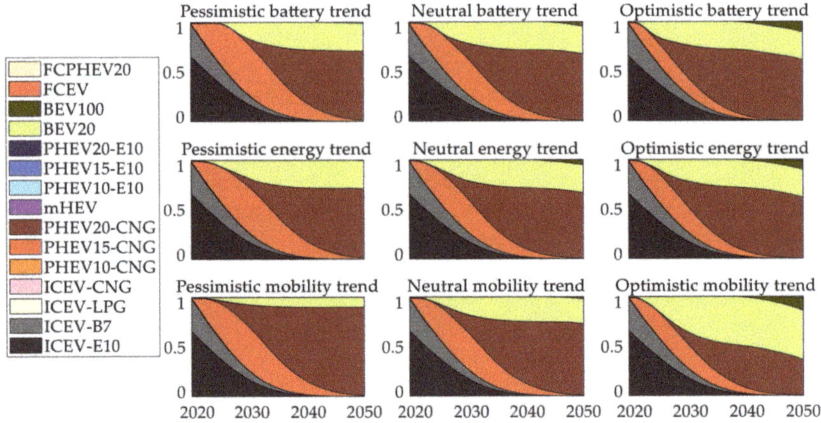

Figure 21. Sensitivity-analysis of battery, energy, and mobility trends on the vehicle fleet transition.

On first sight, we notice the strong similarity with the base, worst-, and best-case scenarios discussed earlier. The trend of the battery production emissions and the trend in the energy sector both have a similar effect on the fleet transition. In more optimistic scenarios, the introduction of BEV20 and PHEV20-CNG takes place earlier. Consequently, the role of the PHEV15-CNG is moderately reduced. In very optimistic cases, BEV100 are introduced into the fleet starting in 2039. Regarding the mobility trend, we observe that in more optimistic scenarios the PHEV15-CNG are out-phased earlier and the PHEV20-CNG are introduced slightly later. This is due to higher user acceptance of short-range vehicles—BEV20—and the fact that it takes several years until the BEV20 reach the maximum share of 50% short-ranged vehicles of the fleet (optimistic case).

4. Conclusions

We presented an optimization model to identify optimal vehicle fleet transition paths for given scenarios with the cumulative amount of GHG emissions caused by all vehicles of the fleet taking into account all life-cycle phases of the vehicles. This way, the relevance of powertrain concepts can be compared in the context of a fleet transition over a 30-year time period. Furthermore, the identified optimal transitions give important insight about suitable timing and rates of introduction of new powertrain concepts. Special focus lies on the interaction of the fleet with the energy sector. Note that, for all powertrain concepts, the vehicles were parametrized in a way that reduces their life-cycle GHG emissions, as in [9]. The results do not represent the most likely fleet transitions, for which different prognosis have been performed as presented in Section 1, but rather reveal the hidden ecological potential of the mobility sector and serve as guideline for decision makers that want to achieve GHG-optimal results.

Aspects that have not been considered in this study are the costs of ownership of the different powertrain concepts and the user convenience, for which the charging infrastructure is essential. These two levers are of major relevance for policy in order to influence the composition of the vehicle fleet. At the same time, these levers will hinder the optimal fleet transition, if disregarded.

In particular, we analyzed the situation in Germany and modeled several parameters with yearly resolution. For four key influence parameters of interest, we modelled several scenarios, followed by an extensive sensitivity-analysis. The battery production emissions trend, the energy sector transition, and the mobility trend highly influence the cumulative GHG emissions of the fleet.

Interestingly, the optimal fleet transitions for different scenario parameters show various similarities, which allow to draw generic conclusions on how the transition of the German vehicle fleet should be organized. Based on our results, short-ranged BEVs with moderate battery capacities—like the BEV20 considered in this study—should soon be introduced into the fleet. As Esser et al. already confirmed in their previous work, this powertrain concept has the lowest life-cycle GHG emissions on short-range driving profiles [6]. Furthermore, we conclude that PHEVs are of central relevance for an optimal fleet transition—bear in mind that this fleet transition optimization aims at showing the ecological potential of different powertrain concepts and therefore, as mentioned earlier, PHEVs are assumed to be fully charged previous to every ride. In general, they represent a great compromise between mostly driving electrically on short distances—like BEV20—whilst having the possibility to drive longer distances as well. This behavior overlaps with the driving habits of most car owners in Germany. More in detail, our optimization finds that as time passes and the GWP_{100} of battery production gets lower, an increase of battery capacity of PHEVs leads to an overall reduction of cumulative GHG emissions. This effect is further strengthened—or conditioned by—the successful transition of the energy sector to more renewable energy sources. If PHEVs are not charged regularly, their good ecological potential will not be reached and their relevance for an optimal fleet transition is reduced. Therefore, regulations or incentives to ensure regular charging of the vehicles are essential.

The PHEVs powered by CNG emit less GHG compared to their petrol-powered counterparts. This is due to lower direct operational emissions of CNG. Nevertheless, deeper analyses of high-pressure tanks for natural gas should be conducted. In this paper, their GHG emissions have not been explicitly determined for the production phase of these vehicles.

When focusing on the time period between 2040 and 2050—assuming the energy transition in Germany is mostly completed—we find that the powertrain concepts PHEV20-CNG, BEV100, and FCPHEV20 show very similar life-cycle emissions. By that time, these three powertrain concepts lead to very similar GHG emissions and result in similar cumulative fleet emissions. Other factors, i.e., synergy effects between the mobility and the energy sector, should be further analyzed to determine the best powertrain concepts and fleet transition. This is because CNG and hydrogen can act as energy buffer and energy carriers between both sectors—globally optimizing the resource utilization.

The developed optimization model for the identification of optimal fleet transitions was applied to the German vehicle fleet. The representativeness of the driving profiles for Germany can be further increased, if the ARTEMIS cycles are substituted by actual driving cycles based on fleet driving data covering Germany as in [9]. This would permit a more precise assessment of the GHG emissions during the operational phase of the vehicles using representative driving cycles resulting in the true environmental impact of GHG emissions during the whole life-cycle of vehicles—the real ecological impact (REI) of considered powertrain concepts. Furthermore, the approach can be generalized and applied on further global regions. Future works on this topic can increase the meaningfulness of the results by integrating the GHG emissions of other means of transportation into a global mobility sector optimization environment that also considers different scenarios regarding policy incentives.

Supplementary Materials: The Supplementary Data Table S1: Meta-Analysis of Lithium-Ion Batteries is available online at http://dx.doi.org/10.25534/tudatalib-144. Supplementary Data Table S2: Meta-Analysis of Fuels is available online at http://dx.doi.org/10.25534/tudatalib-145.

Author Contributions: Conceptualization, B.B.B. and A.E.; methodology, B.B.B. and A.E.; software, B.B.B. and A.E.; validation, B.B.B., A.E. and S.W.; formal analysis, B.B.B. and A.E.; investigation, B.B.B. and A.E.; data curation, B.B.B.; writing—original draft preparation, B.B.B.; writing—review and editing, B.B.B., A.E., S.W. and G.F.; visualization, B.B.B.; supervision, L.S. and S.R.; project administration, S.R. All authors have read and agreed to the published version of the manuscript.

Funding: This research received no external funding.

Conflicts of Interest: The authors declare no conflicts of interest.

References

1. Federal Ministry for Environment, Nature Conservation and Nuclear Safety. *Klimaschutzbericht 2018—Zum Aktionsprogramm Klimaschutz 2020 der Bundesregierung*; Federal Ministry for Environment, Nature Conservation and Nuclear Safety: Bonn, Germany, 2019.
2. Knörr, W.; Heidt, C.; Gores, S.; Bergk, F. *Aktualisierung "Daten- und Rechenmodell: Energieverbrauch und Schadstoffemissionen des Motorisierten Verkehrs in Deutschland 1960–2035" (TREMOD) für die Emissionsberichterstattung 2016 Berichtsperiode 1990–2014*; IFEU—Institut für Energie- und Umweltforschung Heidelberg GmbH: Heidelberg, Germany, 2016.
3. Kickhöfer, B.; Brokate, J. *Die Entwicklung des Deutschen Pkw-Bestandes: Ein Vergleich Bestehender Modelle und die Vorstellung Eines Evolutionären Simulationsansatzes*; Institut für Verkehrsforschung; Institut für Fahrzeugkonzepte; German Aerospace Center: Berlin, Germany, 2017.
4. Harrison, G.; Krause, J.; Thiel, C. Transitions and Impacts of Passenger Car Powertrain Technologies in European Member States. *Transp. Res. Procedia* **2016**, *14*, 2620–2629. [CrossRef]
5. Plötz, P.; Gnan, T.; Wietschel, M. *Total Ownership Cost Projection for the German Electric Vehicle Market with Implications for its Future Power and Electricity Demand*; Fraunhofer Institute for Systems and Innovation Research ISI: Karlsruhe, Germany, 2012.

6. Helms, H.; Jöhrens, J.; Kämper, C.; Giegrich, J.; Liebich, A.; Vogt, R.; Lambrecht, U. Weiterentwicklung und vertiefte Analyse der Umweltbilanz von Elektrofahrzeugen. 2016. Available online: https://www.umweltbundesamt.de/sit es/default/files/medien/378/publikationen/texte_27_2016_umweltbilanz_von_elektrofahrzeugen.pdf (accessed on 20 January 2020).
7. Fraunhofer ISE. *Treibhausgas-Emissionen für Batterie- und Brennstoffzellenfahrzeuge mit Reichweiten über 300 km*; Fraunhofer ISE: Freiburg, Germany, 2019.
8. Wietschel, M.; Kühnbach, M.; Rüdiger, D. *Die Aktuelle Treibhausgas- Emissionsbilanz von Elektrofahrzeugen in Deutschland*; Working Paper Sustainability and Innovation No. S 02/2019; Fraunhofer Institute for Systems and Innovation Research ISI: Karlsruhe, Germany, 2019.
9. Esser, A.; Schleiffer, J.-E.; Eichenlaub, T.; Rinderknecht, S. Development of an Optimization Framework for the Comparative Evaluation of the Ecoimpact of Powertrain Concepts. In *VDI-Berichte, Proceedings of the 19th Internationaler VDI-Kongress "Dritev—Getriebe in Fahrzeugen", Bonn, Germany, 10–11 July 2019*; Springer: Berlin/Heidelberg, Germany, 2019.
10. Wuppertal Institut für Klima, Umwelt, Energie GmbH. *STROMbegleitung—Begleitforschung zu Technologien, Perspektiven und Ökobilanzen der Elektromobilität*; Wuppertal Institut für Klima, Umwelt, Energie GmbH: Wuppertal, Germany, 2015.
11. Blat Belmonte, B.; Esser, A.; Weyand, S.; Franke, G. Meta-Analysis of Lithium-Ion Batteries—Identification of the Optimal Passenger Car Vehicle Fleet Transition for Mitigating the Cumulative Life Cycle Greenhouse Gas Emissions until 2050; Supplementary data table S1; Darmstadt, Germany. 2019. Available online: http://dx.doi.org/10.25534/tudatalib-144 (accessed on 21 January 2020).
12. Blat Belmonte, B.; Esser, A.; Weyand, S.; Franke, G. Meta-Analysis of Fuels—Identification of the Optimal Passenger Car Vehicle Fleet Transition for Mitigating the Cumulative Life Cycle Greenhouse Gas Emissions until 2050; Supplementary data table S2; Darmstadt, Germany. 2019. Available online: http://dx.doi.org/10. 25534/tudatalib-145 (accessed on 20 January 2020).
13. Grote, K.-H.; Bender, B.; Göhlich, D. *Dubbel*; Springer: Berlin/Heidelberg, Germany, 2018.
14. Helms, H.; Jöhrens, J.; Hanusch, J.; Höpfner, U.; Lambrecht, U.; Pehnt, M. *UMBRELA—Wissenschaftlicher Grundlagenbericht Gefördert Durch das Bundesministerium für Umwelt, Naturschutz und Reaktorsicherheit (BMU)*; Federal Ministry for the Environment, Nature Conservation and Nuclear Safety: Heidelberg, Germany, 2011; Available online: https://www.erneuerbar-mobil.de/sites/default/files/publications/abschlussbericht-umbrela _1.pdf (accessed on 20 January 2020).
15. Granovskii, M.; Dincer, I.; Rosen, M.A. Life cycle assessment of hydrogen fuel cell and gasoline vehicles. *Int. J. Hydrogen Energy* **2006**, *31*, 337–352. [CrossRef]
16. Bhandari, R.; Trudewind, C.A.; Zap, P. *Life Cycle Assessment of Hydrogen Production Methods—A Review*; Forschungszentrum Jülich, Institute of Energy and Climate Research—Systems Analysis and Technology Evaluation (IEK-STE): Jülich, Germany, 2012.
17. Edwards, R.; Larivé, J.-F.; Rickeard, D.; Weindorf, W. *Well-to-Tank Report Version 4.a—JEC Well-to-Wheels Analysis*; European Commission Joint Research Centre, Institute for Energy: Ispra, Italy, 2014.
18. International Renewable Energy Agency. *Hydrogen from Renewable Power—Technology Outlook for the Energy Transition*; International Renewable Energy Agency: Abu Dhabi, UAE, 2018.
19. BDEW Bundesverband der Energie- und Wasserwirtschaft e.V. *Gas Kann Grün: Die Potentiale von Biogas/Biomethan—Status Quo, Fakten und Entwicklung*; BDEW: Berlin, Germany, 2019.
20. Braune, M.; Grasemann, E.; Gröngröft, A.; Klemm, M.; Oehmichen, K.; Zech, K. *Die Biokraftstoffproduktion in Deutschland—Stand der Technik und Optimierungsansätze*; Deutsches Biomasseforschungszentrum, DBFZ-ReportNr. 22; Deutsches Biomasseforschungszentrum Gemeinnützige GmbH: Leipzig, Germany, 2016.

21. Federal Ministry for Economic Affairs and Energy. *Marktanalyse Windenergie an Land*. *Federal Ministry for Economic Affairs and Energy*; Federal Ministry for Economic Affairs and Energy: Berlin, Germany, 2014. Available online: https://www.erneuerbare-energien.de/EE/Redaktion/DE/Downloads/bmwi_de/marktanalysen-photovoltaik-windenergie-an-land.pdf?__blob=publicationFile&v=4 (accessed on 20 January 2020).

22. The Federal Government of Germany. *Kommission "Wachstum, Strukturwandel und Beschäftigung"—Abschlussbericht*; The Federal Government of Germany: Berlin, Germany, 2019. Available online: https://www.bmwi.de/Redaktion/DE/Downloads/A/abschlussbericht-kommission-wachstum-strukturwandel-und-beschaeftigung.pdf?__blob=publicationFile (accessed on 20 January 2020).

23. German Federal Environment Agency. *Genehmigte Oder im Genehmigungsverfahren Befindliche Kraftwerksprojekte in Deutschland*; German Federal Environment Agency: Dessau-Roßlau, Germany, 2019; Available online: https://www.umweltbundesamt.de/daten/energie/konventionelle-kraftwerke-erneuerbare-energien#textpart-1 (accessed on 9 April 2019).

24. Burger, B. *Energy Charts*; Fraunhofer-Institut für Solare Energiesysteme ISE: Freiburg, Germany, 2018.

25. BDEW Bundesverband der Energie- und Wasserwirtschaft e.V. *Erneuerbare Energien und das EEG: Zahlen, Fakten, Grafiken (2017)—Anlagen, Installierte Leistung, Stromerzeugung, Marktintegration der Erneuerbaren Energien, EEG-Auszahlungen und Regionale Verteilung der EEG-Anlagen*; BDEW: Berlin, Germany, 2017; Available online: https://www.dieter-bouse.de/app/download/5810146463/BDEW_Erneuerbare+Energien+und+das+EEG+-+Zahlen%2C+Fakten+und+Grafiken+2017%2C+Foliensatz+Juli+2017.pdf (accessed on 20 January 2020).

26. German Federal Environment Agency. *Treibhausgasneutrales Deutschland im Jahr 2050*; German Federal Environment Agency: Dessau-Roßlau, Germany, 2013; Available online: https://www.umweltbundesamt.de/sites/default/files/medien/376/publikationen/treibhausgasneutrales_deutschland_im_jahr_2050_langfassung.pdf (accessed on 20 January 2020).

27. German Aerospace Center; IFNE. *Langfristszenarien und Strategien für den Ausbau Erneuerbarer Energien in Deutschland—Leitszenario 2009*; German Aerospace Center; IFNE: Berlin, Germany, 2009.

28. Fritsche, U.; Leuchtner, J.; Matthes, F.C.; Rausch, L.; Simon, K.-H. *Umweltanalyse von Energie-, Transport- und Stoffsystemen: Gesamt-Emissions-Modell Integrierter Systeme (GEMIS) Version 2.1—Erweiterter und Aktualisierter Endbericht*; Institut für angewandte Ökologie e.V.: Freiburg, Germany, 1995.

29. Wernet, G.; Bauer, C.; Steubing, B.; Reinhard, J.; Moreno-Ruiz, E.; Weidema, B. The ecoinvent database version 3 (part I): Overview and methodology. *Int. J. Life Cycle Assess.* **2016**, *21*, 1218–1230. [CrossRef]

30. Federal Motor Transport Authority. Jahresbilanz des Fahrzeugbestandes am 1. Januar 2019. Federal Motor Transport Authority. 2019. Available online: https://www.kba.de/DE/Statistik/Fahrzeuge/Bestand/b_jahresbilanz.html (accessed on 16 April 2019).

31. Federal Ministry of Transport and Digital Infrastructure; German Aerospace Center; DIW. *Verkehr in Zahlen 2018/2019. 47. Jahrgang*; Federal Ministry of Transport and Digital Infrastructure; German Aerospace Center; DIW: Berlin, Germany, 2018.

32. Plötz, P.; Gnan, T.; Kühn, A.; Wietschel, M. *Markthochlaufszenarien für Elektrofahrzeuge*; Fraunhofer ISI: Karlsruhe, Germany, 2013.

33. Federal Motor Transport Authority. *Erneut Mehr Gesamtkilometer Bei Geringerer Jahresfahrleistung je Fahrzeug*; Kraftverkehrs statistik; Federal Motor Transport Authority: Flensburg, Germany, 2018.

34. Federal Motor Transport Authority. *Verkehr in Kilometern der Deutschen Kraftfahrzeuge im Jahr 2017—Jahresfahrleistung in Millionen Kilometer Nach Fahrzeugarten und Fahrzeugalter im Jahr 2017*; Federal Motor Transport Authority; Flensburg, Germany, 2019; Available online: https://www.kba.de/DE/Statistik/Kraftverkehr/VerkehrKilometer/verkehr_in_kilometern_node.html (accessed on 28 May 2019).

35. André, M. The ARTEMIS European driving cycles for measuring car pollutant emissions. *Sci. Total. Environ.* **2004**, *334*, 73–84. [CrossRef] [PubMed]

36. Federal Ministry of Transport and Digital Infrastructure; Institut für Angewandte Sozialwissenschaft GmbH infas; German Aerospace Center; IVT Research GmbH; infas 360 GmbH. *Mobilität in Deutschland—Ergebnisbericht*; Federal Ministry of Transport and Digital Infrastructure; Institut für angewandte Sozialwissenschaft GmbH infas; German Aerospace Center; IVT Research GmbH; infas 360 GmbH: Bonn, Germany, 2019.
37. MathWorks. Fmincon. Available online: https://www.mathworks.com/help/optim/ug/fmincon.html (accessed on 27 September 2019).

 *vehicles*

Article

Design of an Aftermarket Hybridization Kit: Reducing Costs and Emissions Considering a Local Driving Cycle

Jony Javorski Eckert [1,*], Fabio Mazzariol Santiciolli [1], Ludmila Corrêa de Alkmin e Silva [1], Fernanda Cristina Corrêa [2] and Franco Giuseppe Dedini [1]

[1] Integrated Systems Laboratory, University of Campinas, Campinas, São Paulo 13083-970, Brazil; fabio@fem.unicamp.br (F.M.S.); ludmila@fem.unicamp.br (L.C.d.A.e.S.); dedini@fem.unicamp.br (F.G.D.)

[2] Federal Technological University of Paraná, Ponta Grossa, Paran 84016-210á, Brazil; fernandacorrea@utfpr.edu.br

* Correspondence: javorski@fem.unicamp.br

Received: 22 December 2019; Accepted: 6 March 2020; Published: 11 March 2020

Abstract: For decades, drivers and fleet managers have been impacted by the instability of fuel prices, the need to save resources and the duty to meet and attain environmental regulations and certifications. Aiming to increase performance and efficiency and reduce emissions and mileage costs, plug-in electric vehicles (PHEVs) have been pointed out as a viable option, but there are gaps related to tools that could improve the numerous existing conventional vehicles. This study presents the design of an aftermarket hybridization kit that converts a vehicle originally driven by a combustion engine into a PHEV. To achieve this goal, an optimization was conducted with the objective of decreasing the cost (regarding fuel consumption and battery charging) to perform a local driving cycle, while attenuating the tailpipe emissions and reducing the battery mass. The torque curves of the electric motors, the battery capacity, the parameters for a gear shifting strategy and the parameters for a power split control were the design variables in the optimization process. This study used the Campinas driving cycle, which was experimentally obtained in a real-world driving scenario. The use of a local driving cycle to tune the design variables of an aftermarket optimization kit is important to achieve a customized product according to the selling location. Among the optimum solutions, the best trade-off configuration was able to decrease the mileage cost in 22.55%, and reduce the tailpipe emissions by 28.4% CO, 33.55% NOx and 19.11% HC, with the addition of a 137 kg battery.

Keywords: aftermarket hybridization kit; emissions mitigation; local driving cycle; plug-in hybrid electric vehicles; vehicle efficiency

---

1. Introduction

Local and global environmental issues have resulted in the establishment of regulations to reduce vehicle emissions worldwide [1,2]. On the other hand, the instability of fuel prices leads customers to react by reducing consumption [3] or by changing energy source [4] when it is possible. Plugin hybrid electric vehicles are driven by electric motors (EMs) and internal combustion engines (ICE) and can be connected to the grid [5]. They are one of the solutions for mitigating local emissions [6] and absorbing fuel and electric energy price instabilities.

The relevance of the plug-in electric vehicles (PHEVs) in the current engineering is reflected by the recent scientific activities. Du et al. [7] optimized a plug-in hybrid bus to avoid excessive changes of operation modes, reducing the number of engine startups, the clutch abrasion and the jerk. Golpîra and Khan [8] pointed that the increasing use of PHEVs is an important issue to be taken into account in domestic energy power management, while Li et al. [9,10] investigated the advantages for the power distribution systems due to the optimum coordination of the PHEVs charging times. Furthermore, Mohammadi et al. [11] asserted that if the PHEVs are able to work in a bidirectional power mode (draining power from and supplying power to the grid), they can collaborate to regulate the grid voltage and frequency along a day. Zhang et al. [12] studied the integration between PHEVs, other kinds of electric vehicles and the road to achieve global ecodriving conditions. Liu et al. [13] proposed a mixed mutation strategy for a multiobjective optimization algorithm and demonstrated its effect by applying it on a PHEV problem. Wang et al. [14] optimized a 4-wheel drive PHEV considering the Federal Test Procedure (FTP) 72 driving cycle and achieved good results in three antagonistic objectives: −1.21% electric energy consumption, −6.18% fuel consumption and −5.49% acceleration time. Fu et al. [15] optimized a front traction PHEV under the New European Driving Cycle (NEDC), resulting in reductions about 3.45% fuel consumption, 6.15% NOx, 3.71% CO, and 3.57% HC.

Although in general, research is focused on the production and application of new hybrid vehicles, there are good results regarding the installation of aftermarket hybridization kits on combustion vehicles, turning them into PHEVs. Some of these solutions have already been patented [16–19], and companies are selling and installing such kits [20].

Researchers of the University of Salermo have developed an aftermarket hybridization kit by installing EMs in the rear wheels of a combustion vehicle originally driven only through the front wheels. The batteries of this kit can be loaded by regenerative braking, by plugging the kit to the grid, and by the energy acquired through solar panels installed on the vehicle roof [21]. In the best conditions, the resulting vehicle would have the fuel economy improved by 11% under the NEDC and 20.5% considering the FTP 72 driving cycle [21]. It is expected that 10% of vehicle owners are interested in installing this kit [22]. Furthermore, the kit improves the life cycle assessment of the vehicle in which it is installed, and the overall environmental impact is better than the purchase of a brand-new hybrid vehicle [23].

The Integrated System Laboratory (LabSIn) of the University of Campinas has worked on vehicle longitudinal dynamics issues and on the development of an aftermarket hybridization kit for the Brazilian scenario. The regional market is dominated by conventional vehicles, which are numerous (in 2018 there were 600,000 samples in Campinas City [24]; 18,200,000 in the São Paulo State [25]; and 54,700,000 in the whole country [26]).

Key contributions of LabSIn arose from the addition of two EMs to the rear wheels of a conventional vehicle (front wheel ICE traction); the effects on fuel consumption and on gear shifting strategy were experimentally evaluated [27,28]. Such experimental data were input to virtual analysis regarding the hybridization of a conventional vehicle [29,30]. As a result, suitable configurations were able to decrease by up to 34.18%, the cost of performing the combination of the FTP 75 and US06 driving cycles. However, the mentioned studies did not analyze the hybridization kit's influence on the engine emissions. The increase in the ICE warm-up period is amplified by the addition of the extra electric traction system, which supplies a parcel of the power demand and therefore decreases the ICE torque that can increase the generated emissions if the power split between ICE and EMs is not properly defined [31]. Modifications in the standard driving behavior, e.g., changing the gear shifting strategy, can decrease the ICE fuel consumption and emissions [32,33].

In 2019, LabSIn [34] proposed and simulated an aftermarket hybridization kit shown in Figure 1. The EMs are powered by a battery that is charged by the electrical grid when the vehicle is parked. The optimization was conducted to find out the best gear shifting strategy, EMs torque curves and battery size according to different fuel and electric energy cost scenarios, and the robust solutions regarding cost variations were selected. It also was not considered a local driving cycle that may differ from the standard ones, and may result in further optimized vehicle solutions [35–37].

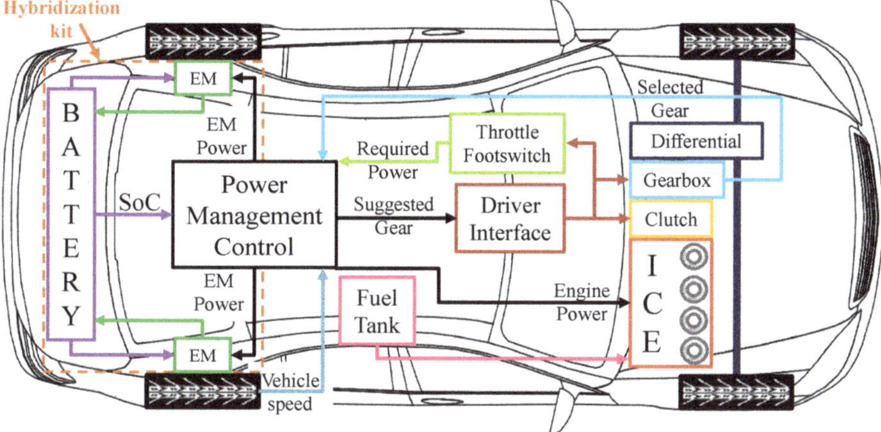

Figure 1. Proposed assembly for the hybridization kit.

Therefore, the aim of the present study is to conduct a multi-objective optimization of the aftermarket hybridization kit. The optimization problem is formulated to decrease the mileage cost (in terms of fuel consumption and battery charging), mitigating the tailpipe emissions and decreasing the battery mass. The selected design variables are the EMs' torque curves, the battery capacity, the parameters of the gear shifting strategy and the parameters of the power split control. Moreover, this study is based on the Campinas driving cycle [38], which was experimentally obtained in a real-world driving scenario. Local driving cycles allow a more realistic analysis of the possible gains of the aftermarket hybridization vehicle, focusing on the local costumer.

2. Simulation Model

In the literature, models of vehicle longitudinal dynamics and vehicle components are often implemented in MATLAB^{TM} [37,39–41]. All the equations presented here have been implemented by the authors by means of MATLAB/Simulink^{TM}block diagrams. In this study, the vehicle longitudinal dynamics are modeled based on the equations presented by Gillespie [42]. However, some modifications were required to include the electric drive train in the model.

The vehicle required traction torque T_{req} (Nm) is defined by Equation (1) as a function of the movement resistance forces caused by the gravitational acceleration g (m/s^2), vehicle mass M (kg) and required acceleration a_{req} (m/s^2), the aerodynamic drag D_A (N), the tires rolling resistance R_x (N) and the climbing resistance that varies according to the road angle α (rad).

$$T_{req} = r(D_A + R_x + M\left(g\sin(\alpha) + a_{req}\right)) \tag{1}$$

The tire dynamic radius r (m) can be defined as proposed by Genta and Morello [43] as 98% of the geometric radius r_g (m), that varies according to the tire standard geometry. The vehicle speed effects are also considered by means of the kv factor [44].

$$r = 0.98r_g(1 + 0.01kv) \tag{2}$$

The rolling resistance R_x (Equation (3)) and the aerodynamic drag D_A (Equation (4)) increase with the vehicle speed V (m/s). Moreover, D_A varies as a function of the air density ρ (kg/m^3), the projection of the vehicle frontal area A (m^2) and the C_D coefficient that represents the vehicle shape.

$$R_x = 0.0981\left(1 + \frac{2.24\,V}{100}\right)M \tag{3}$$

$$D_A = \frac{1}{2}\rho V^2 C_D A \tag{4}$$

The required acceleration a_{req} represents the driver behavior. In this paper, a local driving cycle of the city of Campinas, Brazil [38], is applied (Figure 2). The target speed V_c (m/s) is defined as the cycle speed at one time step Δ_t (s) ahead of the current simulation time. Therefore, the requested acceleration can be calculated by Equation (5).

$$a_{req} = \frac{V_c - V}{\Delta_t} \tag{5}$$

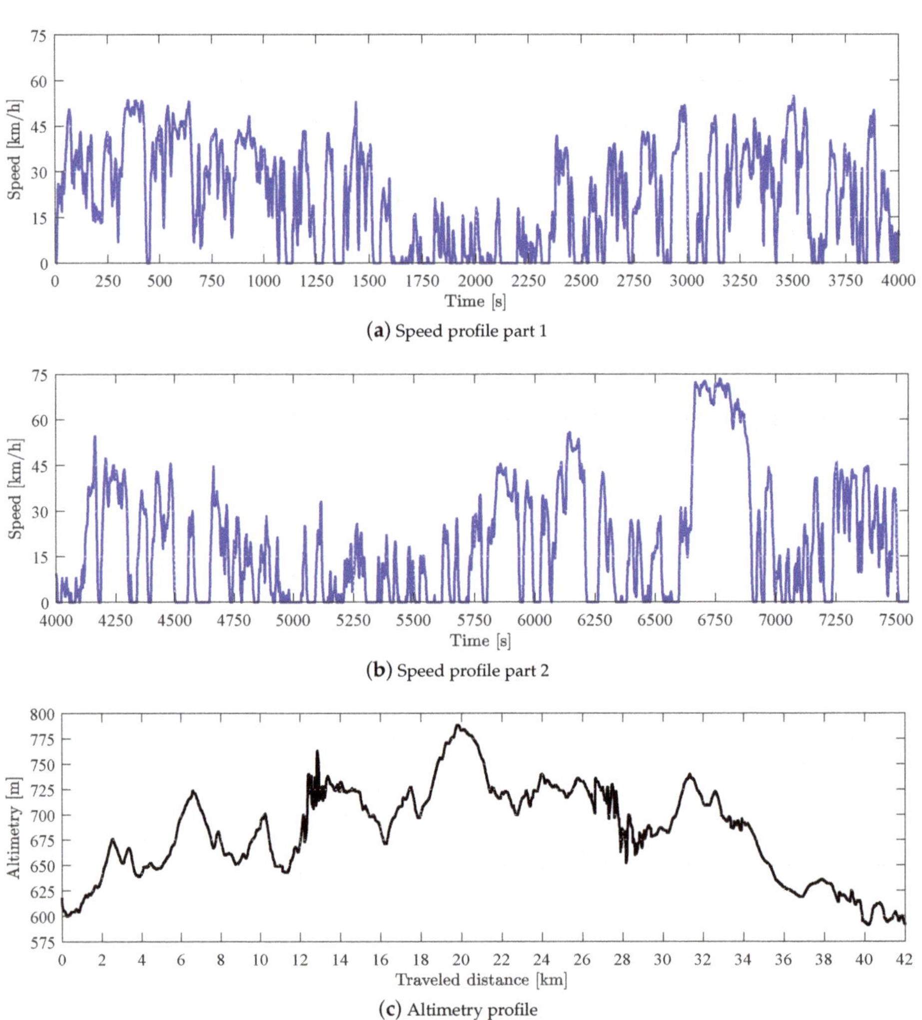

(**a**) Speed profile part 1

(**b**) Speed profile part 2

(**c**) Altimetry profile

Figure 2. Campinas Driving cycle [38].

As aforementioned, the main objective of the aftermarket hybridization kit is to add an extra propelling system in the conventional vehicle. Therefore, the power management control (PMC) defines the traction torque required by the engine T_{eng} (Nm) and the electric motors T_{el} (Nm). In the same way, as presented by Eckert et al. [34], the extra electric motors act only as an auxiliary propelling system. Thus, the throttle pedal signal is acquired and the PMC splits the requested traction torque T_{eng} according to the P_E factor (defined in the optimization problem). The engine torque T_{eng} and torque of the electric motors T_{el} are defined by Equations (6) and (7) respectively, according to the engine inertia I_e (kgm^2); gearbox and differential gear ratios N_t and N_d; inertia I_t (kgm^2) and I_d (kgm^2); the frontal and real wheels inertia I_{wf} (kgm^2) and I_{wr} (kgm^2); and the overall powertrain efficiency η_{td}.

$$T_{eng} = \frac{T_{req}(1 - P_E)}{N_t N_d \eta_{td}} + \left((I_e + I_t)(N_t N_d)^2 + I_d N_d^2 + I_{wf} \right) \frac{a_{req}}{r} \tag{6}$$

$$T_{el} = P_E T_{req} + I_{wr} a_{req} \tag{7}$$

2.1. ICE System Restrictions

Once the required traction torque parcel exerted by the engine is defined, it has to be compared to the available traction torque T_{av} (Nm) that corresponds to the 100% throttle curve (Figure 3a). Moreover, the clutch transmissible torque limit T_{cl} (Nm) (that acts during the gear shifting) also is considered in the definition of the available traction torque.

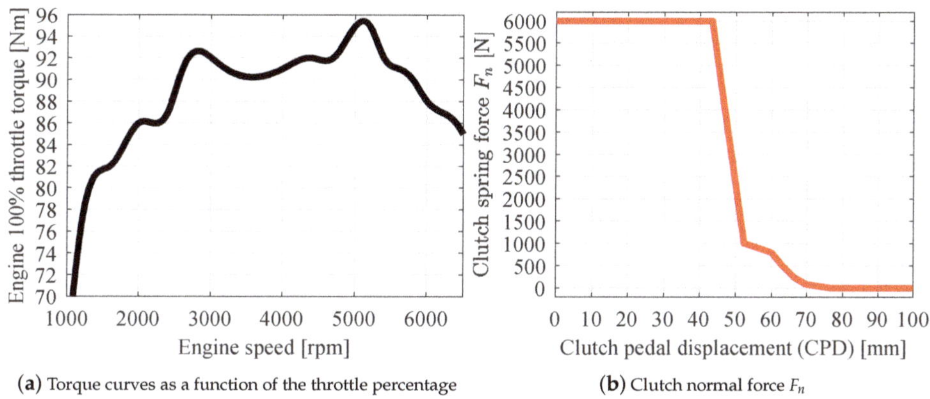

(**a**) Torque curves as a function of the throttle percentage

(**b**) Clutch normal force F_n

Figure 3. Engine torque curve and clutch normal force F_n (N) [45].

The clutch transmissible torque is calculated according to its friction coefficient μ_{cl}, the number of clutch faces N (-) and internal and external disks radii R_i (m) and R_o (m) (see Equation (8)). Moreover, the spring force F_n (N) (Figure 3b) controls the torque transmission according to the clutch pedal displacement (CPD).

$$T_{cl} = \frac{2}{3} \mu_{cl} F_n N \frac{R_o^3 - R_i^3}{R_o^2 - R_i^2} \tag{8}$$

It can be seen in Figure 3b that the clutch maximum F_n value is applied when it is closed (CPD = 0). In this condition, the total engine output torque is transmitted to the gearbox. On the other hand, during the gear shifting process, the clutch disconnects the ICE from the gearbox (CPD = 0 to CPD = 100 mm) in 0.3 s. After the clutch fully opened (CPD = 100 mm), the gear ratio is changed in 0.2 s; then, the ICE is reconnected to the gearbox (CPD = 100 to CPD = 0 mm) in 0.5 s [32]. Therefore, the gearbox input torque T_{gb} (Nm) is defined according to the rules presented in Equation (9) and the available traction torque at the vehicle frontal wheels T_{af} (Nm) is then defined by Equation (10).

$$T_{gb} = \begin{cases} T_{av} \text{ if } T_{cl} > T_{av} \\ T_{cl} \text{ if } T_{cl} \leq T_{av} \end{cases} \tag{9}$$

$$T_{af} = T_{gb}N_tN_d\eta_{td} - \left((I_e + I_t)(N_tN_d)^2 + I_dN_d^2 + I_{wf}\right)\frac{a_{req}}{r} \tag{10}$$

2.2. Electric Drivetrain Restrictions

As the frontal propelling system, the electric drivetrain is also limited by the maximum torque of the EMs curves T_{ev} (Nm) according to their current speed. Therefore, if the required torque T_{el} is higher than the sum of available torques T_{ev} of both in-wheel EMs, there will be performance losses. The available traction torque of the rear-drive system T_{ar} (Nm) is defined by the combination of Equations (11) and (12).

$$T_{rl} = \begin{cases} 2T_{ev} & \text{if } T_{el} > 2T_{ev} \\ T_{el} & \text{if } T_{el} \leq 2T_{ev} \end{cases} \tag{11}$$

$$T_{ar} = T_{rl} - I_{wr}a_{req} \tag{12}$$

2.3. Traction Restrictions

Once the available traction torques T_{af} and T_{ar} are defined, it is possible to verify whether the available torque can be transmitted through the tire/ground contact. The maximum transmissible torque for the frontal $T_{F(max)}$ (Nm) and rear $T_{R(max)}$ (Nm) wheels are defined by Equations (13) and (14) respectively [46]. μ represents the tire-ground peak friction coefficient, and L (m), h (m), b (m) and c (m) correspond respectively to the vehicle's wheelbase; the height of its center of gravity; and the longitudinal distance between its front and rear axles.

$$T_{F(max)} = \mu \left(\frac{Mg\cos\alpha\, c - h\sin\alpha - Mh\, a_{req}}{2L}\right) r \tag{13}$$

$$T_{R(max)} = \mu \left(\frac{Mg\cos\alpha\, b + h\sin\alpha + Mh\, a_{req}}{2L}\right) r \tag{14}$$

Therefore, the effective traction torque of the frontal T_F (Nm) and rear T_R (Nm) propelling system are defined by Equations (15) and (16).

$$T_F = \begin{cases} T_{af} & \text{if } T_{F(max)} \geq T_{af} \\ T_{F(max)} & \text{if } T_{F(max)} < T_{af} \end{cases} \tag{15}$$

$$T_R = \begin{cases} T_{ar} & \text{if } T_{R(max)} \geq T_{ar} \\ T_{R(max)} & \text{if } T_{R(max)} < T_{ar} \end{cases} \tag{16}$$

2.4. Acceleration Iterative Process

With both effective traction torques, it is possible to define the vehicle current acceleration a_x (m/s^2) by means of Equation (17), which is integrated by ODE5 of the Simulink^{TM}, to find out the current vehicle speed V and also the displacement.

$$a_x = \frac{\frac{T_F + T_R}{r} - D_A - R_x - Mg\sin(\alpha)}{M} \tag{17}$$

Due to the many constraints presented, the vehicle acceleration a_x may become lower than the requested a_{req}. Therefore, it is necessary to perform an iterative process among Equations (1), (6), (7) and (9)–(17), considering $a_{req} = a_x$ until the convergence of the a_x value. Moreover, the tire slipping has to be corrected in this iterative process. The coefficient e (Figure 4) estimates the difference between the tire tangential speed and the vehicle displacement speed as a function of the traction torque $T_{F/R}$. This parameter changes the engine and EMs speeds, and therefore the maximum available torques according to the drivetrain respective curves.

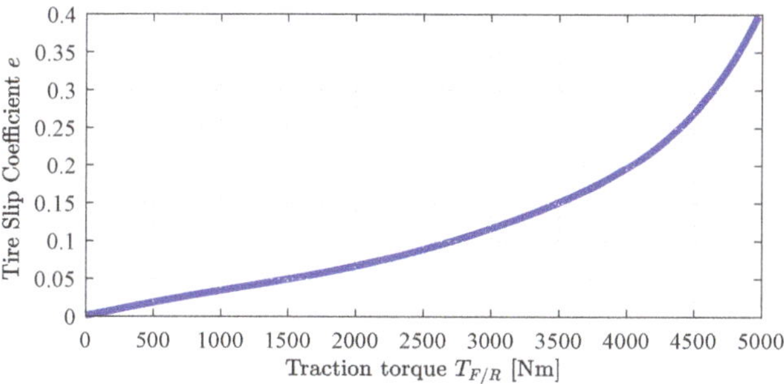

Figure 4. Tire slip coefficient in function of the traction torque $T_{F/R}$.

2.5. Fuel Consumption and Emissions

After the convergence, the engine effective torque T_e (Nm) and speed ω_e are defined by the Equations (18) and (19).

$$T_e = \frac{F_t r}{N_t N_d \eta_{td}} + ((I_e + I_t)(N_t N_d)^2 + I_d N_d^2 + I_w)\frac{a_x}{r} \tag{18}$$

$$\omega_e = \frac{V N_d N_t}{r(1-e)} \tag{19}$$

These values correspond to the engine operation point, and are input to the ADVISOR^{TM} fuel converter block, which defines the ICE fuel consumption F_C (l) and the tailpipe emissions HC (g/km) NOx (g/km) and CO (g/km) according to the ICE maps shown in Figure 5. ADVISOR^{TM} [47] is a free vehicular simulation tool that considers the ICE transient regimes, simulating the combustion heat and its influence in the catalyst efficiency that significantly changes the generated tailpipe emissions [32].

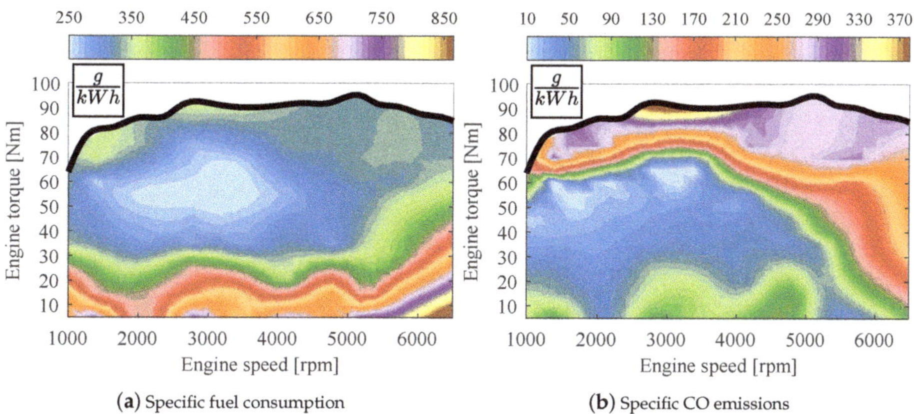

(**a**) Specific fuel consumption

(**b**) Specific CO emissions

Figure 5. *Cont.*

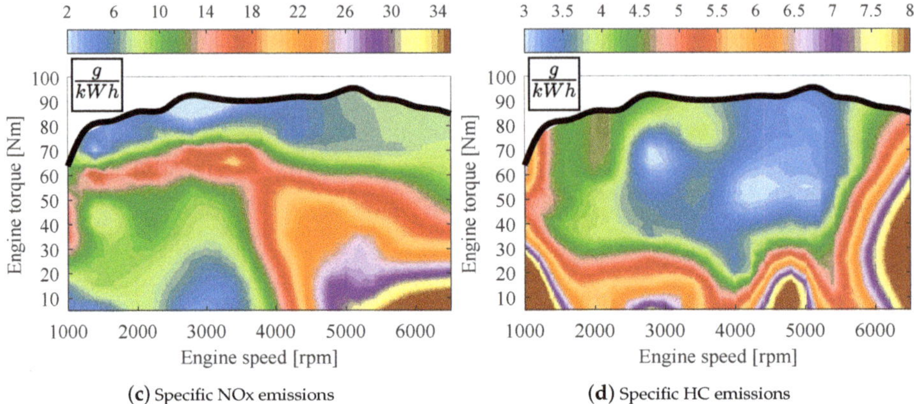

(**c**) Specific NOx emissions   (**d**) Specific HC emissions

Figure 5. Engine maps for fuel consumption and emissions [32,47].

2.6. Battery and Equivalent Fuel Consumption

As initially proposed by Eckert et al. [34], the hybridization kit contains a lead-acid battery pack with a nominal voltage of 48 V. This voltage value was selected because the hybridization kit must be adapted to conventional vehicles. Electrical systems above 60 V can represent a potential danger to the passenger [48], and therefore require a shock protection system [49] that would increase the cost of the kit. Furthermore, the lead-acid batteries were selected due to their maintenance-free, good recycling capacity [50] associated with easy manufacture and high-volume production [51] that decrease the final cost of the kit.

In the simulations, the lead-acid battery block from Simulink^{TM}database [34] is used. The developed control stops the use of the auxiliary electric drive train system when the battery state of charge *SoC* is under 40% to avoid fatigue caused by deep discharge [34]. Aiming to define the best battery pack for the hybridization kit, its capacity B_c (Ah) has to be defined and included as input to the Simulink^{TM}battery block. Moreover, the battery mass M_{bat} (kg) is also calculated by Equation (20) as a function of the lead-acid specific energy S_E = 40 (Wh/kg) and added to the vehicle mass M. Thus, the resulting M_{bat} is added to the vehicle overall mass M.

$$M_{bat} = \frac{V_{bat}B_c}{S_E} \tag{20}$$

Once the driving cycle is simulated, the final battery *SoC* is used as an input to the charge simulation, which infers the cost of recharging the battery from the electric grid. As in Eckert et al. [34], the battery is recharged at constant current [52]—the maximum I (A) value that does not overcome the allowed battery voltage V_{bat} (V). The power consumption P_c (W) to return the battery *SoC* to 100% is defined by Equation (21), considering a charge efficiency η_c that concerns the conversion from AC to DC and the heat losses.

$$P_c = \eta_c I V \tag{21}$$

Differences in the cost of the energy sources may impact the result of the optimization of vehicle solutions [53]. To allow a fair comparison between the conventional vehicles and the hybridized one, the cost of battery charge is converted in a equivalent fuel consumption F_{eq} (l) by Equation (22) according to the charging time C_t (s) and the ratio between the fuel F_{cost} ($/l) and electric energy E_{cost} (\$/Ws) costs. Finally, the F_{eq} is added to the engine fuel consumption F_C, resulting in the total fuel consumption F_T (l) used to compare the results.

$$F_{eq} = P_c C_t \frac{E_{cost}}{F_{cost}} \tag{22}$$

$$F_T = F_C + F_{eq} \tag{23}$$

2.7. Electric Motors

One of the goals of this study was to find out optimum in-wheel EMs to be applied in the aftermarket hybridization kit; a generic torque curve and efficiency map are presented in Figure 6 [34,54,55]. This approach allows for the definition of a theoretical torque curve, based on three main points. The first one is the EM maximum torque T_{max} (Nm) and its respective speed ω_{Tc} (rad/s) at the constant torque phase. The second point is defined at the constant power phase, according to Tong [56], which defines the EM best operating region between $0.1T_{max}$ and $0.3T_{max}$. In this paper, we adopt the upper limit of $0.3T_{max}$, as shown in Equation (24), and the EM speed of this point is defined by Equation (25). Finally, the last torque curve point represents the speed at which the EM reaches null torque, and it is defined by linear progression of the previously defined points (T_{max}, ω_{Tc}) and (T_{Pc}, ω_{Pc}).

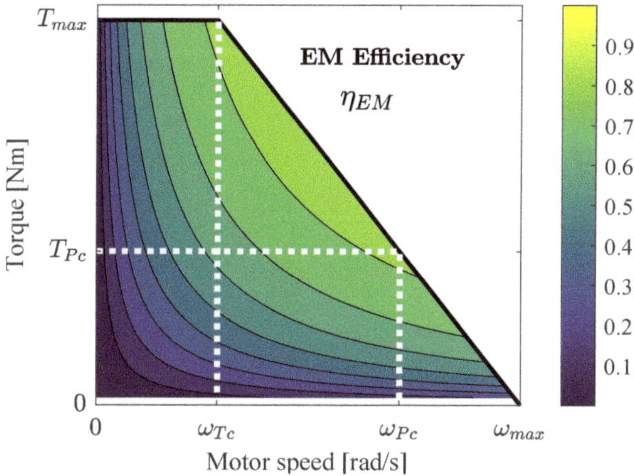

Figure 6. Electric motor efficiency map and torque curve [34,54,55].

$$T_{Pc} = 0.3T_{max} \tag{24}$$

$$\omega_{Pc} = \frac{T_{max}\omega_{Tc}}{T_{Pc}} \tag{25}$$

With the EM torque curve fully defined, the map that defines the EM efficiency η_{EM} is interpolated based on the data presented in Figure 6. Moreover, the EM inertia is interpolated based on some values presented by Corrêa et al. [57]. Finally, the inverter efficiency η_{inv} is obtained from Table 1 and the electric current I (A) results from the Equation (26).

$$I = \frac{T_R V}{r(1-e)\, V_{bat}\, \eta_{EM}\, \eta_{inv}} \tag{26}$$

Table 1. Inverter efficiency η_{inv} map adapted from [58].

T_{EM} [Nm]	ω_{EM} [rad/s]				
	$0.1\omega_{Pc}$	$0.3\omega_{Pc}$	$0.5\omega_{Pc}$	$0.7\omega_{Pc}$	ω_{Pc}
0	0.65	0.84	0.9	0.84	0.83
$0.11T_{max}$	0.74	0.89	0.94	0.91	0.91
$0.33T_{max}$	0.82	0.93	0.96	0.96	0.96
$0.56T_{max}$	0.83	0.94	0.97	0.97	0.97
T_{max}	0.83	0.94	0.97	0.97	0.97

The EMs are able to regenerate some part of the kinetic energy during the braking. In this paper, the regenerative braking is limited to 10% of the EM maximum torque [54,59]. If the required braking torque exceeds the bounded, the remaining parcel is dissipated by the conventional brake of the vehicle [59].

2.8. Vehicle Parameters

The simulated vehicle is based on the 1.0L engine frontal traction Brazilian Chevrolet Celta™. The parameters used to model this conventional vehicle and the aftermarket hybridization kit in the simulations are presented in Table 2.

Table 2. Simulated vehicle parameters [44,45].

Components	Gearbox Position				
	1st	2nd	3rd	4th	5th
Gear ratio (N_t)	4.27	2.35	1.48	1.05	0.8
Gear inertia (I_t) [kgm^2] $\times 10^{-3}$	1.7	2.2	2.9	3.9	5.4
Engine inertia (I_e) [kgm^2]	0.1367				
Differential inertia (I_d) [kgm^2]	9.22×10^{-4}				
Wheels inertia (I_w) [kgm^2]	2				
Differential ratio (N_d)	4.87				
Powertrain efficiency (η_{td})	0.9				
Total vehicle mass (M) [kg]	980				
Vehicle frontal area (A) [m^2]	1.8				
Drag coefficient (C_d)	0.33				
Tires 175/70 R13 radii (r_g) [m]	0.2876				
Tire peak friction coefficient (μ)	0.9				
Wheelbase (L) [m]	2.443				
Gravity center height (h) [m]	0.53				
Rear axle to gravity center (c) [m]	1.460				
Clutch friction coefficient (μ_{cl})	0.27				
Clutch external radius (R_o) [mm]	95				
Clutch internal radius (R_i) [mm]	67				
Number of clutch faces (N)	2				
Vehicle Speed V [m/s]	0	16.67	25	33.33	41.67
$k_v(V)$ Factor	0	0	0.1	0.2	0.4
Electric motors power [kW]	5	7	12	20	30
Electric motors inertia [kgm^2]	0.1	0.13	0.2	0.24	0.3

2.9. Gear Shifting Strategy

Finally, the simulation parameters related to the applied gear shifting strategy have to be defined; they significantly influence the energy management [60], the vehicle performance, the fuel consumption and the emissions [32,33]. The use of an adequate gear shifting strategy improves the gains reached by the vehicle hybridization, once the electric motors decrease the ICE torque demand, allowing for the anticipation of the upshifts, moving the engine to a lower fuel consumption/better efficiency operation point [34]. On the other hand, anticipating the upshifts increases the ICE warm-up period, leading to poor catalyst efficiency and higher tailpipe emissions.

Therefore, the applied shifting strategy has to be carefully defined. The hybridization kit was developed to be adapted to a conventional vehicle a manual gearbox; the most adequate gear will be shown to the driver by means of a user interface adapted to the vehicle. As in previous works [32,34,45], the gear shifting strategy is based on speed limits; in other words, when the vehicle reaches one of these limits, the gear will be changed to the next (upshift) or previous gear (downshift). In the simulation the shifting process is controlled by the vector $\mathbf{V}_{\text{shift}}$ (Equation (27)) that contains the upshift speeds $V_{u(i)}$ (km/h) and the speed interval V_D (km/h) used to define the downshift speeds $V_{d(i)}$ (km/h) (Equation (28)). This difference between $V_{u(i)}$ and $V_{d(i)}$ is necessary to prevent gear shifting instabilities, as highlighted by Xi et al. [61].

$$\mathbf{V}_{\text{shift}} = [V_{u1} \quad V_{u2} \quad V_{u3} \quad V_{u4} \quad V_D] \tag{27}$$

$$V_{d(i)} = V_{u(i)} - V_D \tag{28}$$

The gearbox transmission ratio N_t is then defined according to the shifting rule present in Equation (29).

$$N_t = \begin{cases} \text{Neutral} & \begin{cases} 0 \text{ if } V = 0 \end{cases} \\[2ex] \text{Upshift} & \begin{cases} N_{t1} \ (1^{st} \text{ gear}) \text{ if } 0 < V \le V_{u1} \\ N_{t2} \ (2^{nd} \text{ gear}) \text{ if } V_{u1} < V \le V_{u2} \\ N_{t3} \ (3^{rd} \text{ gear}) \text{ if } V_{u2} < V \le V_{u3} \\ N_{t4} \ (4^{th} \text{ gear}) \text{ if } V_{u3} < V \le V_{u4} \\ N_{t5} \ (5^{th} \text{ gear}) \text{ if } V > V_{u4} \end{cases} \\[5ex] \text{Downshift} & \begin{cases} N_{t1} \ (1^{st} \text{ gear}) \text{ if } 0 < V \le V_{d1} \\ N_{t2} \ (2^{nd} \text{ gear}) \text{ if } V_{d1} < V \le V_{d2} \\ N_{t3} \ (3^{rd} \text{ gear}) \text{ if } V_{d2} < V \le V_{d3} \\ N_{t4} \ (4^{th} \text{ gear}) \text{ if } V_{d3} < V \le V_{d4} \\ N_{t5} \ (5^{th} \text{ gear}) \text{ if } V > V_{d4} \end{cases} \end{cases} \tag{29}$$

3. Optimization

As mentioned before, the hybridization kit is composed of two in-wheel electric motors assembled directly in the vehicle rear wheels and powered by a lead-acid battery. The main objective of this study is, focusing on a local drive cycle (Campinas, Brazil), to define the best configuration for the battery and electric motors and also the power split control and gear shifting strategy associated with each proposed configuration. To do that, a multi-objective optimization problem is formulated and solved by means of the interactive adaptive-weight genetic algorithm (i−AWGA) proposed by Gen, Cheng and Lin [62]. The i−AWGA technique performs a wide search for the best solution, and it is not limited to false minimums, as occurs in some other optimization techniques. Moreover, this method was applied in several previous works, regarding vehicle multi-objective optimization [32–34,59], reaching satisfactory results.

3.1. Problem Formulation

The first optimization criterion $f_1(\mathbf{X})$ is the minimization of the total fuel consumption F_T (Equation (23)) that represents the ICE fuel consumption and the conversion of the battery charge cost in hypothetical fuel (Equation (22)).

The second optimization criterion $f_2(\mathbf{X})$ is the minimization of the engine carbon monoxide (CO (g/km)), nitrogen oxide (NOx (g/km)) and hydrocarbon (HC (g/km)) emissions. These emissions are combined in a single parameter named emission factor E_F [32,33]. The E_F value is defined by the adaptive-weight technique [62] according to Equation (30) as a function of the maximum and minimum values of each gas emission presented in the solution database.

$$E_F(\mathbf{X}) = \frac{CO(\mathbf{X}) - CO_{mim}}{CO_{max} - CO_{mim}} + \frac{NOx(\mathbf{X}) - NOx_{mim}}{NOx_{max} - NOx_{mim}} + \frac{HC(\mathbf{X}) - HC_{mim}}{HC_{max} - HC_{mim}} \tag{30}$$

Finally, the third optimization criterion $f_3(\mathbf{X})$ is the minimization of the battery mass M_{bat} that represents the majority of the extra weight added by the hybridization kit to the vehicle and reduces the available space in the vehicle trunk where it is meant be assembled.

Therefore, the optimization criteria are:

$$f_1(\mathbf{X}) = \min\left(F_T(\mathbf{X})\right) \tag{31}$$

$$f_2(\mathbf{X}) = \min\left(E_F(\mathbf{X})\right) \tag{32}$$

$$f_3(\mathbf{X}) = \min\left(M_{bat}(\mathbf{X})\right) \tag{33}$$

subjected to the constraints C presented in Equation (34).

$$C = \begin{cases} 20\text{ Ah} \leq B_C \leq 150\text{ Ah} \\ 5\% \leq P_E \leq 95\% \\ 250\text{ rpm} \leq \omega_{Tc} \leq 150\text{ rpm} \\ 10\text{ Nm} \leq T_{max} \leq 200\text{ Nm} \\ V_{u1} < V_{u2} < V_{u3} < V_{u4} \\ 1\text{ km/h} \leq V_D \leq 10\text{ km/h} \\ \omega_{min} \leq \omega_e \leq \omega_{max} \\ R \geq 0.9997 \\ CO \leq 40.53\text{ g/km} \\ NOx \leq 6.08\text{ g/km} \\ HC \leq 7.64\text{ g/km} \end{cases} \tag{34}$$

Besides the EMs and battery parameters used to narrow the optimization process, some extra constraints were included to ensure the correct behavior of the vehicle. The first one is the ICE operation range, which cannot be below the idle speed ($\omega_{min} \approx 84$ rad/s) and cannot overcome the maximum allowed speed ($\omega_{max} \approx 680$ rad/s).

The final constraints are included to avoid any configuration that presents poor acceleration performance caused by the fuel consumption minimization, and to prevent configurations that increase the tailpipe emissions due to the ICE warm-up delay. To be considered viable, the PHEV configuration cannot generate more CO, NOx and HC emissions than the conventional vehicle. Moreover, the vehicle performance is evaluated as the correlation coefficient R calculated by the Equation (35) [30,63] in which the simulated speed profile V_s is compared to the standard driving cycle V_{cs} in discrete time steps of 0.1 s, and the analyzed PHEV must reach at least the same performance of the conventional vehicle $R = 0.9997$, using a standard gear shifting strategy proposed by its manufacturer [64].

$$R = \sqrt{\frac{\left(\sum \left(V_{cs} - \bar{V}_c\right)\left(V_s - \bar{V}\right)\right)^2}{\sum \left(V_{cs} - \bar{V}_c\right)^2 \sum \left(V_s - \bar{V}\right)^2}} \tag{35}$$

3.2. Genetic Algorithm

The optimization starts with the definition of chromosomes $\mathbf{X}$ (Equation (36)) that constrain the design variables of a specific potential solution.

$$\mathbf{X} = \begin{bmatrix} B_C & P_E & \omega_{Tc} & T_{max} & \mathbf{V_{shift}} \end{bmatrix} \tag{36}$$

The initial solution database is composed of randomly defined chromosomes (considering the constrains C) that are simulated and included in the population if they reach the minimum required performance.

3.2.1. Selection, Crossover and Mutation

Once an initial solutions population is developed, the results are classified by the i$-$AWGA technique (Equation (37)), which compares the current result of each optimization criterion $f_k(\mathbf{X})$ with the maximum f_k^{max} and minimum f_k^{min} values of the analyzed criterion presented in the population. Moreover, a penalty value $P_p = 0$ is added to the dominated solutions, aiming to decrease their probability of reproduction. On the other hand, the non-dominated solutions (Pareto frontier) receive $P_p = 1$ value, which increases their fitness value Ft($\mathbf{X}$), improving their selection probability as a function of the sum of all $Ft(\mathbf{X})$ values according to the population size PM as shown by Equation (38).

$$Ft(\mathbf{X}) = \sum_{k=1}^{3} \frac{f_k^{max} - f_k(\mathbf{X})}{f_k^{max} - f_k^{min}} + P_p(\mathbf{X}) \tag{37}$$

$$S_P(\mathbf{X}) = \frac{Ft(\mathbf{X})}{\sum_{X=1}^{PM} Ft(\mathbf{X})} \tag{38}$$

The crossover process starts with the selection of two members of the population, which are named Member 1 (M_1) and Member 2 (M_2). The values of the design variables of these selected chromosomes are then randomly combined, generating a new chromosome denominated $\mathbf{X_{cr}}$ which is simulated. It is included in the population if it respects the constraints C and if the minimum performance criteria are reached.

To provide diversity among the values of the design variables present in the population, the mutation process randomly changes some of the chromosomes values of the selected members and of the crossover chromosome $\mathbf{X_{cr}}$, generating three new chromosomes $\mathbf{X_{mt1}}$, $\mathbf{X_{mt2}}$ and $\mathbf{X_{mtc}}$ that are evaluated by the constrains C. Table 3 shows the mutation rules according to the operator $0 \leq Mut \leq 1$ defined by the Matlab^{TM} function *rand*.

Table 3. Mutation operator.

Initial Chromosome	Mutation Operator $0 \geq Mut \geq 1$		Mutated Chromosome
$[\mathbf{X}_{M1}]^T, [\mathbf{X}_{M2}]^T, [\mathbf{X}_{cr}]^T$	$Mut < 0.5$	$Mut \geq 0.5$	$[\mathbf{X}_{mt1}]^T, [\mathbf{X}_{mt2}]^T, [\mathbf{X}_{mtc}]^T$
B_C [Ah]	$B_{mut} = 0$	$-10 \leq B_{mut} \leq 10$	$B_C + B_{mut}$
P_E [%]	$P_{mut} = 0$	$-10\% \leq P_{mut} \leq 10\%$	$P_E + P_{mut}$
ω_{Tc} [rpm]	$\omega_{mut} = 0$	$-100 \leq \omega_{mut} \leq 100$	$\omega_{Tc} + \omega_{mut}$
T_{max} [Nm]	$T_{mut} = 0$	$-20 \leq T_{mut} \leq 20$	$T_{max} + T_{mut}$
$V_{u(i)}$ [km/h]	$V_{mut(i)} = 0$	$-2 \leq V_{mut(i)} \leq 2$	$V_{u(i)} + V_{mut(i)}$
V_D [km/h]	$V_{Dmut} = 0$	$-1 \leq V_{Dmut} \leq 1$	$V_D + V_{Dmut}$

3.2.2. Population Control and Convergence Criterion

The initial population is composed of 100 members (randomly generated chromosomes), which are combined by the crossover and mutation operators. The addition of new solutions increases the population size until it reaches a limit value $P_{lim} = 200$. Once the population limit is reached, the worst solutions (higher Pareto ranking) are eliminated from the population. If a population composed of exclusively non-dominated solutions (Pareto frontier) reaches P_{lim}, the limit is increased $P_{lim} = P_{lim} + 100$ to avoid the elimination of the whole population.

The convergence is defined by the stagnation of the evolution process, which is characterized by the repetition of the Pareto frontier [33,59,65] for over 20 generations of 80 new simulated solutions (crossover and mutation).

Figure 7 shows the optimization flowchart according to the applied method.

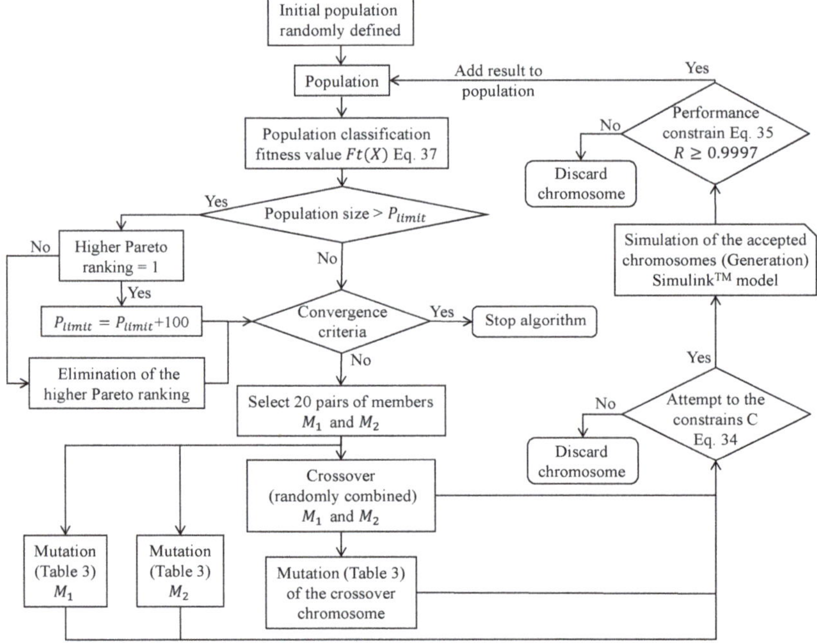

Figure 7. Optimization flowchart.

4. Results and Discussion

After the convergence of the algorithm, the Pareto frontiers of non-dominated solutions are defined as optimum configurations with different compromise among the optimization criteria. Among these solutions, some of them are selected to be analyzed: the minimum energy cost solution min f_1 and the minimum emissions min f_2, the minimum battery min f_3. Moreover, the min f_1 solution presents the best trade-off (max Ft value) among the analyzed criteria.

However, the min f_1 (best trade-off) and min f_2 present large batteries (up to $\approx$144 kg), representing the advantages of the electric powertrain system against the combustion one. On the other hand, the minimum battery min f_3 solution was able to save fuel, but the tailpipe emissions were close to the ones of the conventional vehicle. In addition to these solutions, the Pareto frontier presents several other optimum solutions, and two of them were selected based on the decrease of the battery mass (criterion f_3). These extra solutions were named $Toff_{1.5}$ (making $Ft(f_3)$ 50% more valuable: $1.5 \times Ft(f_3)$) and $Toff_2$ (making $Ft(f_3)$ 100% more valuable: $2 \times Ft(f_3)$), increasing the importance of the Ft value related to the battery mass minimization, selecting the lighter hybridization kit configurations.

Table 4 shows the results of the selected configurations and their respective chromosomes. Moreover, Figure 8 shows the Pareto frontier and highlights the selected configurations.

Table 4. Optimized chromosomes and results.

Solutions		Standard ICE Vehicle	Minimum Cost $(\min(f_1))$	Minimum Emissions $(\min(f_2))$	Minimum Battery $(\min(f_3))$	Trade-Off Solutions 1.5 $Ft(f_3)$ $(Toff_{1.5})$	2 $Ft(f_3)$ $(Toff_2)$
Results	F_C [l]	3.68	2.44	2.46	3.48	2.73	3.04
	F_{eq} [l]	- - -	0.41	0.42	0.06	0.27	0.12
	F_T [l] (f_1)	3.68	2.85	2.88	3.54	2.99	3.16
	CO [g]	40.53	29.02	26.80	39.30	30.89	40.18
	NOx [g]	6.08	4.04	3.93	5.94	5.00	5.74
	HC [g]	7.64	6.18	6.25	7.52	6.79	7.29
	E_F (f_2)	3.11	0.31	0.15	2.87	1.31	2.68
	SoC [%]	- - -	40.06	41.23	49.77	40.03	41.21
	M_{bat} [kg] (f_3)	- - -	137.37	143.97	24.14	89.50	42.55
	R	0.9997	0.9998	0.9997	0.9997	0.9997	0.9998
	Ft	- - -	3.00	2.97	2.24	2.87	2.58
Chromosomes	B_C [Ah]	- - -	114.48	119.97	20.12	74.58	35.45
	P_E [%]	- - -	66.20	55.40	8.22	37.17	16.31
	ω_{Tc} [rpm]	- - -	254.35	253.48	378.9	254.30	255.69
	T_{max} [Nm]	- - -	106.14	139.06	15.42	83.61	43.48
	V_{u1} [km/h]	20	13.65	12.56	11.05	10.24	12.33
	V_{u2} [km/h]	35	23.47	20.71	26.32	23.61	24.14
	V_{u3} [km/h]	70	36.15	37.43	55.00	37.05	37.53
	V_{u4} [km/h]	90	52.88	54.32	57.67	50.89	50.83
	V_D [km/h]	5	8.12	9.41	8.21	9.64	8.77

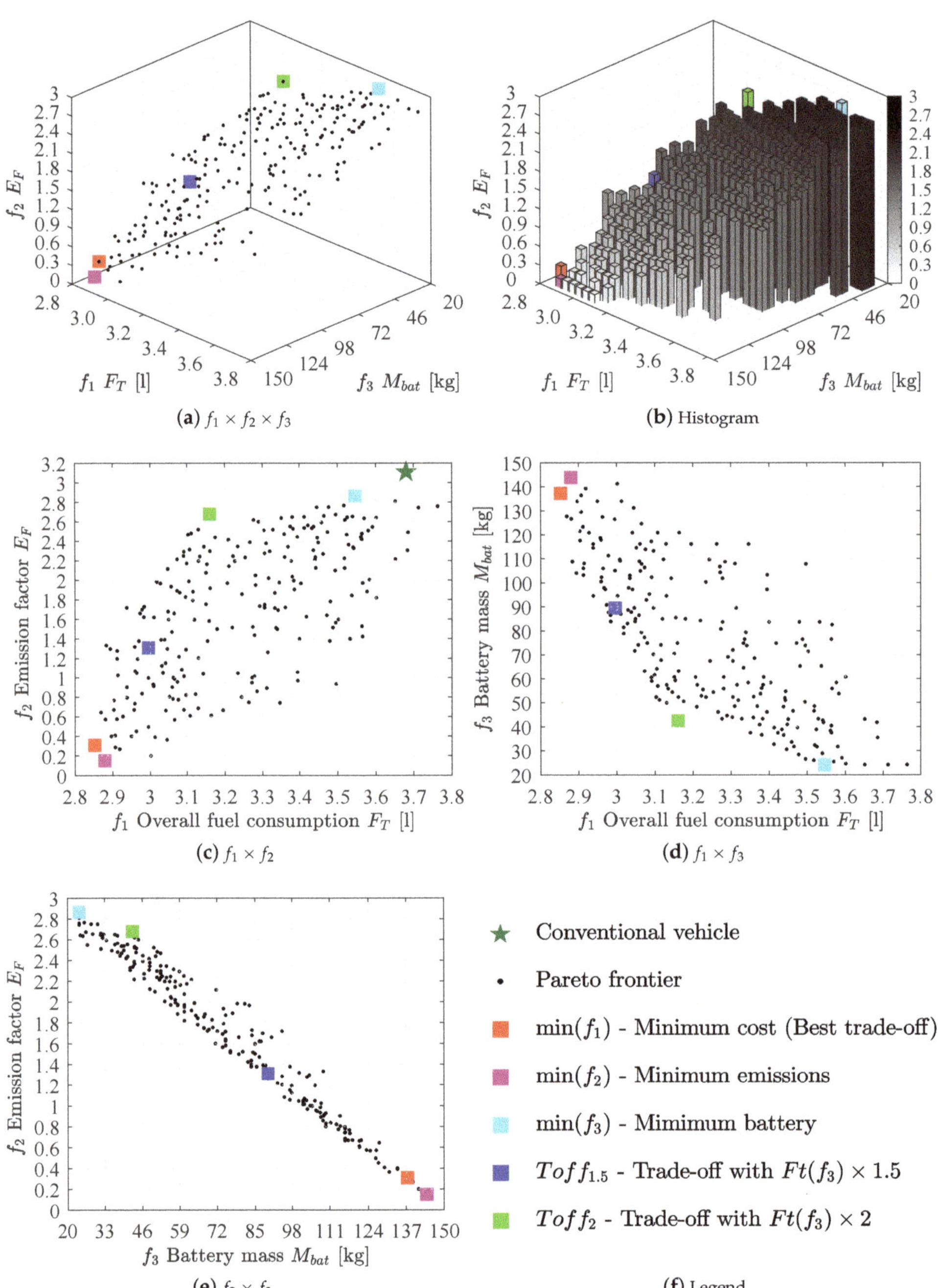

Figure 8. Optimized solutions.

4.1. Standard Vehicle

The standard solution to be compared to the optimized PHEV configurations is the conventional vehicle configuration propelled only by the ICE and using a standard gear shifting strategy proposed by the vehicle manufacturer [64]. Figure 9 shows the ICE running time for each operation point (torque/speed), where it is possible to observe that the ICE presents several points close to the idle/low-speed regime, which is a characteristic of the Campinas driving cycle (Figure 2). The standard gear shifting strategy is shown in Figure 9; it represents the higher speed section of the cycle.

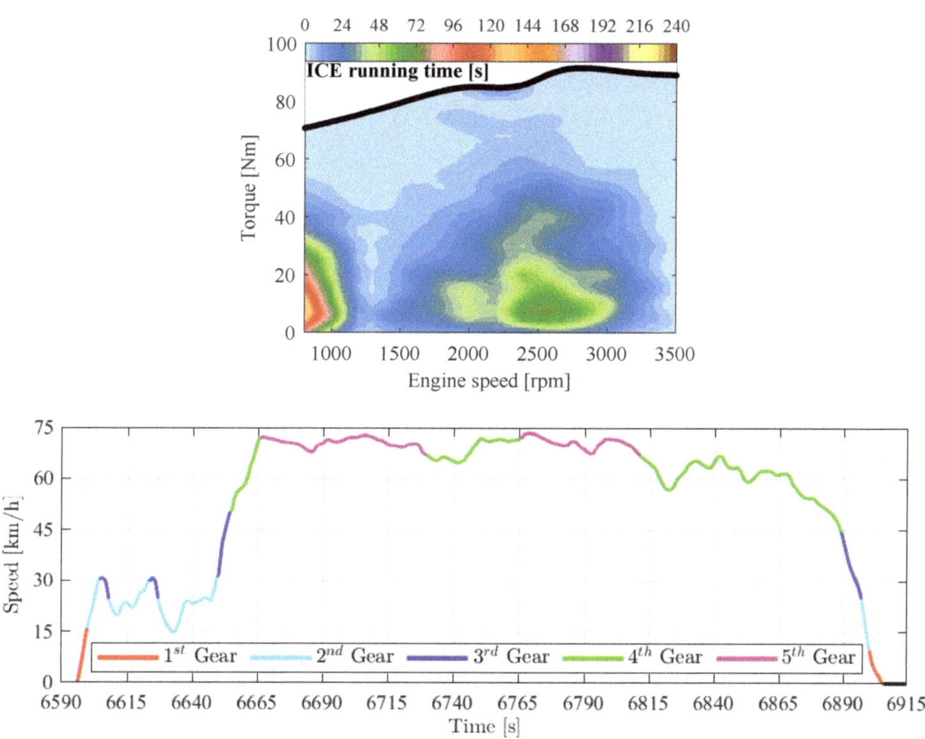

Figure 9. Conventional (internal combustion engine (ICE)) vehicle results.

As it can be seen in Figure 8c and Table 4, the conventional vehicle fuel consumption and resulting emissions are fully dominated by the optimum PHEV configurations.

4.2. Minimum Battery Size Solution min f_3

The minimum battery size solution focuses on minimizing the EMs and battery, resulting in a more compact hybridization kit, with a 24.14 kg battery and low power EMs responsible only for 8.22% of the vehicle power demand. This configuration allows for the upshifting anticipation (Figure 10c) compared to the standard shifting procedure, which saves fuel by moving the ICE operation point to a low-speed/high-torque [32,33], as can be observed in Figure 10a. The EMs operation is presented in Figure 10b where it is possible to observe that in the majority of the time, the electric driving system is used only as a supplementary propelling system, which ensures the vehicle performance even with the upshifting anticipation, and decreases the ICE torque, saving fuel.

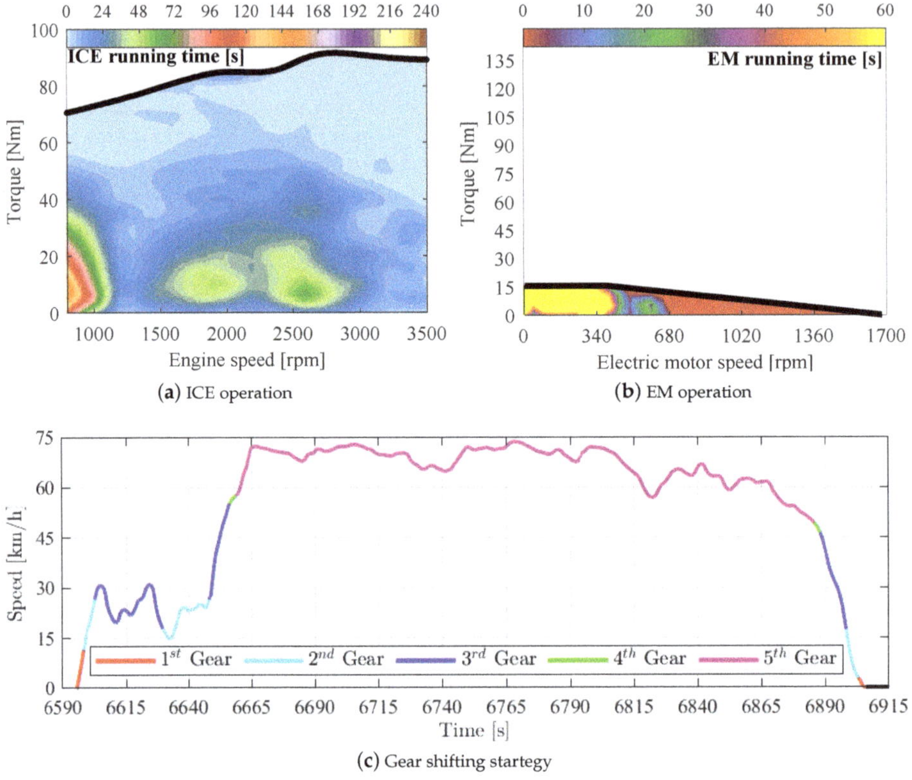

(**a**) ICE operation

(**b**) EM operation

(**c**) Gear shifting startegy

Figure 10. Minimum battery size configuration powertrain behavior.

This configuration was able to reduce the equivalent fuel consumption (F_T) by 3.8%, and present tailpipe emissions close to the standard vehicle results (-3.03% *CO*, -2.3% *NOx* and -1.57% *HC*), as is shown in Table 4.

4.3. Minimum Emissions Solution min f_2

The minimum emission PHEV configuration(min f_2) presents a high level of electrification of the vehicle propelling system, in which the EMs are responsible for 55.4% of the power demand. Therefore, the EMs present high torque, as shown in Figure 11b, in which it is possible to observe that these EMs operate at maximum torque for a large amount of time. Moreover, these high torque EMs also expressively decrease the ICE required torque (Figure 11a), associated with the upshifting anticipation (Figure 11c), keeping the ICE running under ≈ 2700 rpm.

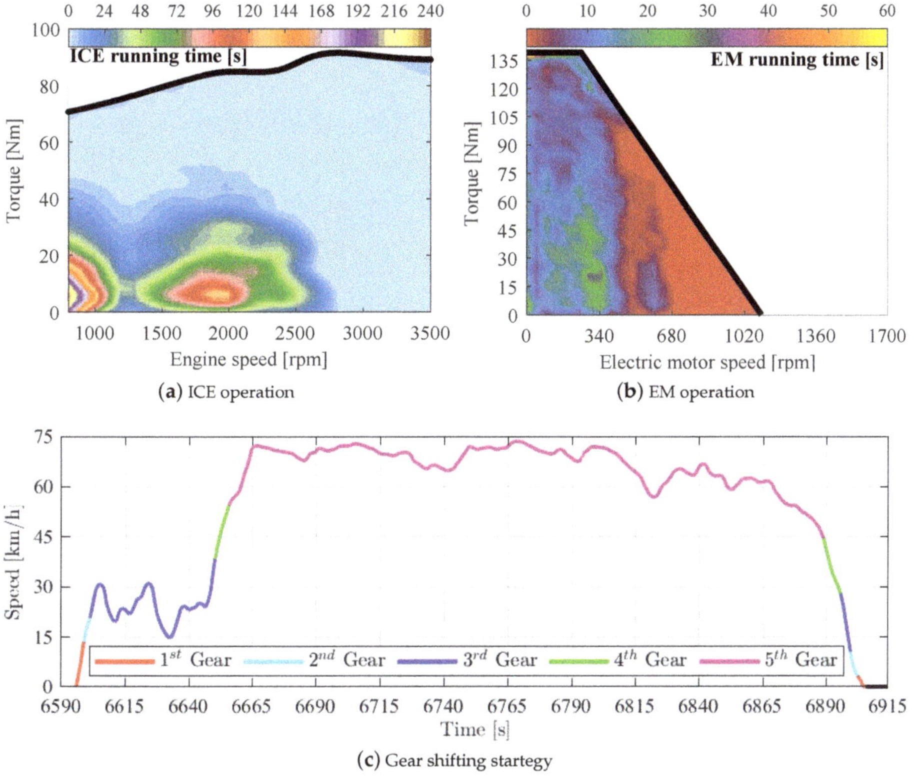

(**a**) ICE operation

(**b**) EM operation

(**c**) Gear shifting startegy

Figure 11. Minimum emissions configuration powertrain behavior.

On the other hand, this configuration needs a large battery of $\approx$144 kg to ensure the availability of the electric propelling system during the 42 km of the analyzed driving cycle. Due to the mentioned minimization of the ICE using, it was possible to save 1.22 L of fuel, compared to the conventional vehicle, which represents 21.74% economy regarding the overall cost to perform the desired path. Due to the heavy use of the electric drivetrain system, this PHEV configuration results in -33.88% CO, -35.36% NOx and -18.19% HC.

Regarding the cost relation between fuel and electricity, the presented results have shown a favorable scenario for vehicle electrification. However, the use of large batteries may lead to other issues, such as a higher cost of the aftermarket hybridization kit, difficulty of assembling the battery in the vehicle and possible interference in vehicle handling [34].

4.4. Minimum Cost Solution min f_1 (Best Trade-Off)

Regarding the optimized scenario, the minimum cost solution min f_1, which represents the minimum equivalent fuel consumption (F_T) among the Pareto frontier is rated as the best-compromised solution (higher fitness value Ft). This PHEV configuration presents EMs that provide up to 105 Nm, which act similarly to the min f_2 EMs, keeping the ICE running at a low speed and torque regime, which together, lead to fuel economy. The electric propelling system is responsible to fulfill 66.20% of the power demand, which results in a high concentration of EM operation points close to the maximum torque curve, as shown in Figure 12b. This configuration was able to decrease the cost to perform the driving cycle in 22.55% and generate -28.4% *CO*, -33.55% *NOx* and -19.11% *HC*—fewer emissions compared to the conventional vehicle.

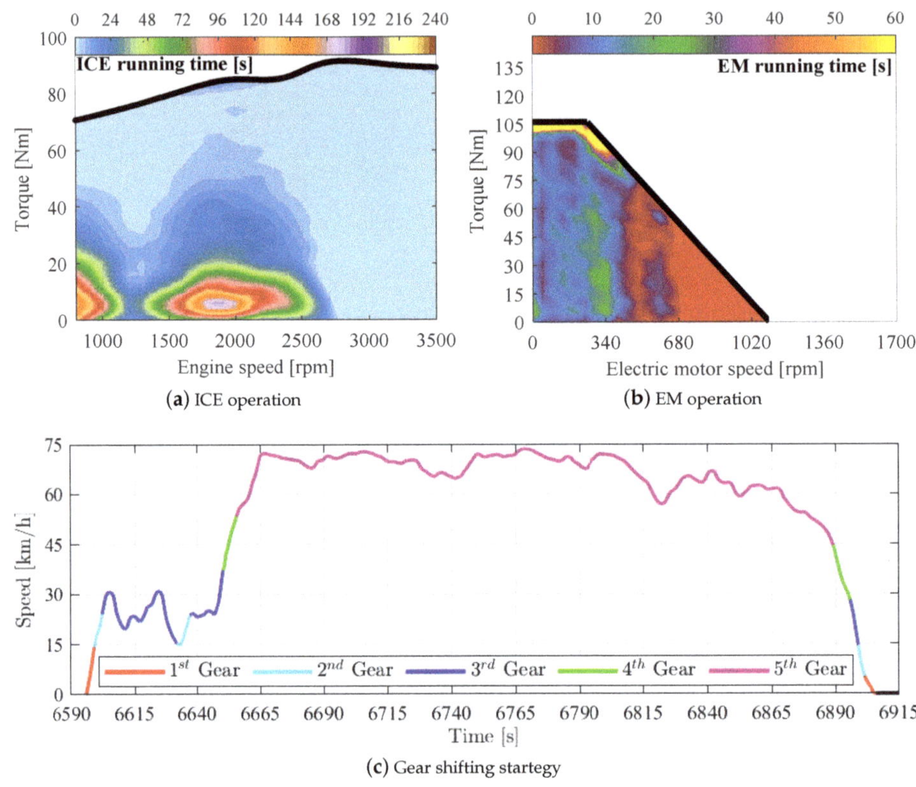

(**a**) ICE operation

(**b**) EM operation

(**c**) Gear shifting startegy

Figure 12. Minimum cost solution powertrain behavior.

Besides the satisfactory reached results regarding cost end emissions of the min f_1 configuration, this aftermarket hybridization kit needs a 137 kg battery, which implies in the same issues listed before regarding large batteries. Therefore, the configurations $Toff_{1.5}$ and $Toff_2$ analyzed below, are included to represent some alternative hybridization kits composed of lighter batteries and EMs, which can be more suitable in cases in which a lower powertrain electrification degree is desired.

4.5. Trade-Off Considering $1.5 \times Ft(f_3)$*: The* $Toff_{1.5}$

The first of the selected trade-off solutions was defined by assigning 50% more weight to the battery mass minimization at Equation (37) that determines the fitness Ft value, used to classify the population. This configuration is powered by a ≈ 90 kg battery and EMs able to perform up to 83 Nm, which operates the majority of the time at maximum torque regime, as shown in Figure 13b. Moreover, the power split control (P_E) was defined as 37.17% due to the lower EMs torque capacity compared to the min f_1 and min f_2 PHEV configurations. Similarly to the previously mentioned configurations, the $Toff_{1.5}$ keeps the ICE running at low torque and speed, which enables 950 mL fuel-saving and an equivalent fuel consumption 18.75% lower compared to the conventional vehicle. Furthermore, the $Toff_{1.5}$ solution also enhanced the ICE emissions, by decreasing 23.78% of the *CO*, 17.76% *NOx* and 11.12% *HC*.

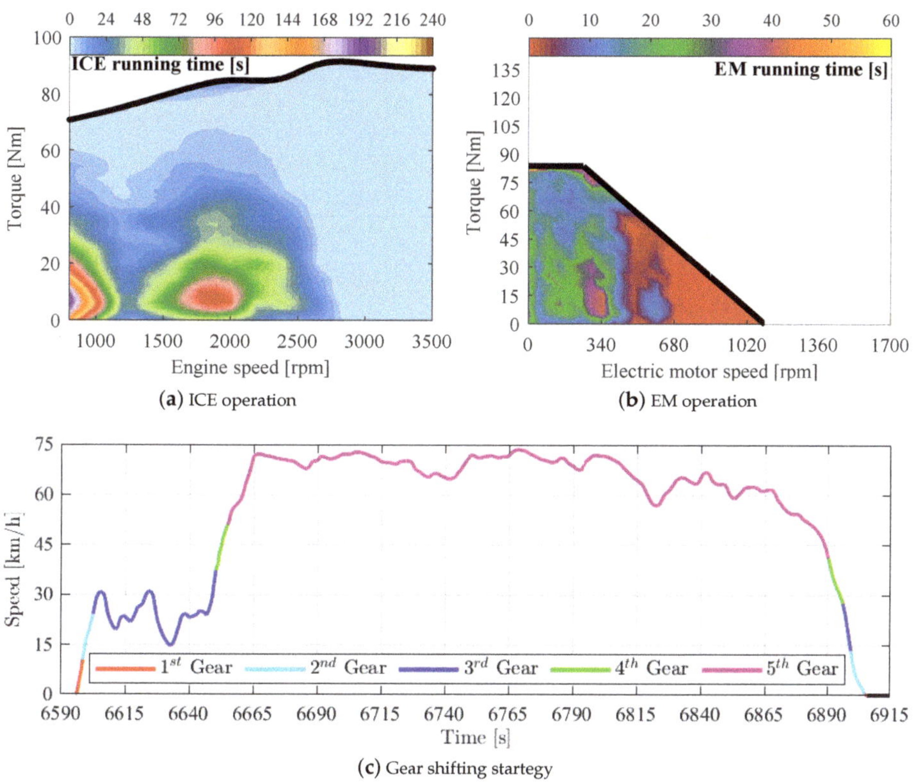

(**a**) ICE operation

(**b**) EM operation

(**c**) Gear shifting startegy

Figure 13. Trade-off configuration ($1.5 \times Ft(f_3)$) powertrain behavior.

4.6. Trade-Off Considering $2 \times Ft(f_3)$: The $To\!f\!f_2$

Finally, the trade-off solution considering the Ft calculated with $2 \times Ft(f_3)$, results in a PHEV configuration with 42.55 kg battery and EMs with 43.48 Nm torque capacity. This configuration decreases the equivalent fuel consumption in 14.13% and the emissions in 0.86% CO, 5.59% NOx and 4.58% HC. The ICE and EMs operation points and the gear shifting strategy are presented in Figure 14.

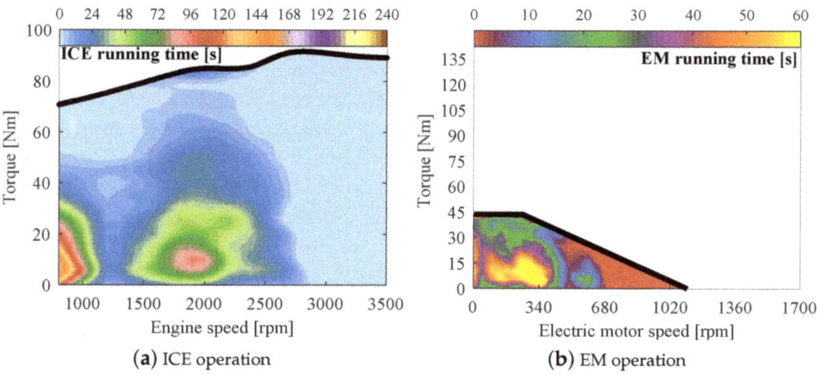

(**a**) ICE operation

(**b**) EM operation

Figure 14. *Cont.*

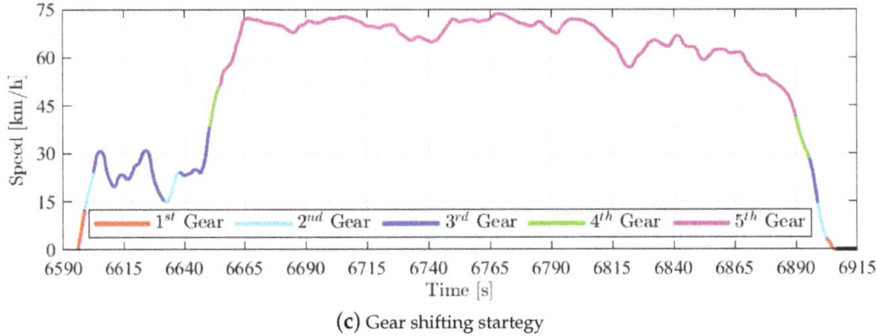

(c) Gear shifting startegy

Figure 14. Trade-off configuration ($2 \times Ft(f_3)$) powertrain behavior.

The $Toff_2$ solution, in the same way as the minimum battery size PHEV configuration ($\min(f_3)$), represents a lower powertrain electrification degree. Even with the lower gains in terms of energy consumption, these configurations may become good alternatives to provide cheaper hybridization kits due to their smaller batteries and electric motors, in addition to their lower interference in the vehicle weight distribution and drivability.

5. Conclusions

In this paper, an aftermarket hybridization kit that converts a vehicle originally driven by a combustion engine into a PHEV was proposed. An optimization problem was formulated, aiming to decrease the overall cost (regarding fuel consumption and battery charging) to perform a local driving cycle, attenuating the tailpipe emissions and reducing the battery mass, which leads to a low price hybridization kit. Some design variables were taken into account; they could change the capacity and the weight of the batteries, the power and the torque of the EMs, the gear shifting strategy and the PMC strategy. Considering the needs of the customer, the optimization of the kit was suitable to a local driving cycle, reflecting the real service loads that it would face.

After the driving cycle was set, an optimization process was conducted by means of the i−AWGA, which was able to find out a Pareto frontier of optimum configurations of the kit. Some of these solutions could be selected, such as the ones that further reduced the travel cost or emissions, resulting in vehicles with a high degree of electrification. These solutions were suitable for customers who were prone to adding more weight to their vehicles and buying a kit with a higher purchase price (due to the cost of large battery packs).

On the other hand, an optimum configuration with the minor battery weight could be found, which would result in a minor purchase price and would be less effective at reducing travel costs and emissions. If battery size was used as a rule, it would be possible to find other optimum solutions in the Pareto frontier to significantly reduce the wanted index but with a non-extravagant additional weight of the kit.

Furthermore, the optimization algorithm led to another aftermarket hybridization kit configuration which presented good trade-off among the optimization criteria. Each of these optimum configurations presented improvements in the overall energy cost (fuel and electricity) and engine emissions according to the applied electrification degree resulting from the assembled electric motors and battery.

Finally, the presented results have shown that it is possible to obtain expressive enhancements by converting a conventional vehicle propelled only by a combustion engine into a PREV. However, this study is only theoretical, and some other steps are necessary to ensure the feasibility of the proposed concept.

Our future work will guide the decisions among all the Pareto solutions. A study regarding the aftermarket hybridization kit cost, the purchase price and the consumer acceptation will contribute to define the kit setup for a certain local. Regarding the emissions, the electric energy mix and the kit material, manufacture and disposal must be taken into account in future analyses to ensure environmental gains. Moreover, the effectiveness of the hybridization kits should be evaluated under a set of different driving scenarios, to ensure the robustness of the designed devices. Finally, the real-time control for vehicle power management control also needs to be developed, aiming to make the hybridization kit feasible.

Author Contributions: Conceptualization, J.J.E. and F.G.D.; methodology, J.J.E. and L.C.d.A.e.S.; software, J.J.E.; formal analysis, J.J.E. and F.M.S.; writing—original draft preparation, J.J.E., F.M.S. and F.C.C.; writing—review and editing, F.M.S., L.C.d.A.e.S. and F.C.C.; visualization, J.J.E.; F.M.S. and F.C.C.; supervision, F.G.D. and L.C.d.A.e.S.; project administration, F.G.D. and L.C.d.A.e.S. All authors have read and agreed to the published version of the manuscript.

Funding: This research received no external funding.

Acknowledgments: The authors wish to thank the State of São Paulo Research Foundation (FAPESP), Federal Technological University of Paraná—UTFPR, Ponta Grossa and the University of Campinas (UNICAMP) for financial support and scholarships.

Conflicts of Interest: The authors declare no conflicts of interest.

References

1. Hooftman, N.; Messagie, M.; Mierlo, J.V.; Coosemans, T. A review of the European passenger car regulations—Real driving emissions vs local air quality. *Renew. Sustain. Energy Rev.* **2018**, *86*, 1–21. [CrossRef]
2. Holjevac, N.; Cheli, F.; Gobbi, M. A simulation-based concept design approach for combustion engine and battery electric vehicles. *Proc. Inst. Mech. Eng. Part D J. Automob. Eng.* **2019**, *233*, 1950–1967. [CrossRef]
3. Shaw, C.; Hales, S.; Edwards, R.; Howden-Chapman, P.; Stanley, J. What can fuel price increases tell us about the air pollution health co-benefits of a carbon price? *J. Transp. Health* **2018**, *8*, 81–90. [CrossRef]
4. Salvo, A. Flexible fuel vehicles, less flexible minded consumers: Price information experiments at the pump. *J. Environ. Econ. Manag.* **2018**, *92*, 194–221. [CrossRef]
5. Mohammadi, F. Design, analysis, and electrification of a solar-powered electric vehicle. *J. Sol. Energy Res.* **2018**, *3*, 293–299.
6. Holjevac, N.; Cheli, F.; Gobbi, M. Multi-objective vehicle optimization: Comparison of combustion engine, hybrid and electric powertrains. *Proc. Inst. Mech. Eng. Part D J. Automob. Eng.* **2020**, *234*, 469–487. [CrossRef]
7. Du, S.; Yang, Y.; Liu, C.; Muhammad, F. Multi-objective real-time optimization energy management strategy for plug-in hybrid electric vehicle. *Proc. Inst. Mech. Eng. Part D J. Automob. Eng.* **2019**, *233*, 1067–1080. [CrossRef]
8. Golpîra, H.; Khan, S.A.R. A multi-objective risk-based robust optimization approach to energy management in smart residential buildings under combined demand and supply uncertainty. *Energy* **2019**, *170*, 1113–1129. [CrossRef]
9. Li, W.; Lin, Z.; Cai, K.; Zhou, H.; Yan, G. Multi-objective optimal charging control of plug-in hybrid electric vehicles in power distribution systems. *Energies* **2019**, *12*, 2563. [CrossRef]
10. Li, W.; Lin, Z.; Zhou, H.; Yan, G. Multi-objective optimization for cyber-physical-social systems: A case study of electric vehicles charging and discharging. *IEEE Access* **2019**, *7*, 76754–76767. [CrossRef]
11. Mohammadi, F.; Nazri, G.A.; Saif, M. A Bidirectional Power Charging Control Strategy for Plug-in Hybrid Electric Vehicles. *Sustainability* **2019**, *11*, 4317. [CrossRef]
12. Zhang, F.; Hu, X.; Langari, R.; Cao, D. Energy management strategies of connected HEVs and PHEVs: Recent progress and outlook. *Prog. Energy Combust. Sci.* **2019**, *73*, 235–256. [CrossRef]
13. Liu, M.; Wang, X.; Sheng, Y.; Wang, L. Improvement of multi-objective differential evolutionary algorithm and its application for Hybrid electric vehicles. In Proceedings of the 2019 Chinese Control And Decision Conference (CCDC), Nanchang, China, 3–5 June 2019; pp. 553–558.
14. Wang, Z.; Cai, Y.; Zeng, Y.; Yu, J. Multi-Objective Optimization for Plug-In 4WD Hybrid Electric Vehicle Powertrain. *Appl. Sci.* **2019**, *9*, 4068. [CrossRef]
15. Fu, X.; Zhang, Q.; Tang, J.; Wang, C. Parameter Matching Optimization of a Powertrain System of Hybrid Electric Vehicles Based on Multi-Objective Optimization. *Electronics* **2019**, *8*, 875. [CrossRef]
16. Kydd, P.H. Electric Hybrid Vehicle Conversion. US Patent 7,681,676, 23 March 2010.
17. Rodriguez, F.; Lukic, S.M.; Wirasingha, S.G.; Emadi, A. Hybrid Electric Conversion Kit for Rear-Wheel Drive, all Wheel Drive, and Four Wheel Drive Vehicles. US Patent 8,011,461, 6 September 2011.
18. Vargas, J. Rechargeable Automobile Electric Power System Configured to Replace the Unpowered Rear Axle of a Front Wheel Drive Vehicle. US Patent 8,118,121, 21 February 2012.

19. Kurdy, T. Electric Vehicle Conversion Kit. US Patent 9,308,810, 5 January 2016.
20. Causton, M.S.; Wu, J. Aftermarket Vehicle Hybridization: Designing a Supply Network for a Startup Company. Ph.D. Thesis, Massachusetts Institute of Technology, Cambridge, MA, USA, 2010.
21. Rizzo, G.; Naddeo, M.; Pisanti, C. Upgrading conventional cars to solar hybrid vehicles. *Int. J. Powertrains* **2018**, *7*, 249–280. [CrossRef]
22. de Luca, S.; Di Pace, R. Aftermarket vehicle hybridization: Potential market penetration and environmental benefits of a hybrid-solar kit. *Int. J. Sustain. Transp.* **2018**, *12*, 353–366. [CrossRef]
23. Tiano, F.A.; Rizzo, G.; De Feo, G.; Landolfi, S. Converting a Conventional Car into a Hybrid Solar Vehicle: A LCA Approach. *IFAC-PapersOnLine* **2018**, *51*, 188–194. [CrossRef]
24. IBGE. Campinas Fleet. Available online: https://cidades.ibge.gov.br/brasil/sp/campinas/pesquisa/22/28120 (accessed on 27 November 2019).
25. IBGE. São Paulo Fleet. Available online: https://cidades.ibge.gov.br/brasil/sp/pesquisa/22/28120 (accessed on 27 November 2019).
26. IBGE. Brazilian Fleet. Available online: https://cidades.ibge.gov.br/brasil/pesquisa/22/28120 (accessed on 27 November 2019).
27. Costa, E.d.S.; Santiciolli, F.M.; Eckert, J.J.; Dionísio, H.J.; Dedini, F.G.; Corrêa, F.C. *Computational and Experimental Analysis of Fuel Consumption of a Hybridized Vehicle*; SAE Technical Paper; SAE International: Warrendale, PA, USA, 2014; doi:10.4271/2014-36-0385. [CrossRef]
28. Costa, E.d.S.; Eckert, J.J.; Santiciolli, F.M.; de Alkmin e Silva, L.C.; Corrêa, F.C.; Dedini, F.G. *Economic and Energy Analysis of Hybridized Vehicle by Means of Experimental Mapping*; SAE Technical Paper; SAE International: Warrendale, PA, USA, 2016; doi:10.4271/2016-36-0368. [CrossRef]
29. Correa, F.C.; Eckert, J.J.; Silva, L.C.; Costa, E.S.; Santiciolli, F.M.; Dedini, F.G. Gear shifting strategy to improve the parallel hybrid vehicle fuel consumption. In Proceedings of the 2015 IEEE Vehicle Power and Propulsion Conference (VPPC), Montreal, QC, Canada, 19–22 October 2015; pp. 1–6.
30. Eckert, J.J.; Santiciolli, F.M.; Silva, L.C.; Costa, E.S.; Corrêa, F.C.; Dedini, F.G. Co-simulation to evaluate acceleration performance and fuel consumption of hybrid vehicles. *J. Braz. Soc. Mech. Sci. Eng.* **2017**, *39*, 53–66. [CrossRef]
31. Eckert, J.J.; Santiciolli, F.M.; de Alkmin, L.C.; dos Santos Costa, E.; Bertoti, E.; Corrêa, F.C.; Dedini, F.G. Fuel consumption and emissions analysis for a hybridized vehicle. *Blucher Eng. Proc.* **2016**, *3*, 580–599.
32. Eckert, J.J.; Santiciolli, F.M.; Bertoti, E.; Costa, E.d.S.; Corrêa, F.C.; Silva, L.C.d.A.e.; Dedini, F.G. Gear shifting multi-objective optimization to improve vehicle performance, fuel consumption, and engine emissions. *Mech. Based Des. Struct. Mach.* **2018**, *46*, 238–253. [CrossRef]
33. Eckert, J.; Santiciolli, F.; Yamashita, R.; Correa, F.; Silva, L.C.A.; Dedini, F. Fuzzy Gear Shifting Control Optimization to Improve Vehicle Performance, Fuel Consumption and Engine Emissions. *IET Control Theory Appl.* **2019**, *13*, 2658–2669. [CrossRef]
34. Eckert, J.J.; Silva, L.C.d.A.e.; Costa, E.d.S.; Santiciolli, F.M.; Corrêa, F.C.; Dedini, F.G. Optimization of electric propulsion system for a hybridized vehicle. *Mech. Based Des. Struct. Mach.* **2019**, *47*, 175–200. [CrossRef]
35. Ho, S.H.; Wong, Y.D.; Chang, V.W.C. Developing Singapore Driving Cycle for passenger cars to estimate fuel consumption and vehicular emissions. *Atmos. Environ.* **2014**, *97*, 353–362. [CrossRef]
36. Pitanuwat, S.; Sripakagorn, A. An Investigation of Fuel Economy Potential of Hybrid Vehicles under Real-World Driving Conditions in Bangkok. *Energy Procedia* **2015**, *79*, 1046–1053, doi:10.1016/j.egypro.2015.11.607. [CrossRef]
37. Jardin, P.; Esser, A.; Givone, S.; Eichenlaub, T.; Schleiffer, J.E.; Rinderknecht, S. The Sensitivity in Consumption of Different Vehicle Drivetrain Concepts Under Varying Operating Conditions: A Simulative Data Driven Approach. *Vehicles* **2019**, *1*, 69–87. [CrossRef]
38. Oliveira, A.d.M.; Bertoti, E.; Eckert, J.J.; Yamashita, R.Y.; dos Santos Costa, E.; e Silva, L.C.d.A.; Dedini, F.G. *Evaluation of Energy Recovery Potential through Regenerative Braking for a Hybrid Electric Vehicle in a Real Urban Drive Scenario*; SAE Technical Paper; SAE International: Warrendale, PA, USA, 2016.
39. Mohammadi, F.; Nazri, G.A.; Saif, M. Modeling, Simulation, and Analysis of Hybrid Electric Vehicle Using MATLAB/Simulink. In Proceedings of the 2019 International Conference on Power Generation Systems and Renewable Energy Technologies (PGSRET), Istanbul, Turkey, 26–27 August 2019; pp. 1–5.

40. Filgueira da Silva, S.; de Moura Fernandes, E.; de Amorim Junior, W.F. *Simulation-Driven Model-Based Approach for the Performance and Fuel Efficiency Trade-Off Evaluation of Vehicle Powertrain*; Automotive Technical Papers; SAE International: Warrendale, PA, USA, 2019.

41. Blagojevic, M.; Djudurovic, M.; Bajic, B. Closed-Form Solution of a Special Case of a Vehicle Longitudinal Motion Model. *Vehicles* **2019**, *1*, 116–126. [CrossRef]

42. Gillespie, T.D. *Fundamentals of Vehicle Dynamics*; SAE International: Warrendale, PA, USA, 1992.

43. Genta, G.; Morello, L. *The Automotive Chassis*; Springer: Berlin, Germany, 2009; Volume 1.

44. Reimpell, J.; Stoll, H.; Betzler, J. *The Automotive Chassis: Engineering Principles*; Butterworth-Heinemann: Oxford, UK, 2001.

45. Eckert, J.J.; Corrêa, F.C.; Santiciolli, F.M.; Costa, E.d.S.; Dionísio, H.J.; Dedini, F.G. Vehicle gear shifting strategy optimization with respect to performance and fuel consumption. *Mech. Based Des. Struct. Mach.* **2016**, *44*, 123–136. [CrossRef]

46. Jazar, R.N. *Vehicle Dynamics*; Springer: Berlin, Germany, 2008.

47. Aaron, B.; Haraldsson, K.; Hendricks, T.; Johnson, V.; Kelly, K.; Kramer, B.; Markel, T.; O'Keefe, M.; Sprik, S.; Wipke, K.; et al. ADVISOR: A systems analysis tool for advanced vehicle modeling. *J. Power Sources*, **2002**, *110*, 255–266.

48. Kerler, M.; Burda, P.; Baumann, M.; Lienkamp, M. A concept of a high-energy, low-voltage EV battery pack. In Proceedings of the 2014 IEEE International Electric Vehicle Conference (IEVC), Florence, Italy, 17–19 December 2014; pp. 1–8.

49. Bojoi, R.; Cavagnino, A.; Cossale, M.; Tenconi, A.; Vaschetto, S. Design trade-off and experimental validation of multiphase starter generators for 48v mini-hybrid powertrain. In Proceedings of the 2014 IEEE International Electric Vehicle Conference (IEVC), Florence, Italy, 17–19 December 2014; pp. 1–7.

50. Pistoia, G. *Electric and Hybrid Vehicles: Power Sources, Models, Sustainability, Infrastructure and the Market*; Elsevier: Amsterdam, The Netherlands, 2010.

51. Jung, J.; Zhang, L.; Zhang, J. *Lead-Acid Battery Technologies: Fundamentals, Materials, and Applications*; CRC Press: Boca Raton, FL, USA, 2015.

52. Young, K.; Wang, C.; Wang, L.Y.; Strunz, K. Electric vehicle battery technologies. In *Electric Vehicle Integration into Modern Power Networks*; Springer: Berlin, Germany, 2013; pp. 15–56.

53. Björnsson, L.H.; Karlsson, S.; Sprei, F. Objective functions for plug-in hybrid electric vehicle battery range optimization and possible effects on the vehicle fleet. *Transp. Res. Part C: Emerg. Technol.* **2018**, *86*, 655–669. [CrossRef]

54. Eckert, J.J.; Silva, L.C.d.A.e.; Santiciolli, F.M.; Costa, E.d.S.; Corrêa, F.C.; Dedini, F.G. Energy storage and control optimization for an electric vehicle. *Int. J. Energy Res.* **2018**, *42*, 3506–3523. [CrossRef]

55. Eckert, J.J.; Silva, L.C.; Costa, E.S.; Santiciolli, F.M.; Dedini, F.G.; Corrêa, F.C. Electric vehicle drivetrain optimisation. *IET Electr. Syst. Transp.* **2017**, *7*, 32–40. [CrossRef]

56. Tong, W. *Mechanical Design of Electric Motors*; CRC Press: Boca Raton, FL, USA, 2014.

57. Correa, F.C.; Eckert, J.J.; Silva, L.C.; Santiciolli, F.M.; Costa, E.S.; Dedini, F.G. Study of Different Electric Vehicle Propulsion System Configurations. In Proceedings of the Vehicle Power and Propulsion Conference (VPPC), Montreal, QC, Canada, 19–22 October 2015; pp. 1–6.

58. Rotering, N.; Ilic, M. Optimal charge control of plug-in hybrid electric vehicles in deregulated electricity markets. *Power Syst. IEEE Trans.* **2011**, *26*, 1021–1029. [CrossRef]

59. Eckert, J.J.; Silva, L.C.A.; Dedini, F.G.; Correa, F.C. Electric Vehicle Powertrain and Fuzzy Control Multi-objective Optimization, considering Dual Hybrid Energy Storage Systems. *IEEE Trans. Veh. Technol.* **2020**, 1. [CrossRef]

60. van Reeven, V.; Hofman, T. Multi-level energy management for hybrid electric vehicles—Part I. *Vehicles* **2019**, *1*, 3–40. [CrossRef]

61. Lu, X.; Xu, X.; Liu, Y. Simulation of gear-shift algorithm for automatic transmission based on matlab. In Proceedings of the 2009 WRI World Congress on Software Engineering, Xiamen, China, 19–21 May 2009; Volume 2, pp. 476–480.

62. Gen, M.; Cheng, R.; Lin, L. *Network Models and Optimization: Multiobjective Genetic Algorithm Approach*; Springer Science & Business Media: Berlin/Heidelberg, Germany, 2008.

63. Barbosa, T.P.; da Silva, L.A.R.; Pujatti, F.J.P.; Gutiérrez, J.C.H. Hydraulic hybrid passenger vehicle: Fuel savings possibilities. *Mech. Based Des. Struct. Mach.* **2020**, 1–19. [CrossRef]

64. General Motors (GM). *Owner Manual Chevrolet Celta 2013*; Technical Report; GM: Saõ Paulo, Brazil, 2013.
65. Lopes, M.V.; Eckert, J.J.; Martins, T.S.; Santos, A.A. Optimizing strain energy extraction from multi-beam piezoelectric devices for heavy haul freight cars. *J. Braz. Soc. Mech. Sci. Eng.* **2020**, *42*, 1–12. [CrossRef]

Review

Non-Volatile Particle Number Emission Measurements with Catalytic Strippers: A Review

Barouch Giechaskiel *, Anastasios D. Melas, Tero Lähde and Giorgio Martini

European Commission—Joint Research Centre (JRC), 21027 Ispra, Italy; anastasios.melas@ec.europa.eu (A.D.M.); tero.lahde@ec.europa.eu (T.L.); giorgio.martini@ec.europa.eu (G.M.)
* Correspondence: barouch.giechaskiel@ec.europa.eu; Tel.: +39-0332-78-5312

Received: 21 May 2020; Accepted: 23 June 2020; Published: 24 June 2020

Abstract: Vehicle regulations include limits for non-volatile particle number emissions with sizes larger than 23 nm. The measurements are conducted with systems that remove the volatile particles by means of dilution and heating. Recently, the option of measuring from 10 nm was included in the Global Technical Regulation (GTR 15) as an additional option to the current >23 nm methodology. In order to avoid artefacts, i.e., measuring volatile particles that have nucleated downstream of the evaporation tube, a heated oxidation catalyst (i.e., catalytic stripper) is required. This review summarizes the studies with laboratory aerosols that assessed the volatile removal efficiency of evaporation tube and catalytic stripper-based systems using hydrocarbons, sulfuric acid, mixture of them, and ammonium sulfate. Special emphasis was given to distinguish between artefacts that happened in the 10–23 nm range or below. Furthermore, studies with vehicles' aerosols that reported artefacts were collected to estimate critical concentration levels of volatiles. Maximum expected levels of volatiles for mopeds, motorcycles, light-duty and heavy-duty vehicles were also summarized. Both laboratory and vehicle studies confirmed the superiority of catalytic strippers in avoiding artefacts. Open issues that need attention are the sulfur storage capacity and the standardization of technical requirements for catalytic strippers.

Keywords: vehicle emissions; particle measurement programme (PMP); portable emissions measurement systems (PEMS); volatile removal efficiency; non-volatiles; solid particle number; catalytic stripper; evaporation tube; artefact

---

1. Introduction

In order to reduce pollution from particulate matter (PM), limits for the PM mass emissions of vehicles are applicable in most countries worldwide [1]. The engine and aftertreatment improvements reduced the emitted PM levels close to the detection limit of the PM methodology [2]. In order to further control the emissions of ultrafine particles, which do not contribute significantly to the PM mass, the European Union (EU) regulations included limits for the non-volatile particle emissions of light-duty, heavy-duty, and non-road mobile machinery since 2011–2013 [3,4].

The particle number methodology is based on sampling from a dilution tunnel where the whole exhaust gas is mixed with ambient air dilution air. In order to measure only non-volatile particles, the sample is pre-conditioned first in a hot diluter at 150 °C and then in an evaporation tube at 350 °C, the so called volatile particle remover [5,6]. After a secondary dilution at ambient temperature, a particle number counter (PNC) counts all (non-volatile) particles above 23 nm. In reality the detection efficiency of the PNC, typically a condensation particle counter (CPC), is around 50% at 23 nm, >90% at 41 nm and 0% at 17 nm. The 23 nm cut-off size was selected in order to include the smallest primary soot particles, and exclude the volatile nucleation mode particles that could survive the pre-conditioning unit. The methodology was extended, but more simplified, to measure directly from the tailpipe during on-road tests with the real-driving emissions (RDE) regulation for both light-duty and heavy-duty vehicles [7]. The portable emissions measurement systems (PEMS) set the volatile particle remover at 300 °C, in order to reduce energy consumption. From the controlled laboratory environment, the measurements can take place under any environmental and driving conditions, thus the measurement equipment can be challenged. Indeed, the survival of volatile particles through the thermal pre-conditioning unit or formation of volatile particles after the unit can lead to false high concentrations of volatiles that will be counted as non-volatile particles. This was quite unlikely with the 23 nm cut-off size of the PNC, but has happened in the sub-23 nm range [8,9], raising concerns as to whether the methodology should be extended below 23 nm.

On the other hand, a major criticism of the particle number methodology is that it excludes an important fraction of non-volatile particles below 23 nm that are emitted by many vehicle technologies [10]. For example, sub-23 nm non-volatile particles have been reported for heavy-duty vehicles [11], especially during urea injection [12], gasoline fueled vehicles [13], both port-fuel and direct injection vehicles. Recently, concerns for CNG (compressed natural gas) fueled vehicles were also raised [3,12]. Mopeds and motorcycles have also a high percentage of sub-23 nm particles. For this reason, the intention is to extend the methodology to include particles from approximately 10 nm. The 10 nm size was selected in order to permit the upgrade of existing PN systems without investment costs and to minimize the higher risk of surviving (or re-nucleating) volatile particles [14]. Furthermore, it will be suitable for the upgrade of the current 23 nm PEMS [15,16]. Developing a methodology to count smaller than 10 nm particles would be extremely challenging due to the uncertainties related to particle losses and calibration procedures.

A proposed methodology to eliminate the volatiles and make the methodology more robust is the catalytic stripper [17]. The idea is based on catalytic converters of vehicles, which remove hydrocarbons (in addition to CO and NO_x). Typically, they utilize a ceramic honeycomb monolithic substrate (support) with many small parallel channels. The monolith walls are coated with an active catalyst layer [18]. The catalytic coating includes noble metal catalysts (such as Pt and Pd) impregnated on a porous high surface area layer of inorganic carrier (support), the washcoat, such as alumina and silica. The support material often contains additives such as promoters and stabilizers. Based on NO_x storage catalysts [19,20], the catalytic strippers include sometimes a sulfur trap to further improve their removal efficiency of sulfuric acid containing aerosol. Catalytic strippers have been used not only in the automotive field [21], but also in the aviation field [22], and for atmospheric measurements [23].

Thermodenuders can also remove volatiles species from the exhaust gas [24,25]. In the thermodenuders, the sample is first heated in an evaporation tube, and then it passes through an unheated section containing adsorbing material, most often activated carbon, which adsorbs most of the evaporated components.

The objective of this study is to review the volatile removal efficiency and sulfur storage capacity of catalytic strippers and their suitability for inclusion in the particle number systems. The review results are compared with the current methodology that uses hot dilution and evaporation tube and with thermodenuders used in research. In order to put the results into the right context, actual volatile emission levels from vehicles are also summarized. Finally, recommendations for technical requirements for catalytic strippers are proposed.

2. Materials and Methods

This review focused on studies that characterized catalytic strippers and evaporation tubes as standalone systems or in volatile particle removers of particle number systems (i.e., downstream of a hot dilution at temperature >150 °C). We searched in Google Scholar with keywords "catalytic stripper", "evaporation tube + particles" and "volatile removal efficiency". The search resulted in 533, 900 and 73 documents, respectively. From the abstracts it was decided whether the documents were relevant or not. The only acceptance criterion was the inclusion in the text of volatile removal efficiency with laboratory aerosols or vehicles' exhaust. Studies that used evaporation tubes or catalytic strippers for measurements of vehicle exhaust were excluded, unless there was discussion of possible artefacts, or comparisons between catalytic strippers and evaporation tubes on the removal of volatile particles. In total, 29 studies were selected, 18 used a laboratory aerosol, and 14 a vehicle's aerosol (3 were in common). Table 1 summarizes the studies included.

Table 1. Number of studies assessing evaporation tubes (ET), catalytic strippers (CS), or thermodenuders (TD). The laboratory studies were conducted with hydrocarbons (HCs) and/or sulfuric acid (H_2SO_4). In brackets the studies using CS with sulfur trap.

System	Laboratory HCs	Laboratory H_2SO_4	Vehicles' Aerosol
ET	13	6	14
CS	15 (9)	7 (5)	8 (7)
CS vs. ET or TD	10 (6)	6 (4)	8 (7)

We used the same search engine for sulfur storage capacity of catalysts and SO_2 to SO_3 conversion rates. Keywords such as "sulfur trap", sulfur storage catalyst", "sulfur absorber", "sulfur poisoning" were used. Only key studies were used because the objective was to put the catalytic stripper assessment results into perspective (i.e., compare with automotive catalytic converters).

We also reviewed emission levels of volatile and semi-volatile components at the tailpipe of the vehicles or at the dilution tunnel. The focus was on two-stroke mopeds, motorcycles, gasoline and diesel light-duty and heavy-duty vehicles mainly of recent technologies. Special attention was paid to emission levels during the cold start of the engines, filter regenerations, or prolonged highway operation, where high levels of volatiles are expected. The search was narrowed to hydrocarbons, SO_2, SO_3 and H_2SO_4, which were the species that were used for the evaluation of the catalytic strippers and evaporation tubes. The research was not extensive, as the target was to find typical maximum levels that the particle number systems can be challenged.

3. Results

The particle number systems measure non-volatile particles >23 nm. Note that "solids" or "non-volatile" particles are by definition the particles that remain after thermal pre-treatment at 350 °C for a residence time of approximately 0.2–0.4 s. The term "volatiles" in this text covers all species that are in gaseous state at temperatures below 350 °C. More appropriate terms such as semi-volatile or intermediate volatility compounds will not be used [26].

3.1. Volatile Particle Remover

The particle number systems consist of a volatile particle remover (VPR) and a particle number counter (PNC). Figure 1 presents a volatile particle remover with an evaporation tube (upper panel) or a catalytic stripper (lower panel).

3.1.1. Evaporation Tube-Based Systems

The exhaust gas at the dilution tunnel is already diluted and cooled down to ambient temperature levels (usually between 20 °C and 50 °C). The diluted aerosol typically consists of soot particles with some species condensed on them, volatile nucleation mode particles, and volatile species in the gas phase. The volatile particles contain high percentages of compounds such as organics and sulfuric acid [27]. The soot particles are aggregates consisting of primary soot particles having diameters between 20 nm and 30 nm [28]. The geometric mean diameter of the soot mode is around 50–70 nm [2], but for modern gasoline engines, means of around 30 nm have also been reported [4]. The nucleation mode size-range is usually below 20 nm.

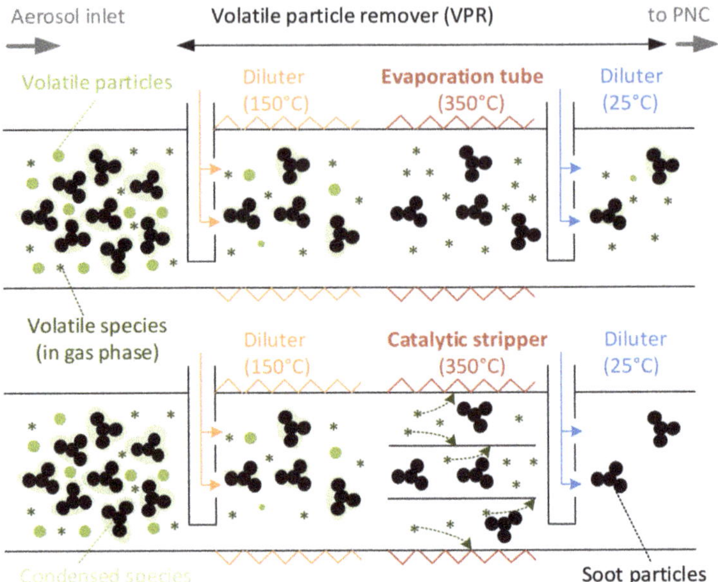

Figure 1. Schematic of volatile particle removers based on evaporation tube (upper panel) and catalytic stripper (lower panel). PNC = particle number counter.

In the primary hot diluter (≥150 °C) some of the condensed material evaporates and the nucleation mode particles evaporate and shrink. Furthermore, the concentration (and partial pressure) of the volatiles species decreases. In the evaporation tube (350 °C) most of the volatile particles and the condensed material evaporate. The cold diluter downstream of the evaporation tube further dilutes the sample to bring the temperature and the soot concentration to an appropriate range for the PNC. During this process it is possible to have re-condensation on the soot particles and/or re-nucleation of volatile species (i.e., formation of new volatile particles) especially if the dilution ratio is not sufficiently high. The nucleated particles due to the low concentrations of available species are unlikely to grow over 23 nm in diameter, and thus they will not be counted by the PNC, which has detection efficiency of 50% for particles of 23 nm diameter. The particles nucleated downstream of the evaporation tube and counted with a PNC are called volatile artefacts [14]. The next sections will summarize the studies that evaluated the concentration levels of volatile species that artefacts can be avoided.

3.1.2. Catalytic Stripper-Based Systems

When a catalytic stripper is used, instead (or in addition to) an evaporation tube, the volatile species in the gas phase are oxidized (Figure 1, lower panel). Then, during dilution and cooling at the secondary diluter the concentrations of nucleating species is too low to form volatile particles and, thus, the possibility for artefacts is minimized.

There are three types of catalytic strippers. Those that comprise:

- Only oxidation catalyst;
- Sulfur trap and oxidation catalyst (in this order);
- Oxidation catalyst and sulfur trap.

The first catalytic stripper for automotive use was mentioned in 1995 [17]. It was constructed from a commercial diesel oxidation catalyst (DOC) and operated at 300 °C. A cooling coil tube downstream of the catalytic stripper reduced the sample temperature to ambient levels. Optimization work in terms of particle losses and volatile removal efficiency was presented in 2007 [29]. In particular, a vortex tube diluter downstream of the catalytic stripper minimized thermophoretic losses. Later the catalytic stripper was installed in a particle number system compliant with the European regulations replacing the evaporation tube [30].

A catalytic stripper with an upstream sulfur trap was presented in 2003 [31]. The sulfur trap prevented poisoning of the oxidation catalyst by trapping the sulfur species and minimized artefacts from the nucleation of SO_3. This concept was compared to a thermodenuder in 2010 [32]. Later the position of the sulfur trap was assessed at a different catalytic stripper and it was found to have higher sulfur storage capacity when placed downstream [33]. The majority of the catalytic strippers in the market for research or regulation purposes consist either only of an oxidation catalyst or oxidation catalyst with a sulfur trap downstream.

3.2. Hydrocarbons

Figure 2 summarizes the volatile removal efficiency of studies that used hydrocarbons (tetracosane C24, octacosane C28, tetracontane C40 and emery oil) for evaporation tube [5,30,32,34–43] and catalytic stripper-based [30,32–41,44–47] systems. Each point is a test. The mass concentration of the challenge aerosol can be seen in the *y*-axis. The *x*-axis is divided in four areas which indicate whether no or some particles remained (or nucleated) downstream of the system under evaluation.

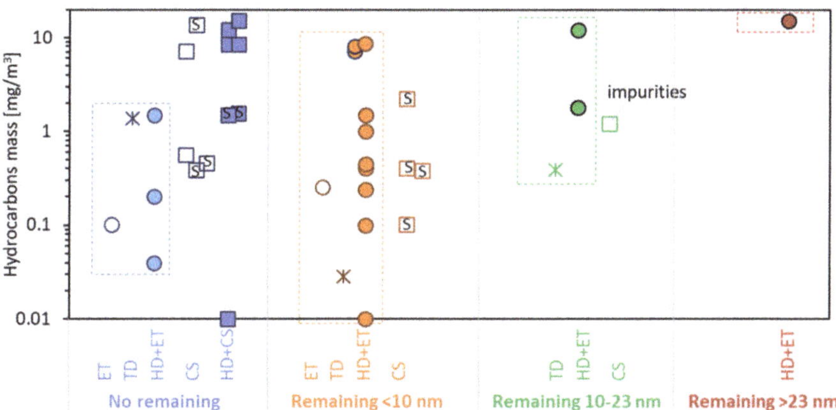

Figure 2. Volatile removal evaluation with hydrocarbons depending on whether none or some particles remained (or re-nucleated) <10 nm, between 10 and 23 nm, and >23 nm. Open circles: evaporation tubes (ET); asterisks: thermodenuders (TD); solid circles: hot dilution and evaporation tubes (HD + ET); open squares: catalytic strippers (CS); solid squares: hot dilution and catalytic strippers (HD + CS). "S" indicates CS with sulfur trap.

- No particles remained downstream of the system under evaluation (blue color). Note that typically the lower detection size of these studies was around 3–6 nm.
- Particles with sizes lower than 10 nm were detected downstream of the system under evaluation (orange color). These cases are of low risk for the future >10 nm regulation but could have artefacts with low cut-off counters.
- Particles in the size region 10–23 nm remained (green color). These cases are risky for the future >10 nm regulation and indicate hydrocarbons mass levels that could lead to "volatile" artefacts.
- Particles larger than 23 nm were detected (red color). These cases would affect also the results of the current >23 nm regulation.

In each area the results are separately plotted for:

- Standalone evaporation tubes (ET) (open circles).
- Thermodenuders (TD) (asterisks).
- Hot dilution (typically 10:1) plus evaporation tubes (HD + ET) (solid circles).
- Standalone catalytic strippers (CS) (open squares).
- Hot dilution plus catalytic strippers (HD + CS) (solid squares).

For better clarity the evaporation tube (stand alone or downstream of a hot dilution) and thermodenuder tests are enclosed in dashed squares. Only the tests with the maximum reported concentrations in each study are plotted.

The hot dilution plus evaporation tube (HD + ET) results resulted in artefacts >23 nm when challenged with 15 mg/m^3 [34], while levels 2–12 mg/m^3 had artefacts in the 10–23 nm range [36,39]. Artefacts <10 nm were detected with masses in a wide range of 0.01 mg/m^3 to 8.5 mg/m^3 [5,30,32,36,37,39–41]. On the other hand, no artefacts were seen with masses up to 1.5 mg/m^3 for the combination of hot dilution and evaporation tube [43], confirming that the results might be system dependent. Standalone evaporation tubes (ET) could have artefacts <10 nm already at low levels (0.25 mg/m^3) [5].

The catalytic stripper systems had no artefacts above 10 nm, with one exception which was probably solid residuals from impurities in the atomized material [45,48]. Masses up to 2 mg/m^3 resulted in some particles remaining <10 nm for some standalone catalytic strippers [36,40,41,44]. For other standalone or with hot-dilution catalytic stripper systems, masses up to 8–15 mg/m^3 had no particles remaining [34,36,39]. There was no difference between CS with or without sulfur trap, as both can remove hydrocarbon masses >10 mg/m^3 without any remaining particles.

3.3. H$_2$SO$_4$

Figure 3 summarizes the results for H$_2$SO$_4$ and H$_2$SO$_4$ with hydrocarbons (C24 or C28) for evaporation tube-based [32,35,38–41] and catalytic stripper-based [32,35,38–41,44] systems. Particles >23 nm remained only with thermodenuder (3.8 mg/m^3) [32] or evaporation tube after hot dilution (0.3 mg/m^3) [39]. When the concentration was 1 mg/m^3, some particles in the 10–23 nm range remained for the thermodenuder, and evaporation tube after hot dilution [40]. With mixture of H$_2$SO$_4$ with hydrocarbons, the evaporation tube had artefacts in the 10–23 nm range even at 0.1 mg/m^3 [41]. Particles <10 nm remained even at very low masses (<0.2 mg/m^3 of H$_2$SO$_4$ or H$_2$SO$_4$ and HCs) for the thermodenuder and the evaporation tube after hot dilution, although in one case, the evaporation tube with hot dilution could handle 2.5 mg/m^3 with artefact <10 nm [38]. Only the catalytic stripper could completely remove sulfuric acid aerosol without any particle remaining at concentrations ranging from 0.1 mg/m^3 [35,39,41] to 1 mg/m^3 [38] or even 9.3 mg/m^3 [44]. At the >10 mg/m^3 range only one standalone catalytic stripper with a sulfur trap downstream of the oxidation catalyst was evaluated [44]. Although it could remove up to 9.3 mg/m^3 H$_2$SO$_4$ without any particles remaining, at 16 mg/m^3 some particles remained below 10 nm, and at 28 mg/m^3 particles remained at the 10–23 nm range. Only one study compared a catalytic stripper with and without a sulfur trap [40]. The version with sulfur trap could handle up to 1 mg/m^3 without significant artefacts while without the trap remaining particles were detected already on H$_2$SO$_4$ concentrations above 0.07 mg/m^3. In general, catalytic strippers with a sulfur trap could handle higher H$_2$SO$_4$ containing aerosol.

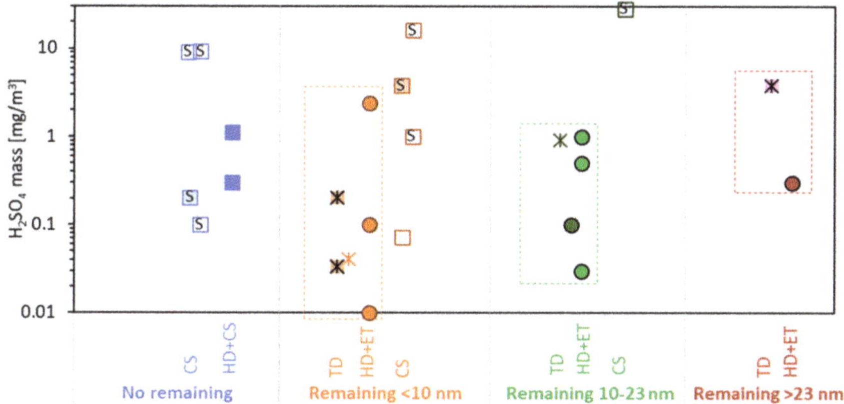

Figure 3. Volatile removal evaluation with sulfuric acid aerosol or sulfuric acid and hydrocarbons (symbols with background) depending on whether none or some particles remained (or re-nucleated) <10 nm, between 10 and 23 nm, and >23 nm. Open circles: evaporation tubes (ET); asterisks: thermodenuders (TD); solid circles: hot dilution and evaporation tubes (HD + ET); open squares: catalytic strippers (CS); solid squares: hot dilution and catalytic strippers (HD + CS). "S" indicates CS with sulfur trap.

3.4. Ammonium Sulfate

A catalytic stripper based on a commercial vehicle aftertreatment DOC, used downstream of a hot dilution, could evaporate and shrink ammonium sulfate particles, but a residual with a mean between 10 nm and 20 nm remained due to the atomization method [29]. In another study, a CS downstream of a hot diluter started to show some breakthrough only at inlet concentration of around 1×10^7 p/cm^3 (8.5 mg/m^3) [39]. The hot diluter and evaporation tube combination had breakthrough at a one order of magnitude lower concentrations. A later study confirmed the findings and showed that the breakthrough particles had a size <23 nm [38].

3.5. Vehicles' Exhaust

Figure 4, which is organized in the same way as the previous figures, summarizes the results with vehicles' exhaust for evaporation tube [3,8,33,34,41,49–57] and catalytic stripper [3,8,33,34,41,50,52,57] systems. The volatile mass concentrations refer to the inlet of the systems, which were sampling from the full dilution tunnel. They were estimated from the nucleation mode size distributions. The horizontal lines indicate the levels of volatiles of the different tested categories: mopeds >10 mg/m^3, diesel with DPF but without diesel oxidation catalyst (DOC) 6 mg/m^3, and all the rest (GDI, DPF regenerations and heavy duty diesel vehicles) <1 mg/m^3. The chemical composition of the volatiles is unknown (i.e., was not measured in any of the studies).

The hot dilution and evaporation tube had cases with remaining particles even at low levels: with 0.05 mg/m^3 there were particles <10 nm [41], while with mass 0.5 mg/m^3 or higher particles 10–23 nm remained (in one case also particles >23 nm) [49]. However, there was a case in which the hot dilution and evaporation tube could handle 15 mg/m^3 without artefacts (primary hot dilution 25:1) [34]. The catalytic stripper (standalone or downstream of a hot dilution) did not have cases where particles >10 nm remained. There were cases that particles <10 nm remained at mass concentrations 0.65 mg/m^3 [52] or 6.1 mg/m^3 [33] (as standalone) or 30–50 mg/m^3 downstream of a hot dilution [8,34]. There were many experiments with CS that no particles remained with concentrations even >10 mg/m^3.

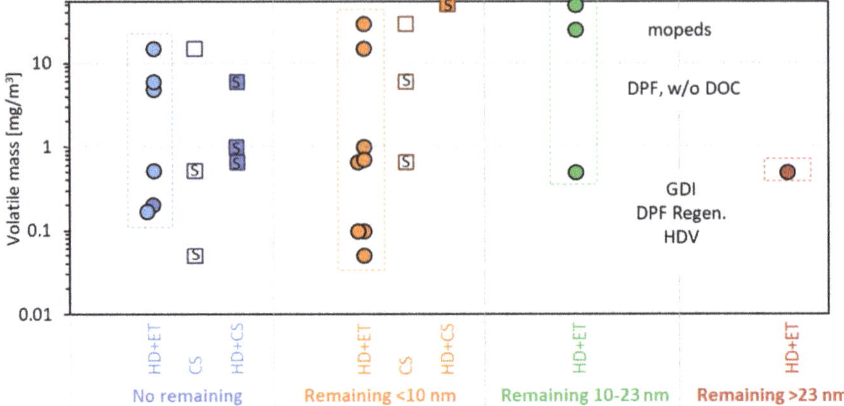

Figure 4. Volatile removal evaluation with vehicles' exhaust depending on whether none or some particles remained (or re-nucleated) <10 nm, between 10 and 23 nm, and >23 nm. Solid circles: hot dilution and evaporation tubes (HD + ET); open squares: catalytic strippers (CS); solid squares: hot dilution and catalytic strippers (HD + CS). "S" indicates CS with sulfur trap. DOC = diesel oxidation catalyst; DPF = diesel particulate filter; GDI = gasoline direct injection; HDV = heavy-duty vehicle.

4. Discussion

This review summarized the studies that evaluated evaporation tubes and catalytic strippers with laboratory or vehicles' exhaust aerosol. The implications of the results will be discussed in the following sections.

4.1. Evaporation Tube

The concept that was introduced in the particle number regulations is very efficient, as summarized in the literature [14]. The hot dilution with the evaporation tube in combination with a counter that measures particles >23 nm is a reliable methodology as artefacts have rarely been reported.

The temperature of the evaporation tube and the residence time is appropriate for the evaporation of volatile particles. The theory estimates necessary time around 10–16 ms for typical commercial systems [5,44], which is much less than the 200–400 ms residence time in the systems, although the available time at the critical temperature region (>250 °C) is shorter. The upstream dilution at 150 °C pre-heats the sample and ensures that the desired temperature will be reached in the evaporation tube. Most importantly, it dilutes the sample and reduces the partial pressures, thus minimizing the possibility of nucleation downstream of the evaporation tube. Nevertheless, nucleation has been reported many times with both laboratory and vehicles' exhaust aerosol. Theoretically, the nucleation of hydrocarbons is unlikely because very high concentrations are required (>3 mg/m³ [5]). However, the experimental data with laboratory hydrocarbon aerosols showed that standalone evaporation tubes could not handle 0.25 mg/m³. Evaporation tubes, even with a hot dilution upstream (dilution 10:1), could not handle 8.5 mg/m³ (0.85 mg/m³ at the inlet of the evaporation tube) as particles <10 nm remained (or nucleated). The case where particles >23 nm remained was atomized emery oil (15 mg/m³) with hot dilution 25:1. In this case it is possible that there were impurities and the evaporated emery oil condensed on the solid residuals downstream of the evaporation tube growing them to the over 23 nm range [48].

When sulfuric acid is available, formation of nucleation particles is very probable. Theoretical calculation and experimental results have shown nucleation possibility even at 0.3 µg/m³ levels [5]. Small amounts of ammonia (<20 ppb) may significantly enhance the binary nucleation rate of sulfuric acid and water [58–60], for example at vehicles with SCR (selective catalytic reduction for NO_x)

systems or gasoline vehicles with three-way catalysts (TWC) [61]. Experimentally, nucleation particles downstream of a VPR with evaporation tube were measured from a heavy-duty vehicle equipped with SCR [62]. However, the size of the H_2SO_4 nuclei (critical clusters) is approximately 1–1.5 nm and high concentrations of H_2SO_4 are needed to grow them to 3 nm in diameter [63,64]. Interestingly, laboratory experiments with sulfuric acid (and humidity) have reported size distributions peaking at 18 nm [65] or even >200 nm [44]. Similarly, experiments with SO_2 (converted to SO_3) gave nucleation modes peaking at 16 nm (1.8 ppm SO_2) [66], 45 nm (61.4 ppm SO_2) [66], or even >70 nm (200 ppm) [33].

Nuclei typically grow to bigger than 3 nm sizes with organics [67], even at low concentrations as experiments with heavy-duty engine have shown [68]. The fuel sulfur was 1 ppm, and the (effective) contribution of the lubricant was 2–9 ppm (estimated sulfuric acid concentration 1.5–5.8 × 10^{12} cm^{-3}) and the final mean size was <10 nm [69]. The nucleation mode peaked at 15 nm when the sulfur (fuel plus effective lubricant) was approximately 50 ppm (sulfuric acid 3.3 × 10^{13} cm^{-3}). Growth rates of 6–24 nm/s were measured for a heavy-duty engine with 8 mg/m^3 organics emissions [70]. Another study found a nucleation mode of 6 nm when organics were available with concentration 10^{11} cm^{-3} [71]. With appropriate residence time (1 s), in the presence of a high amount of organics (10^{14} cm^{-3}) the nucleation mode peaked at 20 nm [72]. Approximately 1 ppm of propane corresponds to 1 × 10^{12} cm^{-3} organics [73]. This is calculated using the appropriate molecular mass and an assumed 10–30% fraction of total organics potentially condensing on the particulates phase [73]. Thus, it can be assumed that with hydrocarbon concentrations >100 ppm at the outlet of the evaporation tube the nucleated sulfuric acid particles can grow to sizes of 20 nm.

The laboratory tests with sulfuric acid (and hydrocarbons in some cases) confirmed the previous studies. In all cases with the evaporation tubes there was re-nucleation. Mass of 0.9 mg/m^3 or lower resulted in nucleation mode particles >10 nm even with a hot dilution of 10:1. However, for the majority of the cases, with high dilution ratios (>100:1 primary hot dilution) there are no reported cases of re-nucleation [74] except with 2-stroke mopeds. Studies comparing evaporation tubes with catalytic strippers gave equivalent results (within 10%) for both 23 nm and 10 nm systems [39,75,76]. Higher differences of the 10 nm systems were attributed to the different cut-off curves of the counters.

4.2. Catalytic Stripper

The studies with catalytic strippers proved that much higher concentrations can be handled for both hydrocarbons and/or sulfuric acid aerosols than with evaporation tubes. For example, no artefacts were seen when catalytic strippers were challenged with hydrocarbons. In a few cases without hot pre-diluter particles <10 nm remained. Similarly, most concentrations of sulfuric acid were efficiently removed. Only in a few cases standalone systems had particles remaining <10 nm. The results with vehicle exhaust were similar to the laboratory aerosols. The standalone catalytic stripper had in some cases particles remaining <10 nm, while the evaporation tube systems often had particles remaining >10 nm at various challenge mass levels.

The conclusion based on the current experimental results is to use a catalytic stripper when measurements of particles >10 nm are conducted. For lower sizes, the catalytic stripper is necessary, but not artefact-free, and a hot pre-diluter is recommended. For lower than 10 nm measurements, the high particle losses of the catalytic stripper have also to be considered.

4.2.1. Oxidation and Sulfur Trap Parts

Figure 5a plots separately examples of the two parts of a catalytic stripper: the oxidation part and the sulfur storage part. In actual catalytic strippers the two parts are continuous. At some catalytic strippers the sulfur trap is upstream of the oxidation part, in some downstream, and in some others it does not exist. As discussed before (Figure 1), the oxidation part evaporates nucleation mode particles and oxidizes the volatile particles (actually the particles should be evaporated earlier in an evaporation tube in order to have more efficient oxidation in the gaseous phase). The condensed material on the soot particles is also evaporated and oxidized. In this oxidation part the SO_2 and SO_3

are considered to remain unaffected; especially catalytic strippers without a sulfur trap rely on this assumption. The actual SO_2 to SO_3 conversion, however, depends on the catalyst formulation as will be discussed in Section 4.2.2. Modern engines equipped with oxidation converters and depending on the fuel and lubricant sulfur content, can have relatively high SO_3 concentrations and this could result in nucleation of sulfuric acid at the exit of the catalytic stripper. The sulfur trap retains sulfur, minimizing this possibility.

At low temperatures (<350 °C) sulfur is stored through adsorption or condensation (physically bounded sulfate), while, at higher temperatures (>400 °C) sulfur is stored as chemically bound metal sulfate [77]. Adhesion of sulfate in mist form on the catalyst is also possible at low temperatures. Under oxygen deficient conditions H_2S is formed, which is poisonous for metal surfaces.

Sulfur traps may include an oxidation catalyst, e.g., platinum, to facilitate the SO_2 to SO_3 reaction. Typical sulfur storage systems are based on alkaline earth or alkali metal oxides and their mixtures on alumina washcoat. Other materials such as zinc, nickel, chromium, and copper are used either as stand-alone scavengers or as promoters, to modify trap performance, strengthen desired reactions, and influence the adsorption/desorption temperatures.

4.2.2. SO_2 to SO_3 Conversion

Figure 5b summarizes SO_2 to SO_3 conversion efficiencies from the literature of commonly used oxidation catalysts (based also on the review in [78]). The space velocity was around 35,000–50,000 h^{-1}. Higher space velocities result in lower conversions [79,80]. Conversion rates for other catalyst materials can be found in the literature (e.g., [81,82]). The extent of the conversion depends mainly on the exhaust gas temperature, the type of noble metal used and the loading, and composition of the washcoat. The SO_2 to SO_3 conversion starts to increase at around 200 °C, reaches a maximum at 450 °C as the reaction kinetics accelerate and then gradually decreases due to thermodynamic equilibrium limitations. The three studies with catalytic strippers are in agreement with the published data and vary from 0% to 37.5% at 275 °C to 300 °C actual aerosol temperature [30,33,36]. The need for a low or high SO_2 to SO_3 conversion depends on the concept of the catalytic stripper (only oxidation part or with sulfur trap).

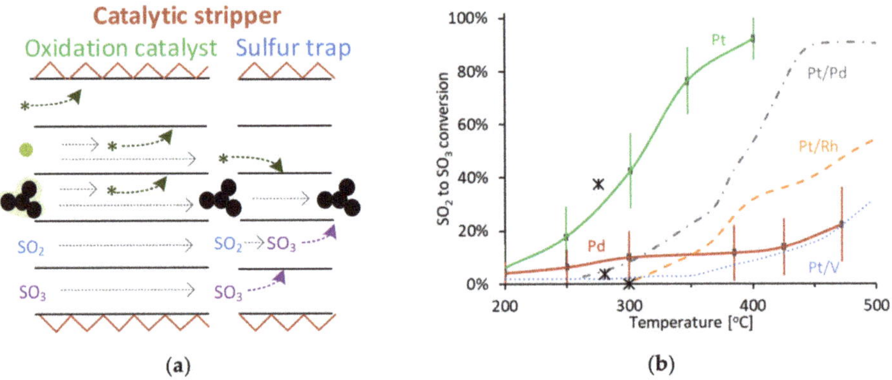

(a)　　　(b)

Figure 5. Catalytic stripper: (**a**) the two parts of a catalytic stripper: oxidation catalyst and sulfur trap. Products of reactions such as CO_2 and H_2O are not plotted; (**b**) SO_2 to SO_3 conversion for various noble metals and combinations. Pt [79,83–86], Pt/Pd [87], Pt/Rh [87], Pt/V [84], Pd [84,87]. Asterisks are experimental data with catalytic strippers [30,33,36].

4.2.3. Storage Capacity

The SO_x (SO_2 and/or SO_3) retention capacity is typically expressed in milligrams of SO_x captured per gram of sorbent. A review of the literature of sorbent systems for the removal of the SO_x from flue gases reported values from a few mg to hundreds of mg per gram of sorbent [88]. In vehicle applications, the catalyst capacity in sulfate storage depends on the noble metal and the washcoat composition. Studies with diesel oxidation catalysts (DOCs) with 1% Pt on alumina found 16–50 mg sulfates per gram of washcoat (1.6–5%) at 350 °C [80,89,90], but 0% for SiC washcoat [89]. A Pd on alumina adsorbed up to 7% SO_2 at 320 °C [91]. At temperatures higher than 400 °C, percentages of up to 4% have been reported for Pt, Pt/Pd, or Pt/V catalysts [84,92–94]. Dedicated studies with sulfur traps have reported SO_2 capacity of 40–70% at temperatures 300–400 °C [95,96].

Materials are generally post-characterized for total sulfur content. However, a substantial concentration of SO_3 can be present in the flue gas well before the total capacity is reached [81]. This means that the total capacity of an absorber or adsorber is a poor indication of actual performance during operation. Therefore, the performance of catalytic strippers is characterized in real time.

Figure 6 plots a hypothetical sulfur storage test based on published similar studies [20,33,36,83,97–103]. At time t_0, SO_2 in O_2 and N_2 is inserted in the catalytic stripper. The SO_2 level is typically between 20 and 150 ppm (100% in the figure) in order to have adequate measurement accuracy with the SO_2 analyzer. Initially, the measured SO_2 outlet concentration is zero because the SO_2 is stored in the trap (either as SO_2 or metal sulfate after conversion to SO_3). At some point (t_b) breakthrough of SO_2 will start. From this point, the trapping efficiency is not 100%. SO_3 will also start to appear at the outlet of the catalytic stripper. For simplification, in this figure, it is assumed that SO_3 appears at the same time with the SO_2, but this is not necessarily true. This SO_3 poses a risk for particles formation. Figure 6b gives an example of a particle size distribution measured at the outlet of the catalytic stripper during the breakthrough period (t_2). Note that during the initial storage phase no particles were emitted (t_1). At point (t_{eq}) the sulfur storage capacity will be saturated and the SO_2 and SO_3 outlet concentrations will be stabilized. From t_{eq} the sulfur trap is behaving like an oxidation catalyst and produces SO_3 with high probability for particle formation (Figure 6b, size distribution t_3). The exact time of appearance and the magnitude of the size distributions depend on many parameters such as SO_2 levels, SO_2 to SO_3 conversion rate, humidity, and availability of hydrocarbons [33,66]. The transition from t_b to t_{eq} was plotted as a linear function for simplicity reasons; however, in reality it is sigmoidal-like curve.

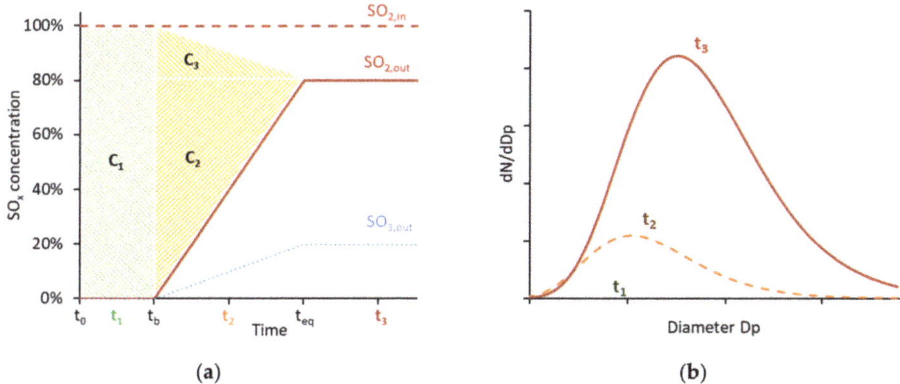

Figure 6. Hypothetical sulfur storage test with SO_2: (**a**) SO_x concentrations over time; (**b**) particle size distributions during the sulfur storage test.

The "total storage capacity" area is $C_1 + C_2 + C_3$. Note that usually there are no SO_3 measurements, so the determination of C_3 is not possible and not reported. Note also that even if the SO_3 breakthrough starts after the SO_2 breakthrough, the definitions of the areas remain the same, because if there is already SO_3 in the aerosol, it might not be completely stored in the sulfur trap already at time t_b. The important area, however, is C_1, which is called "complete storage capacity". The determination of the breakthrough point (t_b) is typically set to 1% SO_2 to cover the uncertainty of the SO_2 measurement, but 0% has also been used for catalytic strippers.

The sulfur mass C_m [mg/s] passing through the catalyst is:

$$C_m = 10^{-6} \times C_v \times \rho_{SO2} \times (Q/60) \times (M_S/M_{SO2}) \times 10^3, \tag{1}$$

where C_v is the SO_2 passing through the catalyst [ppmv], ρ_{SO2} is the density of SO_2 at 0 °C [2.86 g/L], Q is the catalytic stripper flow rate at 0 °C [L/min], M_S is the molecular mass of sulfur [32 g/mol], M_{SO2} is the molecular weight of sulfur dioxide [64 g/mol]. The sulfur capacity C [mg S] can be estimated by integrating C_m for the appropriate time and assuming a retention efficiency R. Typically, the "complete" capacity should be reported (C_1 in Figure 6a), where the retention efficiency R is 100% (integration until t_b). After t_b the retention efficiency can be estimated from the surface area of the graph.

$$C = \Sigma_{0-t} (R_i \times C_m). \tag{2}$$

The composition and the amount of the washcoat of catalytic strippers is not disclosed, so "total" captured sulfur mass is reported (or mass/L of catalyst). Values of approximately 6–23 mg or 0.7–1 g/L have been reported [33,36], with re-nucleation artefacts beginning already at 20–40% of the capacity. It can be assumed that research in this area can further increase the storage capacity. For example, a sulfur trap stored almost up to 15 g/L with no evidence of poisoning of a NO_x catalyst downstream of the sulfur trap [104].

Based on Equations (1) and (2), the time until saturation t [h] of a catalytic stripper with sulfur trap can be estimated based on the following equation:

$$t = 11.7 \times C/Q/E, \tag{3}$$

where E are the expected SO_x emissions [ppmv SO_x]. Assuming 2.5 mg effective (complete) sulfur capacity (C_1 in Figure 6a) and 1 L/min flowrate, for 0.5 ppm SO_x emissions, the time until saturation would be 58 h. Considering the dilution at the dilution tunnel and at the pre-diluter, this time would increase approximately 100 times. Similarly, a system at the tailpipe would also use a high dilution of at least 100:1. It should be highlighted that a catalytic stripper used directly at the tailpipe needs some dilution in order to ensure that enough oxygen will be available for oxidation at rich or stoichiometric modes of engine operation.

4.2.4. Position of Sulfur Trap

The first CS with sulfur trap had the trap upstream of the oxidation catalyst in order to protect the oxidation catalyst from poisoning [31,105], because the sulfates formed inhibit reactants from adsorbing on the active sites [77,106]. Tests showed that the total sulfur capacity was 60% lower compared to the configuration where the sulfur trap was downstream [33]. It was assumed that, the sulfur was primarily stored as SO_2, because no SO_3 was available upstream of the sulfur trap. Further tests showed that poisoning of the oxidation catalyst (saturated with sulfur) did not affect the volatile removal [36]. Nevertheless, the light-off temperatures were lower when the sulfur trap was upstream [33]. As the catalytic stripper is used at a constant high temperature (350 °C), the light-off temperature is not of concern. For these reasons, the downstream position is the preferred one in commercial systems.

4.3. Open Issues

The superiority of the catalytic stripper compared to the evaporation tube for sub-23 nm measurements was shown experimentally in many studies. The addition of a sulfur trap in many cases gave an advantage of retention of sulfuric acid. However, there are no technical requirements or specifications for catalytic strippers, so non-optimized systems might be used resulting in lower than expected performance. The following sections will discuss issues related to particle losses, sulfur storage capacity, and typical vehicles' aerosol.

4.3.1. Particle Losses

The particle losses in a volatile particle remover are due to diffusion and thermophoresis. The thermophoretic losses are around 25% without any dilution downstream of the volatile particle remover, but less than 10% with dilution. The diffusion losses are size-dependent and for catalytic strippers are from 15% to 40% more compared to 100 nm [29,33,36]. The thermophoretic losses are similar for catalytic strippers and evaporation tubes for similar setup. The major drawback of catalytic strippers versus evaporation tubes is the additional size dependent diffusion losses. Tests with a particle number system replacing the evaporation tube with a catalytic stripper gave <20% difference at the 10 nm losses [39]. The design of a catalytic stripper is a compromise between volatile removal efficiency and non-volatile particles' penetrations [29,44]. Particle number systems with optimized catalytic strippers have achieved <20% losses at 10 nm [29,36], but typically the losses are higher [39,107]. The regulation uses an average particle number concentration reduction factor (PCRF) of 30 nm, 50 nm, and 100 nm to take into account the particle losses and the dilution ratio. This average PCRF is based on the fact that vehicles' size distributions have geometric mean diameter around 50 nm. In order to minimize the influence of the small particle losses on the results, restrictions for the sub-23 nm sizes are necessary. The future >10 nm regulation will have a limit for the ratio $PCRF_{15}$ to $PCRF_{100}$ of <2. Another solution is to use two particle number counters with different cut-off sizes (at 23 and 10 nm, approximately) and correct for the losses below 23 nm [8].

For 30 nm (and larger) sizes the differences of the losses between identical systems with evaporation tubes or catalytic strippers are less than 10% [39]. This value is well within the range seen between various commercial systems with evaporation tubes. The average PCRF, which takes into account the losses at 30 nm, 50 nm and 100 nm, will partly offset the size-dependent losses. Furthermore, the penetration of the complete system does not depend so much on the VPR, but mainly on the PNC [76]. Consequently, the differences of the systems with 23 nm PNCs will be within the typical experimental uncertainties (5%). Indeed, most comparisons of >23 nm systems with and without catalytic strippers gave comparable results [39,75,76]. With the current regulations, catalytic strippers are not allowed in the volatile particle removers, because it is required that the materials do not react with exhaust gas components. However, assuming that a system with catalytic stripper fulfils all the rest requirements (e.g., $PCRF_{30}$ to $PCRF_{100}$ ratio <1.3), and based on the aforementioned findings, there is no reason to exclude catalytic strippers for >23 nm measurements. Thus, future regulation should permit the use of catalytic strippers even for measurements of particles >23 nm.

4.3.2. SO_3 Conversion Risks

The catalytic strippers with only the oxidation part do not convert SO_2 to SO_3, but this needs to be checked, because at the temperature required by the regulation (350 °C) some conversion may take place (as discussed in Figure 5b). This could result in higher risk for re-nucleation compared to an evaporation tube (where no SO_2 to SO_3 conversion takes place). In this case it is necessary that most of the hydrocarbons are removed, so the nuclei will not grow to the measurement range of the counting instruments.

Catalytic strippers with sulfur trap showed efficient removal even for high concentrations of sulfuric acid. However, the risk with these systems is that they may be saturated if proper attention is not given. At the moment there is no alarm/warning for such cases. Due to their high SO_2 to SO_3 conversion, when they are saturated they may result in higher nucleation probability compared to the simple oxidation catalytic strippers or evaporation tubes. For modern light-duty vehicles with low sulfur fuel and lubricant, the time until saturation is hundreds of hours (with some dilution) [33]. For applications using high sulfur content fuels (e.g., marine engines), however, they could saturate only in a couple of hours [35]. Maximum SO_2 concentrations are given in Table 2 for recent technologies (and fuels). Higher concentrations could be measured when the fuel sulfur content is >15 ppm, e.g., in India, Central and South America or in Africa [108]. For example, the expected SO_2 concentrations are <1 ppm for modern vehicles [35], and peaks of 20 ppm may appear, and in extreme cases (e.g., regenerations) even higher (150 ppm) [109]. However, measurement campaigns in countries with high sulfur content fuel have reported average SO_2 concentrations of 20–120 ppm [110] or even higher when there were maintenance issues with the vehicles [111]. As it was discussed in Figure 6b, high SO_2 or SO_3 concentrations can result in sulfuric acid nucleation particles peaking at sizes >10 nm [33,66].

In addition to the SO_2 to SO_3 reaction, it is not well understood how the catalytic strippers react with the exhaust aerosol. In a few cases non-volatile particle formation has been found in evaporation tubes, but also in catalytic strippers [32,41]. In another case, a catalytic stripper measured higher non-volatile particle concentrations than an evaporation tube [50]. This is an area that needs further research.

4.3.3. Vehicles' Volatile Aerosol

Table 2 gives the mass of volatiles reported as nucleation mode (NM) particles in the literature. These are the actual (maximum) levels that particle number systems may be challenged with. What is not always known is the chemical composition, and when detailed analysis is undertaken, hundreds of compounds can be found [112]. Furthermore, it is not possible to define easily a "typical" composition as it depends on the engine out emissions, the aftertreatment devices, the fuel, the lubricant, and the driving style. For example, for a heavy-duty engine without aftertreatment devices, branched alkanes and alkyl-substituted cycloalkanes from unburned fuel and/or lubricating oil contributed >95% of the nanoparticle mass, and sulfuric acid <5% [113]. Another study indicated that the organic component of diesel nanoparticles comprised compounds with carbon numbers in the C24–C32 range, derived almost entirely from unburned oil [114]. For a heavy-duty diesel vehicle with a catalyzed continuously regenerating trap (CRT) and 420 ppm sulfur fuel the majority of the NM was sulfates [115]. Similarly, for light-duty NM particles, the sulfate to organic ratios were from 0.2 to 20 depending on the engine load (and exhaust gas temperature) for 350 ppm fuel sulfur [116]. Chemical analysis of a heavy-duty DPF having mainly nucleation mode particles (1.2×10^{11} p/km) showed a high amount of BTEX compounds (benzene, toluene, ethylbenzene, m,p-xylenes, and o-xylene) and PAHs (polycyclic aromatic hydrocarbons) [117]. Similar results were noted for light-duty vehicles and, additionally, alkanes had high concentrations [118]. Toluene, isopentane, m,p-xylene, 1,2,4-trimethylbenzene, and o-xylene were the dominant volatile organic compounds emitted from the tailpipe exhaust of the motorcycles [119].

Table 2. Peak concentrations during cold start, during the test cycle or regenerations. HCs and SO_2 concentrations refer to the tailpipe, NM to the dilution tunnel. HC = hydrocarbons; HD = heavy-duty; LD = light-duty; NM = nucleation mode.

Technology	Period	NM [mg/m³]	Gas HCs [ppm]	SO_2 [ppm]	References
Moped	Cold start	>10	>50,000	60	[8,34,49]
Motorcycle	Cold start		10,000–35,000	12–20	[120–122]
Motorcycle	Cycle	0.4	300–5000	9–20	[123,124]
LD Gasoline	Cold start	5	12,000	10	[125–130]
LD Diesel	Cold start	low	350	10	[128,131]
LD Diesel	Regeneration	1	200	25	[57,132–134]
HD Diesel	Regeneration	1–12	250	150	[109,115,135–139]

For completeness, examples of measured gaseous hydrocarbons are given in Table 1. They are measured according to the regulated procedure (with heated line at 191 °C), thus they represent "light" hydrocarbons and they do not necessarily correlate with the NM reported in the same table. Furthermore, typically less than 30% of total HC is condensed on soot particles. The catalytic strippers remove the hydrocarbons more efficiently in the gas phase than in the particle phase [33]. Thus, the interpretation of these concentrations as requirements for catalytic strippers should be done with precaution. Moreover, the gas oxidation efficiency check may not be necessary if the volatile particles' removal efficiency is checked. Nevertheless, it could be a useful and easy check for the long term volatile removal efficiency stability, especially as the catalytic stripper is poisoned by sulfur compounds over time. At the temperatures prescribed in the regulations (300–400 °C) the oxidation efficiency of CH_4 is low, thus a gas such as propane is more suitable and also typically available in most laboratories that do vehicle emissions testing (for calibration of other gas analyzers).

4.3.4. Catalytic Stripper Technical Requirements

Table 3 gives suggestions for future tests that could be required for the characterization of catalytic strippers (standalone or in a VPR system). Examples and citations where more details can be found are also given. Regarding particle losses, the requirements of the regulation should be fulfilled, i.e., $PCRF_{15}$ to $PCRF_{100}$ ratio <2. The wall temperature is also defined in the requirements, usually between 300 °C to 400 °C. For catalytic strippers, it seems that the lower temperature limit is more advantageous because the SO_2 to SO_3 conversion is lower. This conversion ratio should be reported, especially for systems without a sulfur trap, which should have very low conversion. Systems with a sulfur trap should report the sulfur storage capacity. Two values should be reported: the complete (i.e., until no SO_2 is measured downstream of the catalytic stripper) and the total (i.e., until the SO_2 concentration at the outlet of the catalytic stripper is stabilized). The measurement of SO_3 (and H_2SO_4) is challenging. In the presence of water vapor in particular, sulfuric acid will be formed, that is corrosive and there are concerns regarding potential damage to the analytical equipment or even artefacts [81]. The particle removal efficiency should also be given. At the moment the preferred generation method is evaporation-condensation technique with tetracontane particles, as the atomization method of oils can result in residual particles from impurities in the atomized material. Even though the oxidation efficiency (e.g., of propane) is not so relevant for the specific application of catalytic strippers, it could be evaluated as a simple check of the long-term stability of the oxidation efficiency.

Table 3. Suggestions for technical requirements of catalytic strippers. The references give the experimental setups of the proposed tests. HC = hydrocarbons; VRE = volatile removal efficiency.

Test	Comment	Example	Reference
Particle losses	As required in the regulation		[33,36]
Wall temperature	As required in the regulation	300 °C to 350 °C	Regulation
SO_2 to SO_3 conversion	To be declared	%	[33,36]
Sulfur storage capacity	Based on SO_2	mg	[33,36]
VRE H_2SO_4	Feasibility to be assessed	>99.9%; >1 mg/m^3	[39,44]
VRE gaseous HCs	Long term efficiency C_3H_8	>99.9%; >10,000 ppm	[33]
VRE particle HCs	Tetracontane particles	>99.9%; >1 mg/m^3	[30,33,36]

5. Conclusions

This review summarized the volatile removal efficiency of evaporation tubes (ET), thermodenuders (TD), and catalytic strippers (CS). Some of the catalytic strippers had only an oxidation part and some had, additionally, a sulfur trap upstream or downstream of the oxidation part.

Theoretically, CS oxidize the hydrocarbons resulting in smaller (if any) growth of any volatile nucleation mode particles, compared to evaporation tubes that reduce hydrocarbons only by dilution. CS with sulfur traps store sulfur compounds, thus, further minimizing any possibility of nucleating particles. On the other hand, if the sulfur trap is saturated, any existing SO_3 or SO_3 formed in the CS might enhance nucleation. At very high SO_2 and/or SO_3 concentrations the nucleated particles may grow to the measuring range of the instruments. Considering that modern vehicles emit some SO_3 (and thus there is potential for nucleation) and the ET does not oxidize the hydrocarbons, the

CS is a safer option for avoiding volatile artefacts (i.e., measuring re-nucleated volatile particles as non-volatiles).

The summary of the published experimental results showed that ET and TD could handle up to 2 mg/m^3 of hydrocarbons without artefacts, but in some cases particles <10 nm remained. Masses >2 mg/m^3 resulted in artefacts in the 10–23 nm range or even >23 nm (15 mg/m^3) on some occasions. CS did not have any artefacts >10 nm in any of the studies (with one exception that were presumably solid residuals). Inlet masses up to 13 g/m^3 could be handled with CS in most cases without any particles remaining, but in some cases particles <10 nm were detected after CS even at hydrocarbon compound masses below 2 mg/m^3.

With sulfuric acid containing aerosol, particles were always detected downstream of ET and TD: they were <10 nm with <0.2 mg/m^3 upstream concentrations (with one exception of hot dilution and evaporation tube that could handle up to 2.5 mg/m^3). The CS could efficiently handle the sulfuric acid aerosol. However, at concentrations >1 mg/m^3 some particles below 10 nm remained. The limited data show that with a sulfur trap higher masses can be handled. In two cases, CS with a sulfur trap could handle at least 9 mg/m^3 without any particles remaining.

The vehicle exhaust aerosol had remaining particles larger than 10 nm for masses >0.5 mg/m^3 and <10 nm even for masses of 0.05 mg/m^3 for the ET systems. The CS had no cases with remaining particles >10 nm. There were cases with particles <10 nm when the mass was 0.65 mg/m^3 (standalone) or 50 mg/m^3 (downstream of hot dilution).

In general, the experimental results of the reviewed studies showed that the CS is the safest option for minimum artefacts for measurements >10 nm. Studies below 10 nm still need attention, even with CS, both regarding the volatile removal efficiency, but also the high particle losses. The above results also confirm that the CS should be allowed in place or in addition to the ET of the existing systems measuring >23 nm.

Due to the substantial variability of the results and the different removal efficiencies of the various systems, in order to standardize the evaluation of the systems, it is recommended to include a volatile removal efficiency test. Realistic levels of challenge aerosol should be in the order of 1 mg/m^3 (except mopeds that exceed 10 mg/m^3). At the moment, it is not clear what is a representative chemical composition of the volatile particles, but it seems that sulfuric acid plays a major role in their formation and should also be included in the volatile removal assessment of the systems. However, H_2SO_4 is corrosive and there are concerns regarding potential damage to the analytical equipment or even artefacts. Catalytic strippers with sulfur traps should report their storage capacity and those without, should confirm that no (or minimum) SO_2 to SO_3 conversion takes place.

Author Contributions: Conceptualization, B.G.; formal analysis, B.G.; writing—original draft preparation, B.G. and A.D.M.; writing—review and editing, T.L. and G.M.; All authors have read and agreed to the published version of the manuscript.

Funding: This research received no external funding.

Acknowledgments: The authors would like to acknowledge the comments of A. Mamakos and Y. Otsuki at earlier draft versions. B.G. appreciates the discussions with L. Ntziachristos.

Conflicts of Interest: The authors declare no conflict of interest.

References

1. Giechaskiel, B.; Maricq, M.; Ntziachristos, L.; Dardiotis, C.; Wang, X.; Axmann, H.; Bergmann, A.; Schindler, W. Review of motor vehicle particulate emissions sampling and measurement: From smoke and filter mass to particle number. *J. Aerosol Sci.* **2014**, *67*, 48–86. [CrossRef]
2. Giechaskiel, B.; Mamakos, A.; Andersson, J.; Dilara, P.; Martini, G.; Schindler, W.; Bergmann, A. Measurement of automotive nonvolatile particle number emissions within the European legislative framework: A review. *Aerosol Sci. Technol.* **2012**, *46*, 719–749. [CrossRef]

3. Giechaskiel, B.; Lähde, T.; Suarez-Bertoa, R.; Clairotte, M.; Grigoratos, T.; Zardini, A.; Perujo, A.; Martini, G. Particle number measurements in the European legislation and future JRC activities. *Combust. Engines* **2018**, *174*, 3–16. [CrossRef]
4. Giechaskiel, B.; Joshi, A.; Ntziachristos, L.; Dilara, P. European regulatory framework and particulate matter emissions of gasoline light-duty vehicles: A review. *Catalysts* **2019**, *9*, 586. [CrossRef]
5. Giechaskiel, B.; Drossinos, Y. Theoretical investigation of volatile removal efficiency of particle number measurement systems. *SAE Int. J. Engines* **2010**, *3*, 1140–1151. [CrossRef]
6. Burtscher, H. Physical characterization of particulate emissions from diesel engines: A review. *J. Aerosol Sci.* **2005**, *36*, 896–932. [CrossRef]
7. Giechaskiel, B.; Bonnel, P.; Perujo, A.; Dilara, P. Solid particle number (SPN) portable emissions measurement systems (PEMS) in the European legislation: A review. *IJERPH* **2019**, *16*, 4819. [CrossRef]
8. Giechaskiel, B.; Vanhanen, J.; Väkevä, M.; Martini, G. Investigation of vehicle exhaust sub-23 nm particle emissions. *Aerosol Sci. Technol.* **2017**, *51*, 626–641. [CrossRef]
9. Johnson, K.C.; Durbin, T.D.; Jung, H.; Chaudhary, A.; Cocker, D.R.; Herner, J.D.; Robertson, W.H.; Huai, T.; Ayala, A.; Kittelson, D. Evaluation of the European PMP methodologies during on-road and chassis dynamometer testing for dpf equipped heavy-duty diesel vehicles. *Aerosol Sci. Technol.* **2009**, *43*, 962–969. [CrossRef]
10. Giechaskiel, B.; Manfredi, U.; Martini, G. Engine exhaust solid sub-23 nm particles: I. literature survey. *SAE Int. J. Fuels Lubr.* **2014**, *7*, 950–964. [CrossRef]
11. Rönkkö, T.; Virtanen, A.; Kannosto, J.; Keskinen, J.; Lappi, M.; Pirjola, L. Nucleation mode particles with a nonvolatile core in the exhaust of a heavy duty diesel vehicle. *Environ. Sci. Technol.* **2007**, *41*, 6384–6389. [CrossRef]
12. Giechaskiel, B. Solid particle number emission factors of Euro VI heavy-duty vehicles on the road and in the laboratory. *IJERPH* **2018**, *15*, 304. [CrossRef] [PubMed]
13. Mayer, A.; Czerwinski, J.; Kasper, M.; Ulrich, A.; Mooney, J.J. *Metal Oxide Particle Emissions from Diesel and Petrol Engines*; SAE: Warrendale, PA, USA, 2012.
14. Giechaskiel, B.; Martini, G. Engine exhaust solid sub-23 nm particles: II feasibility study for particle number measurement systems. *SAE Int. J. Fuels Lubr.* **2014**, *7*, 935–949. [CrossRef]
15. Giechaskiel, B.; Mamakos, A.; Woodburn, J.; Szczotka, A.; Bielaczyc, P. Evaluation of a 10 nm particle number portable emissions measurement system (PEMS). *Sensors* **2019**, *19*, 5531. [CrossRef] [PubMed]
16. Giechaskiel, B.; Lähde, T.; Gandi, S.; Keller, S.; Kreutziger, P.; Mamakos, A. Assessment of 10-nm particle number (PN) portable emissions measurement systems (PEMS) for future regulations. *IJERPH* **2020**, *17*, 3878. [CrossRef] [PubMed]
17. Abdul-Khalek, I.S.; Kittelson, D.B. Real time measurement of volatile and solid exhaust particles using a catalytic stripper. *SAE Trans.* **1995**, *104*, 462–478. [CrossRef]
18. Twigg, M.V. Catalytic control of emissions from cars. *Catal. Today* **2011**, *163*, 33–41. [CrossRef]
19. Roy, S.; Baiker, A. NOx storage–reduction catalysis: From mechanism and materials properties to storage–reduction performance. *Chem. Rev.* **2009**, *109*, 4054–4091. [CrossRef]
20. Liu, G.; Gao, P.-X. A review of NOx storage/reduction catalysts: Mechanism, materials and degradation studies. *Catal. Sci. Technol.* **2011**, *1*, 552–568. [CrossRef]
21. Xu, F.; Chen, L.; Stone, R. Effects of a catalytic volatile particle remover (VPR) on the particulate matter emissions from a direct injection spark ignition engine. *Environ. Sci. Technol.* **2011**, *45*, 9036–9043. [CrossRef]
22. Lobo, P.; Durdina, L.; Brem, B.T.; Crayford, A.P.; Johnson, M.P.; Smallwood, G.J.; Siegerist, F.; Williams, P.I.; Black, E.A.; Llamedo, A.; et al. Comparison of standardized sampling and measurement reference systems for aircraft engine non-volatile particulate matter emissions. *J. Aerosol Sci.* **2020**, *145*, 105557. [CrossRef]
23. Lin, Y.; Bahreini, R.; Zimmerman, S.; Fofie, E.A.; Asa-Awuku, A.; Park, K.; Lee, S.-B.; Bae, G.-N.; Jung, H.S. Investigation of ambient aerosol effective density with and without using a catalytic stripper. *Atmos. Environ.* **2018**, *187*, 84–92. [CrossRef]
24. Burtscher, H.; Baltensperger, U.; Bukowiecki, N.; Cohn, P.; Hüglin, C.; Mohr, M.; Matter, U.; Nyeki, S.; Schmatloch, V.; Streit, N.; et al. Separation of volatile and non-volatile aerosol fractions by thermodesorption: Instrumental development and applications. *J. Aerosol Sci.* **2001**, *32*, 427–442. [CrossRef]

25. Huffman, J.A.; Ziemann, P.J.; Jayne, J.T.; Worsnop, D.R.; Jimenez, J.L. Development and characterization of a fast-stepping/scanning thermodenuder for chemically-resolved aerosol volatility measurements. *Aerosol Sci. Technol.* **2008**, *42*, 395–407. [CrossRef]

26. Donahue, N.M.; Kroll, J.H.; Pandis, S.N.; Robinson, A.L. A two-dimensional volatility basis set—Part 2: Diagnostics of organic-aerosol evolution. *Atmos. Chem. Phys.* **2012**, *12*, 615–634. [CrossRef]

27. Maricq, M.M. Chemical characterization of particulate emissions from diesel engines: A review. *J. Aerosol Sci.* **2007**, *38*, 1079–1118. [CrossRef]

28. Choi, S.; Myung, C.L.; Park, S. Review on characterization of nano-particle emissions and PM morphology from internal combustion engines: Part 2. *Int. J. Autom. Technol.* **2014**, *15*, 219–227. [CrossRef]

29. Khalek, I.A. Sampling system for solid and volatile exhaust particle size, number, and mass emissions. *SAE Trans.* **2007**, *116*, 122–133. [CrossRef]

30. Khalek, I.A.; Bougher, T. Development of a solid exhaust particle number measurement system using a catalytic stripper technology. *SAE Int. J. Engines* **2011**, *4*, 610–618. [CrossRef]

31. Kittelson, D.B.; Stenitzer, M. A new catalytic stripper for removal of volatile particles. In Proceedings of the 7th ETH Conference on Combustion Generated Nanoparticles, Zurich, Switzerland, 18–20 August 2003.

32. Swanson, J.; Kittelson, D. Evaluation of thermal denuder and catalytic stripper methods for solid particle measurements. *J. Aerosol Sci.* **2010**, *41*, 1113–1122. [CrossRef]

33. Amanatidis, S.; Ntziachristos, L.; Giechaskiel, B.; Katsaounis, D.; Samaras, Z.; Bergmann, A. Evaluation of an oxidation catalyst ("catalytic stripper") in eliminating volatile material from combustion aerosol. *J. Aerosol Sci.* **2013**, *57*, 144–155. [CrossRef]

34. Giechaskiel, B.; Zardini, A.; Martini, G. Particle emission measurements from L-category vehicles. *SAE Int. J. Engines* **2015**, *8*, 2322–2337. [CrossRef]

35. Amanatidis, S.; Ntziachristos, L.; Karjalainen, P.; Saukko, E.; Simonen, P.; Kuittinen, N.; Aakko-Saksa, P.; Timonen, H.; Rönkkö, T.; Keskinen, J. Comparative performance of a thermal denuder and a catalytic stripper in sampling laboratory and marine exhaust aerosols. *Aerosol Sci. Technol.* **2018**, *52*, 420–432. [CrossRef]

36. Melas, A.D.; Koidi, V.; Deloglou, D.; Daskalos, E.; Zarvalis, D.; Papaioannou, E.; Konstandopoulos, A.G. Development and evaluation of a catalytic stripper for the measurement of solid ultrafine particle emissions from internal combustion engines. *Aerosol Sci. Technol.* **2020**, *54*, 704–714. [CrossRef]

37. Giechaskiel, B.; Riccobono, F.; Bonnel, P. *Feasibility Study on the Extension of the Real-Driving Emissions (RDE) Procedure to Particle Number (PN): Chassis Dynamometer Evaluation of Portable Emission Measurement Systems (PEMS) to Measure Particle Number (PN) Concentration: Phase II*; Publications Office: Luxembourg, 2015; ISBN 978-92-79-51003-8.

38. Otsuki, Y.; Tochino, S.; Kondo, K.; Haruta, K. *Portable Emissions Measurement System for Solid Particle Number Including Nanoparticles Smaller than 23 nm*; SAE International: Warrendale, PA, USA, 2017.

39. Otsuki, Y.; Takeda, K.; Haruta, K.; Mori, N. *A Solid Particle Number Measurement System Including Nanoparticles Smaller than 23 Nanometers*; SAE International: Warrendale, PA, USA, 2014.

40. Kiwull, B.; Wolf, J.C.; Niessner, R. *Evaluation of Volatile Particle Remover Devices for Exhaust Particle Quantification*; Technische Universität München: Munich, Germany, 2015.

41. Zheng, Z.; Johnson, K.C.; Liu, Z.; Durbin, T.D.; Hu, S.; Huai, T.; Kittelson, D.B.; Jung, H.S. Investigation of Solid Particle Number Measurement: Existence and Nature of Sub-23nm Particles Under PMP Methodology. *J. Aerosol Sci.* **2011**, *42*, 883–897. [CrossRef]

42. Kasper, M. *Characterisation of the Second Generation PMP "Golden Instrument"*; SAE International: Warrendale, PA, USA, 2008.

43. Giechaskiel, B.; Carriero, M.; Martini, G.; Krasenbrink, A.; Scheder, D. Calibration and validation of various commercial particle number measurement systems. *SAE Int. J. Fuels Lubr.* **2009**, *2*, 512–530. [CrossRef]

44. Swanson, J.; Kittelson, D.; Giechaskiel, B.; Bergmann, A.; Twigg, M. A miniature catalytic stripper for particles less than 23 nanometers. *SAE Int. J. Fuels Lubr.* **2013**, *6*, 542–551. [CrossRef]

45. Emery Oil Removal Efficiency. Available online: http://catalytic-instruments.com/wp-content/uploads/2019/09/Application-note0007.pdf (accessed on 24 June 2020).

46. Crayford, A.P.; Johnson, M.; Marsh, R.; Sevcenco, Y.; Walters, D.; Williams, P.; Christie, S.; Chung, W.; Petzold, A.; Ibrahim, A.; et al. *SAMPLE III (SC01): Contribution to Aircraft Engine PM Certification Requirement and Standard First Specific Contract—Final Report*; EASA: Cologne, Germany, 2011.

47. Kim, S.; Kondo, K.; Otsuki, Y.; Haruta, K. *A New On-Board PN Analyzer for Monitoring the Real-Driving Condition*; SAE International: Warrendale, PA, USA, 2017.

48. Ho, J.; Kournikakis, B.; Gunning, A.; Fildes, J. Submicron aerosol characterization of water by a differential mobility particle sizer. *J. Aerosol Sci.* **1988**, *19*, 1425–1428. [CrossRef]

49. Giechaskiel, B.; Chirico, R.; DeCarlo, P.F.; Clairotte, M.; Adam, T.; Martini, G.; Heringa, M.F.; Richter, R.; Prevot, A.S.H.; Baltensperger, U. Evaluation of the particle measurement programme (PMP) protocol to remove the vehicles' exhaust aerosol volatile phase. *Sci. Total Environ.* **2010**, *408*, 5106–5116. [CrossRef]

50. Giechaskiel, B. Differences between tailpipe and dilution tunnel sub-23 nm nonvolatile (solid) particle number measurements. *Aerosol Sci. Technol.* **2019**, *53*, 1012–1022. [CrossRef]

51. Mamakos, A.; Martini, G.; Marotta, A.; Manfredi, U. Assessment of different technical options in reducing particle emissions from gasoline direct injection vehicles. *J. Aerosol Sci.* **2013**, *63*, 115–125. [CrossRef]

52. Ntziachristos, L.; Amanatidis, S.; Samaras, Z.; Giechaskiel, B.; Bergmann, A. Use of a catalytic stripper as an alternative to the original PMP measurement protocol. *SAE Int. J. Fuels Lubr.* **2013**, *6*, 532–541. [CrossRef]

53. Mamakos, A.; Martini, G.; Manfredi, U. Assessment of the legislated particle number measurement procedure for a Euro 5 and a Euro 6 compliant diesel passenger cars under regulated and unregulated conditions. *J. Aerosol Sci.* **2013**, *55*, 31–47. [CrossRef]

54. Giechaskiel, B.; Riccobono, F.; Mendoza-Villafuerte, P.; Grigoratos, T. *Particle Number (PN)—Portable Emissions Measurement Systems (PEMS) Heavy Duty Vehicles Evaluation Phase at the Joint Research Centre (JRC)*; Publications Office: Luxembourg, 2016.

55. Zheng, Z.; Durbin, T.D.; Karavalakis, G.; Johnson, K.C.; Chaudhary, A.; Cocker, D.R.; Herner, J.D.; Robertson, W.H.; Huai, T.; Ayala, A.; et al. Nature of sub-23-nm particles downstream of the European particle measurement programme (PMP)-compliant system: A real-time data perspective. *Aerosol Sci. Technol.* **2012**, *46*, 886–896. [CrossRef]

56. Ntziachristos, L.; Giechaskiel, B.; Pistikopoulos, P.; Samaras, Z. *Comparative Assessment of Two Different Sampling Systems for Particle Emission Type-Approval Measurements*; SAE International: Warrendale, PA, USA, 2005.

57. Giechaskiel, B. Particle number emissions of a diesel vehicle during and between regeneration events. *Catalysts* **2020**. [CrossRef]

58. Yu, F. Effect of ammonia on new particle formation: A kinetic H_2SO_4-H_2O-NH_3 nucleation model constrained by laboratory measurements. *J. Geophys. Res.* **2006**, *111*. [CrossRef]

59. Lemmetty, M.; Vehkamäki, H.; Virtanen, A.; Kulmala, M.; Keskinen, J. Homogeneous ternary H_2SO_4–NH_3–H_2O nucleation and diesel exhaust: A classical approach. *Aerosol Air Qual. Res.* **2007**, *7*, 489–499. [CrossRef]

60. Benson, D.R.; Yu, J.H.; Markovich, A.; Lee, S.-H. Ternary homogeneous nucleation of H_2SO_4, NH_3, and H_2O under conditions relevant to the lower troposphere. *Atmos. Chem. Phys.* **2011**, *11*, 4755–4766. [CrossRef]

61. Suarez-Bertoa, R.; Pechout, M.; Vojtíšek, M.; Astorga, C. Regulated and non-regulated emissions from Euro 6 diesel, gasoline and CNG vehicles under real-world driving conditions. *Atmosphere* **2020**, *11*, 204. [CrossRef]

62. Czerwinski, J.; Zimmerli, Y.; Mayer, A.; Heeb, N.; Lemaire, J.; D'Urbano, G.; Bunge, R. *Testing of Combined DPF+SCR Systems for HD-Retrofitting—VERTdePN*; SAE International: Warrendale, PA, USA, 2009.

63. Sipila, M.; Berndt, T.; Petaja, T.; Brus, D.; Vanhanen, J.; Stratmann, F.; Patokoski, J.; Mauldin, R.L.; Hyvarinen, A.-P.; Lihavainen, H.; et al. The role of sulfuric acid in atmospheric nucleation. *Science* **2010**, *327*, 1243–1246. [CrossRef]

64. Uhrner, U.; von Löwis, S.; Vehkamäki, H.; Wehner, B.; Bräsel, S.; Hermann, M.; Stratmann, F.; Kulmala, M.; Wiedensohler, A. Dilution and aerosol dynamics within a diesel car exhaust plume—CFD simulations of on-road measurement conditions. *Atmos. Environ.* **2007**, *41*, 7440–7461. [CrossRef]

65. Olin, M.; Alanen, J.; Palmroth, M.R.T.; Rönkkö, T.; Dal Maso, M. Inversely modeling homogeneous H_2SO_4—H_2O nucleation rate in exhaust-related conditions. *Atmos. Chem. Phys.* **2019**, *19*, 6367–6388. [CrossRef]

66. Karjalainen, P.; Rönkkö, T.; Pirjola, L.; Heikkilä, J.; Happonen, M.; Arnold, F.; Rothe, D.; Bielaczyc, P.; Keskinen, J. Sulfur driven nucleation mode formation in diesel exhaust under transient driving conditions. *Environ. Sci. Technol.* **2014**, 2336–2343. [CrossRef] [PubMed]

67. Du, H.; Yu, F. Nanoparticle formation in the exhaust of vehicles running on ultra-low sulfur fuel. *Atmos. Chem. Phys.* **2008**, *8*, 4729–4739. [CrossRef]

68. Vaaraslahti, K.; Keskinen, J.; Giechaskiel, B.; Solla, A.; Murtonen, T.; Vesala, H. Effect of lubricant on the formation of heavy-duty diesel exhaust nanoparticles. *Environ. Sci. Technol.* **2005**, *39*, 8497–8504. [CrossRef]
69. Lemmetty, M.; Rönkkö, T.; Virtanen, A.; Keskinen, J.; Pirjola, L. The effect of sulphur in diesel exhaust aerosol: Models compared with measurements. *Aerosol Sci. Technol.* **2008**, *42*, 916–929. [CrossRef]
70. Khalek, I.A.; Kittelson, D.B.; Brear, F. *Nanoparticle Growth During Dilution and Cooling of Diesel Exhaust: Experimental Investigation and Theoretical Assessment*; SAE International: Warrendale, PA, USA, 2000.
71. Arnold, F.; Pirjola, L.; Rönkkö, T.; Reichl, U.; Schlager, H.; Lähde, T.; Heikkilä, J.; Keskinen, J. First online measurements of sulfuric acid gas in modern heavy-duty diesel engine exhaust: Implications for nanoparticle formation. *Environ. Sci. Technol.* **2012**, *46*, 11227–11234. [CrossRef]
72. Vouitsis, E.; Ntziachristos, L.; Samaras, Z. Modelling of diesel exhaust aerosol during laboratory sampling. *Atmos. Environ.* **2005**, *39*, 1335–1345. [CrossRef]
73. Vouitsis, E.; Ntziachristos, L.; Samaras, Z. Theoretical investigation of the nucleation mode formation downstream of diesel after-treatment devices. *Aerosol Air Qual. Res.* **2008**, *8*, 37–53. [CrossRef]
74. Yamada, H.; Funato, K.; Sakurai, H. Application of the PMP methodology to the measurement of sub-23 nm solid particles: Calibration procedures, experimental uncertainties, and data correction methods. *J. Aerosol Sci.* **2015**, *88*, 58–71. [CrossRef]
75. Giechaskiel, B.; Martini, G. JRC sub-23 nm update. In Proceedings of the 37th PMP meeting, Brussels, Belgium, 7 October 2015.
76. AVL Comments on GTR15 Ammendment. Available online: https://wiki.unece.org/display/trans/PMP+Web+Conference+02+April (accessed on 24 June 2020).
77. Neyestanaki, A.K.; Klingstedt, F.; Salmi, T.; Murzin, D.Y. Deactivation of postcombustion catalysts, a review. *Fuel* **2004**, *83*, 395–408. [CrossRef]
78. Giechaskiel, B.; Ntziachristos, L.; Samaras, Z.; Casati, R.; Scheer, V.; Vogt, R. *Effect of Speed and Speed-Transition on the Formation of Nucleation Mode Particles from a Light Duty Diesel Vehicle*; SAE International: Warrendale, PA, USA, 2007.
79. Cooper, B.J.; Thoss, J.E. Role of NO in diesel particulate emission control. *SAE Trans.* **1989**, *98*, 612–624. [CrossRef]
80. Kröcher, O.; Widmer, M.; Elsener, M.; Rothe, D. Adsorption and desorption of SOx on diesel oxidation catalysts. *Ind. Eng. Chem. Res.* **2009**, *48*, 9847–9857. [CrossRef]
81. Li, L.; King, D.L. Method for determining performance of sulfur oxide adsorbents for diesel emission control using online measurements of SO_2 and SO_3 in the effluent. *Ind. Eng. Chem. Res.* **2004**, *43*, 4452–4456. [CrossRef]
82. Koutsopoulos, S.; Rasmussen, S.B.; Eriksen, K.M.; Fehrmann, R. The role of support and promoter on the oxidation of sulfur dioxide using platinum based catalysts. *Appl. Catal. A Gen.* **2006**, *306*, 142–148. [CrossRef]
83. Xue, E.; Seshan, K.; van Ommen, J.G.; Ross, J.R.H. Catalytic control of diesel engine particulate emission: Studies on model reactions over a EuroPt-1 (Pt/SiO_2) catalyst. *Appl. Catal. B Environ.* **1993**, *2*, 183–197. [CrossRef]
84. Wyatt, M.; Manning, W.A.; Roth, S.A.; D'Aniello, M.J.; Andersson, E.S.; Fredholm, S.C.G. The design of flow-through diesel oxidation catalysts. *SAE Trans.* **1993**, *102*, 211–223. [CrossRef]
85. Horiuchi, M.; Saito, K.; Ichihara, S. The effects of flow-through type oxidation catalysts on the particulate reduction of 1990's diesel engines. *SAE Trans.* **1990**, *99*, 1268–1278. [CrossRef]
86. Hamzehlouyan, T.; Sampara, C.; Li, J.; Kumar, A.; Epling, W. Experimental and kinetic study of SO_2 oxidation on a Pt/γ-Al_2O_3 catalyst. *Appl. Catal. B Environ.* **2014**, *152*, 108–116. [CrossRef]
87. Koltsakis, G. Catalytic automotive exhaust aftertreatment. *Prog. Energy Combust. Sci.* **1997**, *23*, 1–39. [CrossRef]
88. Mathieu, Y.; Tzanis, L.; Soulard, M.; Patarin, J.; Vierling, M.; Molière, M. Adsorption of SOx by oxide materials: A review. *Fuel Process. Technol.* **2013**, *114*, 81–100. [CrossRef]
89. Smedler, G.; Ahlström, G.; Fredholm, S.; Frost, J.; Lööf, P.; Marsh, P.; Walker, A.; Winterborn, D. *High Performance Diesel Catalysts for Europe Beyond 1996*; SAE International: Warrendale, PA, USA, 1995.
90. Henk, M.G.; Williamson, W.B.; Silver, R.G. *Diesel Catalysts for Low Particulate and Low Sulfate Emissions*; SAE International: Warrendale, PA, USA, 1992.
91. Lampert, J.; Kazi, M.; Farrauto, R. Palladium catalyst performance for methane emissions abatement from lean burn natural gas vehicles. *Appl. Catal. B Environ.* **1997**, *14*, 211–223. [CrossRef]

92. Kärkkäinen, M.; Honkanen, M.; Viitanen, V.; Kolli, T.; Valtanen, A.; Huuhtanen, M.; Kallinen, K.; Vippola, M.; Lepistö, T.; Lahtinen, J.; et al. Deactivation of diesel oxidation catalysts by sulphur in laboratory and engine-bench scale aging. *Top Catal.* **2013**, *56*, 672–678. [CrossRef]

93. Taylor, K.C. Sulfur storage on automotive catalysts. *Ind. Eng. Chem. Prod. Res. Dev.* **1976**, *15*, 264–268. [CrossRef]

94. Wang, Q.; Zhu, J.; Wei, S.; Chung, J.S.; Guo, Z. Sulfur poisoning and regeneration of NOx storage–reduction Cu/K2Ti2O5 catalyst. *Ind. Eng. Chem. Res.* **2010**, *49*, 7330–7335. [CrossRef]

95. Liu, X.; Chen, L.; Qi, G. Enhanced SO_2 capture performance of MnO2 by doping with alkali metal ions for diesel emission control. *Chem. Eng. Technol.* **2018**, *41*, 1675–1681. [CrossRef]

96. Li, L.; King, D.L. High-capacity sulfur dioxide absorbents for diesel emissions control. *Ind. Eng. Chem. Res.* **2005**, *44*, 168–177. [CrossRef]

97. Horiuchi, M.; Saito, K.; Ichihara, S. *Sulfur Storage and Discharge Behavior on Flow-Through Type Oxidation Catalysts*; SAE International: Warrendale, PA, USA, 1991.

98. Centi, G.; Passarini, N.; Perathoner, S.; Riva, A. Combined DeSOx/DeNOx reactions on a copper on alumina sorbent-catalyst. 1. Mechanism of sulfur dioxide oxidation-adsorption. *Ind. Eng. Chem. Res.* **1992**, *31*, 1947–1955. [CrossRef]

99. Limousy, L.; Mahzoul, H.; Brilhac, J.F.; Gilot, P.; Garin, F.; Maire, G. SO_2 sorption on fresh and aged SOx traps. *Appl. Catal. B Environ.* **2003**, *42*, 237–249. [CrossRef]

100. Dawody, J.; Skoglundh, M.; Olsson, L.; Fridell, E. Sulfur deactivation of Pt/SiO$_2$, Pt/BaO/Al$_2$O$_3$, and BaO/Al$_2$O$_3$ NOx storage catalysts: Influence of SO_2 exposure conditions. *J. Catal.* **2005**, *234*, 206–218. [CrossRef]

101. Dawody, J.; Skoglundh, M.; Olsson, L.; Fridell, E. Kinetic modelling of sulfur deactivation of Pt/BaO/Al$_2$O$_3$ and BaO/Al$_2$O$_3$ NOx storage catalysts. *Appl. Catal. B Environ.* **2007**, *70*, 179–188. [CrossRef]

102. Kylhammar, L.; Carlsson, P.-A.; Ingelsten, H.H.; Grönbeck, H.; Skoglundh, M. Regenerable ceria-based SOx traps for sulfur removal in lean exhausts. *Appl. Catal. B Environ.* **2008**, *84*, 268–276. [CrossRef]

103. Hamzehlouyan, T.; Sampara, C.S.; Li, J.; Kumar, A.; Epling, W.S. Kinetic study of adsorption and desorption of SO_2 over γ-Al$_2$O$_3$ and Pt/γ-Al$_2$O$_3$. *Appl. Catal. B Environ.* **2016**, *181*, 587–598. [CrossRef]

104. Yoshida, K.; Asanuma, T.; Nishioka, H.; Hayashi, K.; Hirota, S. *Development of NOx Reduction System for Diesel Aftertreatment with Sulfur Trap Catalyst*; SAE International: Warrendale, PA, USA, 2007.

105. Englund, J.; Xie, K.; Dahlin, S.; Schaefer, A.; Jing, D.; Shwan, S.; Andersson, L.; Carlsson, P.-A.; Pettersson, L.J.; Skoglundh, M. Deactivation of a Pd/Pt bimetallic oxidation catalyst used in a biogas-powered Euro VI heavy-duty engine installation. *Catalysts* **2019**, *9*, 1014. [CrossRef]

106. Argyle, M.; Bartholomew, C. Heterogeneous catalyst deactivation and regeneration: A review. *Catalysts* **2015**, *5*, 145–269. [CrossRef]

107. Chasapidis, L.; Melas, A.D.; Tsakis, A.; Zarvalis, D.; Konstandopoulos, A. *A Sampling and Conditioning Particle System for Solid Particle Measurements Down to 10 nm*; SAE International: Warrendale, PA, USA, 2019.

108. Miller, J.; Jin, L. *Global Progress toward Soot-Free Diesel Vehicles*; International Council Clean Transportation (ICCT) Report; ICCT: Washington, DC, USA, 2019.

109. Yamada, H.; Inomata, S.; Tanimoto, H. Mechanisms of increased particle and VOC emissions during DPF active regeneration and practical emissions considering regeneration. *Environ. Sci. Technol.* **2017**, *51*, 2914–2923. [CrossRef] [PubMed]

110. Yasar, A.; Haider, R.; Tabinda, A.B.; Kausar, F.; Khan, M. A comparison of engine emissions from heavy, medium, and light vehicles for CNG, diesel, and gasoline fuels. *Pol. J. Environ. Stud.* **2013**, *22*, 1277–1281.

111. Tavares, J.R.; Sthel, M.S.; Campos, L.S.; Rocha, M.V.; Lima, G.R.; da Silva, M.G.; Vargas, H. Evaluation of pollutant gases emitted by ethanol and gasoline powered vehicles. *Proc. Environ. Sci.* **2011**, *4*, 51–60. [CrossRef]

112. Gentner, D.R.; Worton, D.R.; Isaacman, G.; Davis, L.C.; Dallmann, T.R.; Wood, E.C.; Herndon, S.C.; Goldstein, A.H.; Harley, R.A. Chemical composition of gas-phase organic carbon emissions from motor vehicles and implications for ozone production. *Environ. Sci. Technol.* **2013**, *47*, 11837–11848. [CrossRef]

113. Tobias, H.J.; Beving, D.E.; Ziemann, P.J.; Sakurai, H.; Zuk, M.; McMurry, P.H.; Zarling, D.; Waytulonis, R.; Kittelson, D.B. Chemical analysis of diesel engine nanoparticles using a nano-DMA/thermal desorption particle beam mass spectrometer. *Environ. Sci. Technol.* **2001**, *35*, 2233–2243. [CrossRef]

114. Sakurai, H.; Tobias, H.J.; Park, K.; Zarling, D.; Docherty, K.S.; Kittelson, D.B.; McMurry, P.H.; Ziemann, P.J. On-line measurements of diesel nanoparticle composition and volatility. *Atmos. Environ.* **2003**, *37*, 1199–1210. [CrossRef]

115. Swanson, J.J.; Kittelson, D.B.; Watts, W.F.; Gladis, D.D.; Twigg, M.V. Influence of storage and release on particle emissions from new and used CRTs. *Atmos. Environ.* **2009**, *43*, 3998–4004. [CrossRef]

116. Schneider, J.; Hock, N.; Weimer, S.; Borrmann, S.; Kirchner, U.; Vogt, R.; Scheer, V. Nucleation particles in diesel exhaust: Composition inferred from in situ mass spectrometric analysis. *Environ. Sci. Technol.* **2005**, *39*, 6153–6161. [CrossRef]

117. Karavalakis, G.; Gysel, N.; Schmitz, D.A.; Cho, A.K.; Sioutas, C.; Schauer, J.J.; Cocker, D.R.; Durbin, T.D. Impact of biodiesel on regulated and unregulated emissions, and redox and proinflammatory properties of PM emitted from heavy-duty vehicles. *Sci. Total Environ.* **2017**, *584*, 1230–1238. [CrossRef]

118. Alves, C.A.; Lopes, D.J.; Calvo, A.I.; Evtyugina, M.; Rocha, S.; Nunes, T. Emissions from light-duty diesel and gasoline in-use vehicles measured on chassis dynamometer test cycles. *Aerosol Air Qual. Res.* **2015**, *15*, 99–116. [CrossRef]

119. Tsai, J.-H.; Huang, P.-H.; Chiang, H.-L. Air pollutants and toxic emissions of various mileage motorcycles for ECE driving cycles. *Atmos. Environ.* **2017**, *153*, 126–134. [CrossRef]

120. Ntziachristos, L.; Vonk, W.A.; Papadopoulos, G.; van Mensch, P.; Geivanidis, S.; Mellios, G.; Papadimitriou, G.; Steven, H.; Elstgeest, M.; Ligterink, N.E.; et al. *Effect Study of the Environmental Step Euro 5 for L-Category Vehicles*; European Commission: Brussels, Belgium, 2017; ISBN 978-92-79-70203-7.

121. Kontses, A.; Ntziachristos, L.; Zardini, A.A.; Papadopoulos, G.; Giechaskiel, B. Particulate emissions from L-Category vehicles towards Euro 5. *Environ. Res.* **2020**, *182*, 109071. [CrossRef]

122. Costagliola, M.A.; Murena, F.; Prati, M.V. Exhaust emissions of volatile organic compounds of powered two-wheelers: Effect of cold start and vehicle speed. Contribution to greenhouse effect and tropospheric ozone formation. *Sci. Total Environ.* **2014**, *468*, 1043–1049. [CrossRef]

123. Liu, X.; Deng, B.; Fu, J.; Xu, Z.; Liu, J.; Li, M.; Li, Q.; Ma, Z.; Feng, R. The effect of air/fuel composition on the HC emissions for a twin-spark motorcycle gasoline engine: A wide condition range study. *Chem. Eng. J.* **2019**, *355*, 170–180. [CrossRef]

124. Szymlet, N.; Lijewski, P.; Sokolnicka, B.; Siedlecki, M.; Domowicz, A. Analysis of research method, results and regulations regarding the exhaust emissions from two-wheeled vehicles under actual operating conditions. *J. Ecol. Eng.* **2020**, *21*, 128–139. [CrossRef]

125. Kašpar, J.; Fornasiero, P.; Hickey, N. Automotive catalytic converters: Current status and some perspectives. *Catal. Today* **2003**, *77*, 419–449. [CrossRef]

126. Jaworski, A.; Mądziel, M.; Lejda, K. Creating an emission model based on portable emission measurement system for the purpose of a roundabout. *Environ. Sci. Pollut. Res.* **2019**, *26*, 21641–21654. [CrossRef]

127. Zhu, G.; Liu, J.; Fu, J.; Xu, Z.; Guo, Q.; Zhao, H. Experimental study on combustion and emission characteristics of turbocharged gasoline direct injection (GDI) engine under cold start new European driving cycle (NEDC). *Fuel* **2018**, *215*, 272–284. [CrossRef]

128. Yang, Z.; Liu, Y.; Wu, L.; Martinet, S.; Zhang, Y.; Andre, M.; Mao, H. Real-world gaseous emission characteristics of Euro 6b light-duty gasoline- and diesel-fueled vehicles. *Transp. Res. Part D Transp. Environ.* **2020**, *78*, 102215. [CrossRef]

129. Ma, C.; Wu, L.; Mao, H.; Fang, X.; Wei, N.; Zhang, J.; Yang, Z.; Zhang, Y.; Lv, Z.; Yang, L. Transient characterization of automotive exhaust emission from different vehicle types based on on-road measurements. *Atmosphere* **2020**, *11*, 64. [CrossRef]

130. Simonen, P.; Kalliokoski, J.; Karjalainen, P.; Rönkkö, T.; Timonen, H.; Saarikoski, S.; Aurela, M.; Bloss, M.; Triantafyllopoulos, G.; Kontses, A.; et al. Characterization of laboratory and real driving emissions of individual Euro 6 light-duty vehicles—Fresh particles and secondary aerosol formation. *Environ. Pollut.* **2019**, *255*, 113175. [CrossRef]

131. Ko, J.; Son, J.; Myung, C.-L.; Park, S. Comparative study on low ambient temperature regulated/unregulated emissions characteristics of idling light-duty diesel vehicles at cold start and hot restart. *Fuel* **2018**, *233*, 620–631. [CrossRef]

132. Bikas, G.; Zervas, E. Regulated and non-regulated pollutants emitted during the regeneration of a diesel particulate filter. *Energy Fuels* **2007**, *21*, 1543–1547. [CrossRef]

133. Jung, S.; Lim, J.; Kwon, S.; Jeon, S.; Kim, J.; Lee, J.; Kim, S. Characterization of particulate matter from diesel passenger cars tested on chassis dynamometers. *J. Environ. Sci.* **2017**, *54*, 21–32. [CrossRef]
134. R'Mili, B.; Boréave, A.; Meme, A.; Vernoux, P.; Leblanc, M.; Noël, L.; Raux, S.; D'Anna, B. Physico-chemical characterization of fine and ultrafine particles emitted during diesel particulate filter active regeneration of Euro 5 diesel vehicles. *Environ. Sci. Technol.* **2018**, *52*, 3312–3319. [CrossRef]
135. Ruehl, C.; Smith, J.D.; Ma, Y.; Shields, J.E.; Burnitzki, M.; Sobieralski, W.; Ianni, R.; Chernich, D.J.; Chang, M.-C.O.; Collins, J.F.; et al. Emissions during and real-world frequency of heavy-duty diesel particulate filter regeneration. *Environ. Sci. Technol.* **2018**, *52*, 5868–5874. [CrossRef] [PubMed]
136. Giechaskiel, B.; Gioria, R.; Carriero, M.; Lähde, T.; Forloni, F.; Perujo, A.; Martini, G.; Bissi, L.M.; Terenghi, R. Emission factors of a Euro VI heavy-duty diesel refuse collection vehicle. *Sustainability* **2019**, *11*, 1067. [CrossRef]
137. Doozandegan, M.; Hosseini, V.; Ehteram, M.A. Solid nanoparticle and gaseous emissions of a diesel engine with a diesel particulate filter and use of a high-sulphur diesel fuel and a medium-sulphur diesel fuel. *Proc. Inst. Mech. Eng. Part. D J. Automob. Eng.* **2017**, *231*, 941–951. [CrossRef]
138. Rothe, D.; Knauer, M.; Emmerling, G.; Deyerling, D.; Niessner, R. Emissions during active regeneration of a diesel particulate filter on a heavy duty diesel engine: Stationary tests. *J. Aerosol Sci.* **2015**, *90*, 14–25. [CrossRef]
139. Smith, J.D.; Ruehl, C.; Burnitzki, M.; Sobieralski, W.; Ianni, R.; Quiros, D.; Hu, S.; Chernich, D.; Collins, J.; Huai, T.; et al. Real-time particulate emissions rates from active and passive heavy-duty diesel particulate filter regeneration. *Sci. Total Environ.* **2019**, *680*, 132–139. [CrossRef] [PubMed]

 *vehicles*

Article

On the Impact of Maximum Speed on the Power Density of Electromechanical Powertrains

Daniel Schweigert [1,*]**, Martin Enno Gerlach** [2]**, Alexander Hoffmann** [2]**, Bernd Morhard** [1]**,
Alexander Tripps** [1]**, Thomas Lohner** [1]**, Michael Otto** [1]**, Bernd Ponick** [2] **and Karsten Stahl** [1]

[1]   Gear Research Centre (FZG), Technical University of Munich, Boltzmannstraße 15,
     85748 Garching near Munich, Germany; morhard@fzg.mw.tum.de (B.M.); alexander.tripps@tum.de (A.T.);
     lohner@fzg.mw.tum.de (T.L.); otto@fzg.mw.tum.de (M.O.); stahl@fzg.mw.tum.de (K.S.)
[2]   Institute for Drive Systems and Power Electronics, Leibniz University Hannover, 30167 Hannover, Germany;
     martin.gerlach@ial.uni-hannover.de (M.E.G.); alexander.hoffmann@ial.uni-hannover.de (A.H.);
     ponick@ial.uni-hannover.de (B.P.)
*   Correspondence: schweigert@fzg.mw.tum.de; Tel.: +49-89-289-15774

Received: 30 May 2020; Accepted: 17 June 2020; Published: 25 June 2020

Abstract: In order to achieve the European Commission's ambitious climate targets by 2030, BEVs (Battery Electric Vehicles) manufacturers are faced with the challenge of producing more efficient and ecological products. The electromechanical powertrain plays a key role in the efficiency of BEVs, which is why the design parameters in the development phase of electromechanical powertrains must be chosen carefully. One of the central design parameters is the maximum speed of the electric machines and the gear ratio of the connected transmissions. Due to the relationship between speed and torque, it is possible to design more compact and lighter electric machines by increasing the speed at constant power. However, with higher speed of the electric machines, a higher gear ratio is required, which results in a larger and heavier transmission. This study therefore examines the influence of maximum speed on the power density of electromechanical powertrains. Electric machines and transmissions with different maximum speeds are designed with the state-of-the-art for a selected reference vehicle. The designs are then examined with regard to the power density of the overall powertrain system. Compared to the reference vehicle, the results of the study show a considerable potential for increasing the power density of electromechanical powertrains by increasing the maximum speed of the electric machines.

Keywords: E-Mobility; powertrain design; high-speed; electric machine design; transmission design; gearbox

---

1. Introduction

In 2011, the EU Commission for Energy, Climate change and Environment published ambitious plans for the future climate and energy policy framework. By 2030, greenhouse gas emissions have to be lowered by at least 40% in comparison to 1990. Furthermore, the energy efficiency of products should, at the same time, increase by at least 32.5%. Limit values for CO_2 emissions of newly registered cars of each manufacturer are also part of this initiative. The average emissions of a manufacturer's fleet in 2021 must, for example, be lower than the limit of 95 grams of CO_2 per 100 km [1]. If manufacturers exceed the limit values for CO_2 emissions, they are expected to pay substantial fines. In order to achieve the ambitious goals, set for the protection of the environment and to avoid heavy fines, the mobility behavior has to change significantly.

The electrification of the automotive powertrain is expected to play an important role in this change. A combination of battery electric vehicles (BEVs) and electricity from renewable energy sources can enable a CO_2 neutral and largely pollutant-free usage phase and thus make a significant contribution to reducing greenhouse gas emissions in the future. Furthermore, the tank-to-wheel efficiency of electromechanical powertrains in BEVs, which is significantly higher than that of internal combustion engines (ICE), is intended to make a contribution to the increase of efficiency of mobility [1].

The electromechanical powertrain of a BEV typically consists of power electronics, a drive motor or electric machine and a transmission. While the power electronics are responsible for the conversion of electrical power to control the electric motor, the transmission's gear ratio is used to adapt the electric machine's speed and torque to the required values at the drive axle. An important parameter in the design process of an electromechanical powertrain is the maximum speed of the electric machines in combination with the gear ratio of the transmission to reach the desired maximum speed of the vehicle. A higher speed of the electric machine leads to a lower required torque at constant power, which means that more compact and lighter electric machines can be designed. However, a transmission with an increased input speed needs a higher gear ratio to reach the same output speed and must therefore be designed larger and heavier. Nevertheless, there is a trend towards increasing speeds of electric machines in a BEV. The positive influence of increased speeds of the electric machine on the power density of the whole powertrain has been proven in various studies [2,3]. This positive influence may play an important role in the achievement of ultimate efficiency and climate goals by 2030.

This study presents a detailed analysis of the influence of maximum speed of the electric machine on the power density of an electromechanical powertrain. For this purpose, conceptual designs of electric machines with various maximum speeds and transmissions with suitable overall gear ratios have been developed. The considered speeds are set from $n = 12,000 \text{ min}^{-1}$ to $n = 50,000 \text{ min}^{-1}$. While the speed increases and thus the torque of the electric machines decreases, the gear ratio has to increase to remain a constant output torque and vehicle speed. These boundary conditions are based on the parameters of the BMW i3 (120Ah) as reference vehicle. Table 1 shows the parameters of the electric machine and transmission relevant for this study. The voltage level of the battery is increased to satisfy the demands of the high-speed electric machine.

Table 1. Relevant parameters and view of the BMW i3 reference vehicle (120Ah). Data from [4].

BMW i3 (120 Ah)	
Empty weight	1245 kg
Top speed	150 km/h
Rated power	75 kW
Peak power	135 kW
Battery voltage	660 V
Rated torque (EM)	150 Nm
Peak torque (EM)	250 Nm
Gear ratio	9.665
Max. speed (EM)	$11,400 \text{ min}^{-1}$
Max. speed (drive axle)	1179.5 min^{-1}
Rated torque (drive axle)	1449.75 Nm
Max. torque (drive axle)	2416.25 Nm

The electric machine of the reference vehicle is designed as a permanent magnet synchronous machine (PMSM) with a rated power of 75 kW and a peak power of 135 kW. Furthermore, the electric machine has a rated torque of 150 Nm and a peak torque of 250 Nm and reaches a maximum speed of 11,400 1/min. The gear ratio determines the corresponding parameters on the drive axle. All of the designed electric machines and transmissions for every considered maximum speed and overall gear ratio result in a large number of electromechanical powertrains. All of these designs represent potential drivetrains for the reference vehicle and will be analyzed with regard to mass, volume, volumetric and gravimetric power density. This is intended to make a statement regarding the potential of increase of the maximum speed in electromechanically powertrains to increase power density.

After applying the state of the art to design parameters of high-speed electric machines and transmissions in Sections 2–4 present the conceptual design process of the electric machines and transmissions used in this study. The results of the study will be shown in Section 5, followed by a discussion and an outlook of the results in Section 6.

2. State of the Art Considerations for Higher Shaft Speeds

The following section will cover the key aspects of current developments in the field of electrical machines and transmissions with regard to high speeds. After discussing the current technological driver in this field, this section will close with a summary of current and future generation powertrains.

2.1. Electric Machines

In the context of electrical machines, the term high-speed refers to the surface speed of the rotor v_s and not to the synchronous shaft speed

$$n_0 = \frac{f_1}{p},$$

(1)

where f_1 is the fundamental electric stator frequency and p is the number of pole pairs. It is therefore important to understand the relation

$$v_{\text{rot}} = 2\pi r_{0,2} n,$$

(2)

where $r_{0,2}$ is the outer radius of the rotor, because the surface speed v_{rot} is in fact the variable on which the mechanical boundaries depend in the first place. The variable on which the electrical boundaries depend in the first place is the fundamental electric stator frequency f_1. With the importance of these two quantities in mind, subjects such as rotor topologies, winding technologies and materials will be discussed in the context of PMSMs [5].

2.1.1. Rotor Topologies

Rotors for PMSMs are subdivided into two categories, surface type rotors and interior type rotors. The definition is based on the location of the permanent magnet relative to the outer contour of the rotor iron, as shown in Figure 1. Depending on the desired maximum surface speed, surface PM rotors are preferable to interior PM rotors [6].

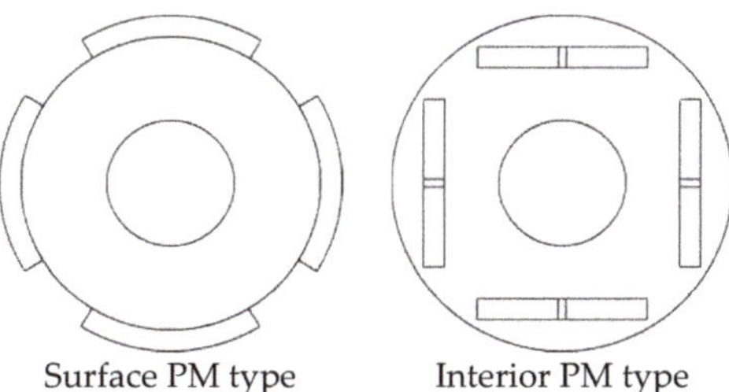

Figure 1. Two different rotor topologies for PMSMs.

The mechanical stresses introduced on the rotor iron due to rotation are proportional to the square of the rotational speed and the square of the outer radius $r_{0,2}$. Due to this strongly non-linear relationship, the mechanical stress on the rotor iron requires special consideration. The introduction of so-called pockets for the permanent magnets has the consequence of weakening the material, which further increases the mechanical stresses [7]. For fast rotating electric machines, it is common to reinforce the rotor with a so-called bandage or sleeve. Such reinforcement is applied on the outer radius and prevents the expansion of the rotating rotor. The insertion of a bandage thus has the consequence of increasing the magnetically effective air gap. A larger magnetic air gap acts like an increased magnetic resistance and reduces the magnetic flux, which has an influence on a number of important parameters of the electric machine. The use of a bandage must therefore be weighed against other possibilities and the influence on all requirements must be checked. This includes the selection of the bandage itself. Different materials will be discussed in a later section. Decisions are usually made by evaluating results from finite element simulations [8].

2.1.2. Rotor Materials

To go into more detail, an overview of materials is presented for the three basic components (besides the shaft) of a PMSM rotor, electrical steel sheets, permanent magnets and bandage material. An example for a high-speed PMSM rotor is shown in Figure 2.

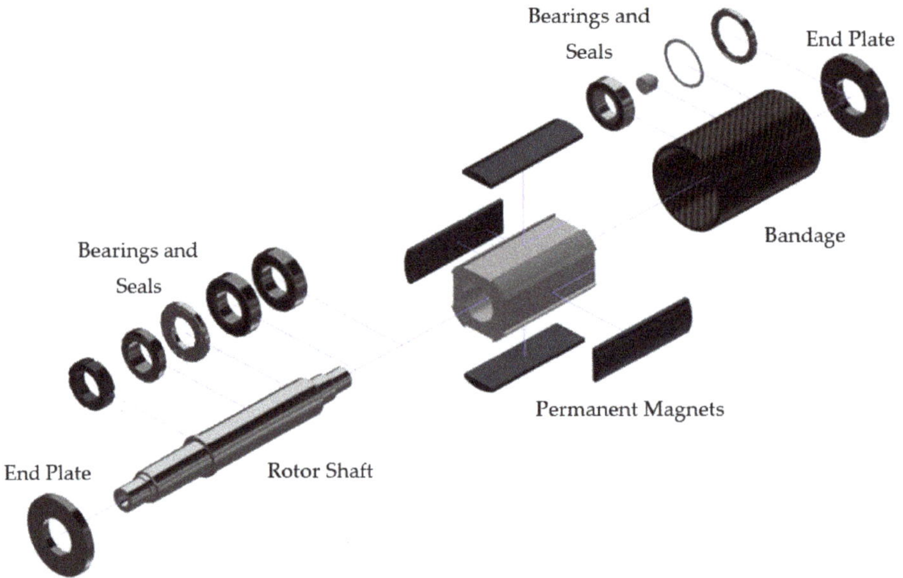

Figure 2. Exploded view of an PMSM rotor assembly.

2.1.3. Electrical Steel Sheets

While industrially used induction machines usually have a laminated rotor, PMSMs do not require a laminated rotor. However, the use of electric sheet metal has a positive effect on the losses of the rotor due to eddy currents and therefor on the rotor temperature. Electrical steel is the most common material seen in electrical machines; with increasing demands on efficiency and power density, electrical steel with cobalt is gaining interest. However, the cost of cobalt electrical steel is considerably higher. High strength steel is particularly interesting for the rotors electrical machines with high speeds [9]. The yield strength of such sheets is higher than that of electrical steel sheets but returns less favorable magnetic properties. Recent developments in electrical steel have resulted in high-strength electrical steel, which aims to close the gap between conventional electrical steel and structural optimized steel [10].

2.1.4. Permanent Magnets

In addition to the different materials for permanent magnets (AlNiCo, NdFeB and SmCo), a variety of different segmentation strategies exist. Segmentation of permanent magnets is performed in order to reduce eddy currents inside of the permanent magnets and to avoid overheating of the rotor assembly [11]. If the rotor in a PMSM reaches unconsidered temperatures, a loss of torque at the shaft and permanent damage due to demagnetization can occur. Currently, permanent magnet segmentation can be performed with a layer thickness of as low as 500 μm in thickness and a <20 μm insulation between the layers [12]. In general, segmentation can be in axial or in tangential direction of the rotor. The selection of suitable permanent magnet shapes and the placement within the rotor are most often aided by magnetostatic simulations using the finite-element-method [13].

2.1.5. Sleeve and Bandage Materials

Materials for sleeves and bandages can be differentiated into steel-based sleeves and composite-based sleeves. Materials for steel sleeves are high strength steels with alloying elements such as chromium, molybdenum and nickel. For steel-based sleeves, the highest tensile strength is given at 700 N/mm^2 for 2.461 NiMo16Cr16Ti; in comparison, the well-known 1.7225 42CrMo4 achieves a tensile strength of 550 N/mm^2 according to the manufacturer's data [14,15]. Both have a good strength to weight ratio and are considerably stronger and harder than standard steels. A side-by-side comparison of the values for tensile strength, density and electrical resistivity shows that the composite-based sleeves are more suitable for high-speed electrical machines than sleeves made of steel. It should be said, that the properties of the fiber and the finished composite sleeve depend heavily on the fabrication process. Therefore, the given data in Table 2 is for bare fiber and is presented to provide a starting point for further analysis.

Table 2. Bare fibers for composite-based sleeve materials. Data from [16].

Material	Glass Fiber	Aramid Fiber	Carbon Fiber
Tensile strength in N/mm^2	3400	2880	3950
Density in kg/m^3	2600	1450	1760
Electrical resistivity in Ωmm^2/m at 20 °C	1×10^{20}	1×10^{20}	1.6×10^7
Typical operating temperature in °C	below 100	below 200	below 1000
Thermal expansion coefficient in 10^{-6} 1/K	5	−3.5	−0.1

2.1.6. Winding Technologies

In the field of traction motors, two different types of distributed windings can be observed for electrical machines. The random winding, consisting of randomly distributed round shaped conductors, and the form wound winding, which is composed of so-called hairpins. Hairpins in the context of windings are solid copper conductors with rectangular shape, are formed into the shape of a conventional hairpin by bending. By connecting several hairpins trough welding, the actual winding is created. Hairpin windings are favored by car manufacturers because the manufacturing process can be fully automated [17]. But this does not mean that all traction motors for electric cars are equipped with hairpin windings. Both winding technologies are still on the market. In the context of high-speed electric machines, hairpin windings have concrete disadvantages. The cross-sectional area of a hairpin must not fall below a certain value, otherwise processing becomes more difficult. Since the winding resistance has a great influence on efficiency, and since the phase current in high-speed electrical machines can reach higher frequencies than in conventional designs, the frequency dependence of the winding resistance has to be considered for the selection of the winding technology. Finally, Figure 3 shows an example of the phase resistance for the two mentioned winding technologies as a function of frequency. The displayed data is acquired via direct measurements at electrical machines. It can be seen that the resistance of the hairpin wound winding is already greater by a factor of 10 at a frequency of 1000 Hz due to current displacement effects. This factor has a direct linear effect on the power loss in the winding [18].

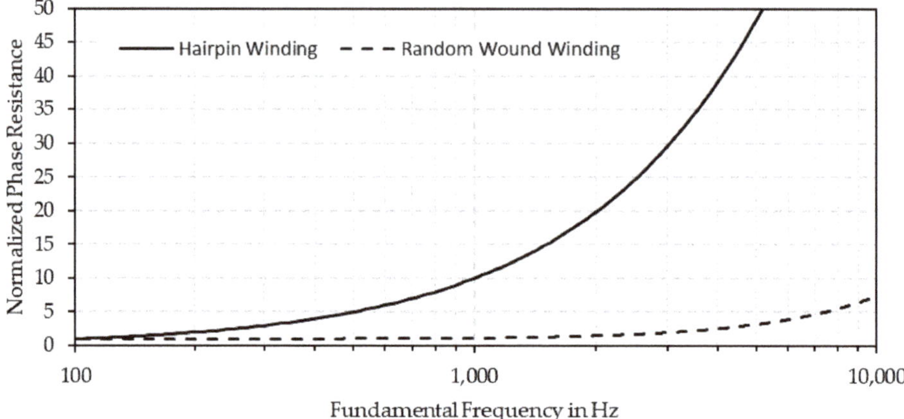

Figure 3. Comparison of the phase resistance of a hairpin winding with a conductor diameter of 0.5 mm and a random wound winding with a conductor width of 3.2 mm and a height of 1.6 mm (Phase resistance is related to DC resistance).

2.2. Technologies for High-Speed Transmissions

High-speed electromechanical powertrains as scope of recent research history [19–24] promise an increased power density of the whole powertrain. Indeed, the required torque and speed at the wheels cause the transmission to be designed with a higher total gear ratio that on the first glance causes increased weight, installation space and cost of the transmission itself. Still, the conglomerate of high-speed electrical machines and high-speed transmissions is able to demonstrate its benefits. To meet the high-level requirements of BEVs concerning range, comfort and operability new technologies have to support and enable transmissions to operate within this high-speed approach.

2.2.1. Improved Efficiency for Maximized Range

As the available range of BEVs is reduced by the power losses of the whole powertrain, the efficiency of the powertrain in general and of the transmission in particular is of prime importance. The power loss of a transmission and hence the efficiency is determined by gear power losses, bearing power losses, sealing power losses and other power losses, e.g., caused by oil pumps. Both, gear power losses and bearing power losses are separated into load-dependent and no-load power losses. In order to reduce the power losses of the transmission, and more precisely the load-dependent gear power losses, low-loss gear geometries and water-containing gear fluids with coefficients of friction smaller than 0.01 [25] (which is referred to as superlubricity [26]) can be used in electromechanical powertrains. As both technologies reduce the frictional losses, they can also be seen as a compensation for the rising sliding speeds with higher input speeds, usually causing a higher risk of scuffing. The low-loss gear geometry concentrates the path of contact to a minimum around the pitch point. As sliding speed rises with increasing distance to the pitch point, high sliding speeds are avoided this way, resulting in reduced load-dependent gear power losses [27]. Hinterstoisser et al. [28] outline in experiments the significant impact of the low-loss gear geometry on the efficiency. With an extreme low-loss gear design, the power losses can be reduced by about 79% compared to a conventional gear design. In order to reduce the mean gear coefficient of friction, water-containing gear fluids promise a significant improvement of the efficiency. Yilmaz et al. [29] demonstrate on the FZG gear efficiency test rig the reduction of the mean gear coefficient of friction and hence of the load-dependent gear power losses by water-containing fluids (cf. Figure 4). Using those fluids, the mean gear coefficient of friction is reduced by up to 82%.

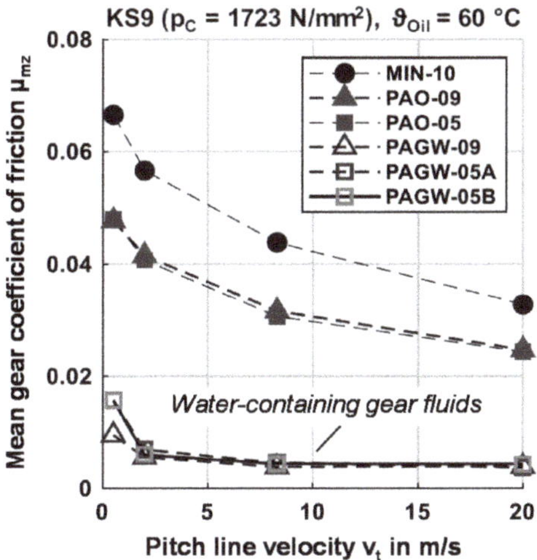

Figure 4. Mean gear coefficient of friction μ_{mz} determined on the FZG gear efficiency test rig at a load p_C of 1723 N/mm^2 for a mineral oil (MIN), polyalphaolefine oils (PAO) and water-containing polyalkylenglycols (PAGW) according to Yilmaz et al. [25].

Concerning bearings, investigations with water-containing gear fluids on the FZG bearing power loss test rig show reduced no-load bearing losses and increased load-dependent bearing losses with higher rotational speeds of roller bearings [30] which additionally supports the approach of high speeds and low loads on the input of high-speed transmissions. The authors mention that hybrid bearings with Si_3N_4 ceramic cylindrical rollers, cronidur races, and polyether ether ketone (PEEK) cages were used to avoid incompatibilities with the investigated water-containing gear fluids [30]. As the presence of hydrogen is suspected to cause white edge cracks (WECs), or at least to support its formation, causing premature failure of the bearings [31–33], WECs have to be considered when using water-containing fluids. As the water content strongly improves its caloric properties, water-containing gear fluids are possible to use as coolant. By this, the whole powertrain including power electronics, electrical machines and the transmission can be cooled and lubricated by one circuit. This holistic thermal management promises further improvement of efficiency of the powertrain as e.g., injection lubrication in the transmission can be used without additional oil pumps.

2.2.2. Improved Acoustics for Increased Comfort

As the internal combustion engine is cut in BEVs, its masking sound is not present anymore. Consequently, the customer faces noise of the transmission in BEVs which can be felt as uncomfortable, particularly because of its tonal character. Furthermore, in comparison to conventional powertrains in ICE, the drive speeds are already higher in BEV series solutions, which leads to new spectral compositions of the noise emissions. In particular, the high-frequency components of the spectrum can be perceived as disturbing by the human ear. The increase in speed in comparison to BEV series solutions, which is carried out in the context of this study to achieve higher power densities, further aggravates this problem [34].

In addition, increasing speeds make it more difficult to operate the transmission subcritically over the entire operating range. The subcritical operating range is the range with a mesh frequency f lower than the torsional natural frequency f_n of the meshing. Especially in the critical operating range with reference speeds (f / f_n) from 0.85 to 1.15, significant additional dynamic forces and thus vibrations and noise emissions occur in the meshing, as stated in Figure 5. Therefore, the critical operating range is typically avoided in ICE by operating the transmission in the subcritical range with low dynamic forces [31,35,36].

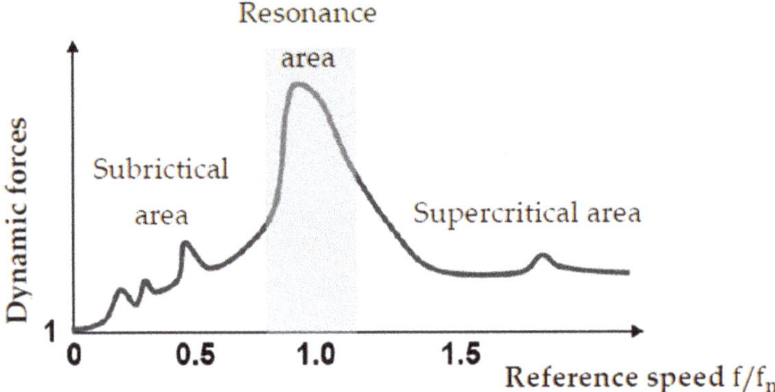

Figure 5. Dynamic factor as a function of resonance ratio.

In case of the first stage of the high speed transmissions of this study, subcritical operating will not be possible over the entire operating range with increasing speed of the electric machines. Hence, special precautions are needed to optimize the acoustical behavior of high-speed transmissions. The basis for this optimization is a precise knowledge of acoustically critical operating ranges of all meshes. For powertrains with double-e-architectures, this knowledge can be used to split the required power and torque between two non-identical power-paths and therefore avoid those areas. If there is only one power path available, resonance areas should be positioned in an area which can be passed [35].

2.2.3. Operability of High-Speed Transmission

The high circumferential speeds caused by the high rotational speeds on the rotor and input shaft cause stricter requirements the bearings and sealings must fulfill in order to ensure the reliability needed.

That is why the bearings of the input shaft and the rotor of high-speed electrical machines need to be designed for high-speeds. The limiting speed of bearings is given by the speed factor ndm consisting of the maximum speed n [min^{-1}] and the mean diameter of the bearing dm [mm]. In the work of Deiml et al. [37] it is mentioned that common SKF bearings are designed with a maximum speed factor of 0.7×10^6 mm/min, but for their high-speed approach sealed bearings with a speed factor of 1.6×10^6 were designed. Usually, spindle bearings (super precision angular ball bearings) according to DIN 628-1 [38] are used in machine tools, reaching highest rotational speeds. With oil lubrication, the bearing manufacturer SCHAEFFLER mentions that spindle bearings are suitable for a speed factor of up to over 3.0×10^6 mm/min [39]. The bearing manufacturer SKF outlines, that hybrid bearings with ceramic balls are superior to all steel bearings in terms of operating speed, as ceramic balls are lighter which causes reduced centrifugal forces leading to reduced losses and heat development [40].

To ensure the operability of the transmission fluid leakage and dirt entry via the input and output shaft have to be avoided by appropriate sealing. Rotary shaft lip seals according to ISO 6194-1 [41], as a part of contact seals and commonly used in automotive sector, operate at circumferential speeds beneath 40 m/s as the frictional losses cause unnecessary high temperatures harming those seals at higher speeds. Contactless seals like labyrinth seals overcome the limiting circumferential speeds as only negligible frictional losses are acting, enabling the operation at highest circumferential speeds [42]. In terms of sealing itself, contactless seals, being more precisely labyrinth seals, are disadvantageous as they are not able to completely seal the transmission, causing leakage [43]. In order to overcome these disadvantages, new sealing technologies have to be used within the high-speed approach.

One promising sealing technology is the gas-lubricated mechanical face seal applied on transmissions. The working principle is that the primer ring and mating ring move apart at rotation of the shaft as a consequence of gas flow (usually air) caused by the aerodynamically optimized surface structure of the mating ring [44]. These seals are characterized by reduced frictional losses, enabling the operation at highest rotational speeds.

2.2.4. Research on the High-Speed Transmission of Speed4E

In the joint research project Speed4E [45] a hyper-high-speed powertrain for electric vehicles is developed, designed, and investigated. The goals of Speed4E include the development of an innovative powertrain for BEVs with rotational speeds of the electrical machines of up to $50,000$ min^{-1}, the integration into a test vehicle, and a holistic thermal management based on a water-containing gear fluid also used for lubrication of the transmission. Different aspects are considered, such as efficiency optimization, mass and cost reduction as well as the power density increase by the high input speeds. In the transmission of the research project Speed4E among other things, the mentioned low-loss gear geometry, NVH-optimized gears, water-containing gear fluids, holistic thermal management, high-speed spindle and hybrid bearings as well as contactless, lifting seals are investigated with respect to input speeds.

2.3. Series and Future Axle Drive Systems

A survey on current single axle drive systems for personal BEV shows, that the peak power is around 150 kW and the peak speed is below $n = 20,000$ min^{-1}. Depending on which components are included in the axle drive system the weight is at around 80 kg. All shown drive systems in Table 3 aim for a high level of integration and combine the main major parts of a drive system: the electric machine, the inverter of power electronics and the transmission.

Table 3. Overview of current axle drive systems from automotive suppliers.

	Continental (3rd gen.) [46]	Bosch eAxle [47]	Nidec Corporation eAxle [48]	BorgWarner iDM eAxle [49]
Primary use	Personal BEV	Personal BEV	Personal BEV	Personal BEV
Peak power	150 kW	300 kW	150 kW	125 kW
Peak speed	n/a	16,000 min^{-1}	15,000 min^{-1}	10,600 min^{-1}
Date and status	In production since 2019	In production since 2019	In production since 2018	In production since 2019
Package	PMSM, Inverter and single speed transmission	Electric machine, transmission and Inverter	Electric machine, Inverter and single speed (10.4:1) transmission	Electric motor. single speed transmission (9.4:1)
Weight	<80 kg	90 kg	83 kg	n/a

3. Conceptual Design of High-Speed Electric Machines

In this section, the design process is introduced for electric machines. Four machines with a rated power of $P_N = 75$ kW and different maximum speeds are designed, as it can be seen in Table 4. The PMSM-B1 and PMSM-B2 machines have interior bar magnets and the PMSM-S1 and PMSM-S2 machines have surface mounted magnets with a bandage. The field weakening behavior of machines with interior magnets is better than with surface mounted magnets. Thus, the rated speed of the PMSM-B1 and PMSM-B2 machines is chosen smaller in relation to the maximum speed, than for the PMSM-S1 and PMSM-S2 machines. The geometry, mass, volume, volumetric and gravimetric power density are determined and presented later.

Table 4. Speeds and power requirements for the machine designs.

	PMSM-B1	PMSM-B2	PMSMS-S1	PMSM-S2
Rated power P_N		75 kW		
Maximum power P_{max}		135 kW		
Rated speed n_N	4000 min^{-1}	6500 min^{-1}	15,000 min^{-1}	25,000 min^{-1}
Maximum speed n_{max}	12,000 min^{-1}	20,000 min^{-1}	30,000 min^{-1}	50,000 min^{-1}

3.1. Design Process of Active Parts

To design an electric machine, different fundamental equations can be used to estimate the size and the geometry of the machine. Therefore, input parameters such as power, speed and voltage have to be provided and assumptions regarding certain machine parameters and boundary conditions have to be specified to start the design process.

Each machine is designed for one specific operating point in the torque speed diagram, i.e., the rated operation, as can be seen in Figure 6. In this case, this point is set to be the point of rated power P_N and rated speed n_N at which the machine's supply voltage U_1 reaches the maximum output voltage of the inverter U_{max} and the maximum continuous torque M_{cont} is provided. The machine's induced voltage

$$U_1 \propto w, \hat{\phi}, n, p, \tag{3}$$

which almost matches the supply voltage, is proportional to the magnetic flux $\hat{\phi}$, the speed n, the number of pole pairs p and the number of series turns per phase w. The torque is proportional to the torque-forming part I_q of the current I_1. Using the dq-plane, the current I_1 can be displayed as the q-axis current I_q and the magnetizing d-axis current I_d [50]. Since the induced voltage reaches the voltage limit U_{max}, for speeds above the rated speed, the current I_d is increased to provide field weakening and decrease the magnetic flux in the machine. Dividing the current I_1 by the number of parallel branches per phase a, the number of conductors per slot z and the conductors cross sectional area of copper A_{co}, the current density

$$S_1 = \frac{I_1}{A_{co} \cdot a} \tag{4}$$

in the slot can be calculated [18]. Depending on the cooling system, the current density $S_{1,cont}$ at which the machine can be operated continuously and the maximum current density $S_{1,max}$ is given. The machine geometry is therefore mainly determined by the magnetic flux $\hat{\phi}$, the current density $S_{1,cont}$, and the maximum voltage U_{max}.

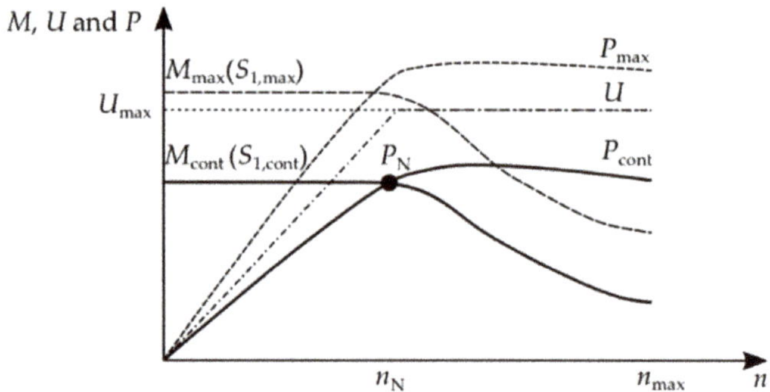

Figure 6. Schematic torque speed diagram.

The design process can be seen in Figure 7. First, the main dimensions of the machine such as bore diameter and length are specified. Then the electric quantities and the main parameters of the machine are calculated and determined. From thereon, the geometry of the lamination and the magnets is calculated, based on the magnetic circuit in the stator and the rotor lamination. The machine design is completed, specifying the winding quantities and the slot geometry. The design process of the machine is followed by a general design of the housing. Finally, the volume and mass of the housing and the machine are calculated and the volumetric and gravimetric power density are determined.

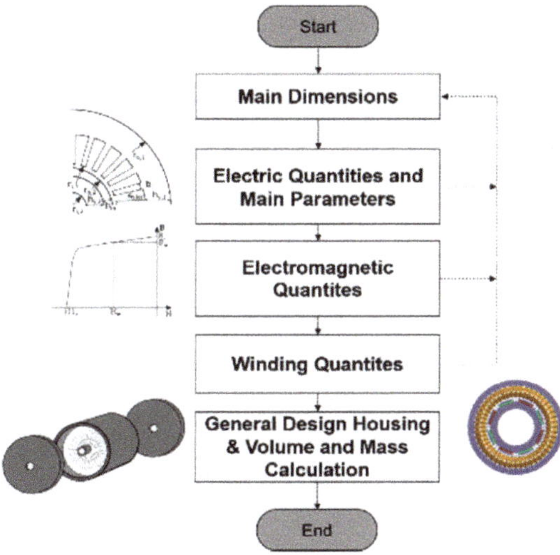

Figure 7. Design process of electric machines.

3.1.1. Main Dimensions

The first step of the design process is a first assessment of the main dimensions of the machine. The bore diameter and the length determine the provided torque of the machine. The torque

$$T = \frac{2\pi}{8} \, d_{i,1}^2 \, l_1 \hat{B}_{\mathrm{mp}} \hat{A}_p \tag{5}$$

is calculated using the spatial fundamental of the electric loading $\hat{A}_p$ and the spatial fundamental of the magnetic flux density of the magnet in the rotor $\hat{B}_{\mathrm{mp}}$ [18]. Estimating these two quantities, $\hat{A}_p$ and $\hat{B}_{\mathrm{mp}}$, the bore volume

$$\frac{2\pi}{8} \, d_{i,1}^2 \, l_i = \frac{P_\mathrm{N}}{2\pi \, n_\mathrm{n} \, \hat{B}_{\mathrm{mp}} \hat{A}_p} \tag{6}$$

can be calculated, if the rated power P_N and speed n_N are given. By setting the ratio of length and bore diameter to $\frac{l_i}{d_{i,1}} = 1.3$, the length and the diameter are determined. This ratio is set accordingly to other traction drives. The electric loading $\hat{A}_p$ and magnetic flux density $\hat{B}_{\mathrm{mp}}$ are chosen according to the maximum speed, (cf. Table 4). For higher speeds, the electric loading $\hat{A}_p$ is chosen smaller. The calculated lengths and diameters of the four machine designs are shown in Table 5.

The parameters need to be validated concerning their maximum surface velocity v_{rot}. The PMSM-B1 and PMSM-B2 machine designs have interior magnets. Their surface velocity should not exceed $v_{\mathrm{rot,max}} = 120\,\mathrm{m/s}$ [6]. The surface mounted magnets of PMSM-S1 and PMSM-S2 are kept by a bandage. The bandage leads to a bigger magnetic air gap, but allows to increase the maximum surface velocity to $v_{\mathrm{rot,max}} = 250\,\mathrm{m/s}$ [5]. The air gap for the designs with interior magnets is assumed to be $\delta = 1\,\mathrm{mm}$. For the PMSM-S1 surface mounted rotor, the bandage height is set to 2 mm and, for the PMSM-S2, the bandage height is set to 4 mm.

The surface velocity of the machine designs

$$v_{rot} = 1.1 \cdot \pi \cdot n_{max} \cdot d_{o,2}$$

(7)

is calculated for an overspeed of 10% of the maximum speed. The results for these machines are listed in Table 5 and do not exceed the limits of the surface velocity. The PMSM-B1 and PMSMS-S1 machine designs show low mechanical utilization compared to the limits. By increasing the bore radius and though the surface velocity, the machine would become larger and heavier.

Table 5. Main dimensions, electric loading and magnetic flux density of the magnet.

	PMSM-B1	PMSM-B2	PMSMS-S1	PMSM-S2
Magnetic flux density of the magnet $\hat{B}_{mp}$	0.95 T	0.9 T	0.8 T	0.8 T
Electric loading $\hat{A}_p$	115 kA/m	100 kA/m	85 kA/m	70 kA/m
Core length l	152 mm	138 mm	115 mm	103 mm
Bore diameter $d_{i,1}$	117 mm	106 mm	88 mm	79.5 mm
Mechanical air gap δ_m	1 mm	1 mm	0.5 mm	0.5 mm
Magnetic air gap δ	1 mm	1 mm	2.5 mm	4.5 mm
Surface velocity v_{rot}	72 m/s	109 m/s	151 m/s	226 m/s

3.1.2. Electric Quantities

To distinguish the rated current I_N of the machine at rated power P_N, the apparent power

$$S_N = \frac{P_N}{\eta_N \ \cos(\phi_N)}$$

(8)

is calculated. The rated phase current

$$I_{N,str} = \frac{S_N}{m_1 \cdot U_{str,max}}$$

(9)

is then determined using the maximal phase voltage $U_{str,max}$ provided by the inverter. m_1 is the number of phases. To determine the current and the apparent power the efficiency η_N and the power factor $\cos(\phi_N)$ at rated operation must be assumed reasonably. In this case η_N and $\cos(\phi_N)$ are set to $\eta_N = 0.95$ and $\cos(\phi_N) = 0.9$. With the maximum phase voltage of $U_{str,max} = 230$ V, the rated current of $I_{N,str} = 127.67$ A is calculated. The maximum current deliverable by the inverter is $I_{max,str} = 205$ A.

The maximum operating frequency

$$f_{max} = \frac{n_{max}}{p}$$

(10)

is determined based on the number of pole pairs p. The number of pole pairs is kept small to keep the hysteresis and eddy current losses low. The number of stator slots N_1 is chosen to feature a high fundamental winding factor and a good compromise between reasonable slot dimensions and a low harmonic leakage factor. The resulting number of slots per pole and phase is accordingly set to $q = 2$. The pole pitch $\tau_p = \frac{d_{i,1}\pi}{2p}$ and the stator slot pitch $\tau_N = \frac{d_{i,1}\pi}{N_1}$ are then defined. The results are shown in Table 6.

Table 6. Electric quantities and main parameters.

	PMSM-B1	PMSM-B2	PMSMS-S1	PMSM-S2
Number of pole pairs p	4	3	2	2
Maximum frequency f_{max}	0.88 kHz	1.1 kHz	1.1 kHz	1.83 kHz
Number of slots N_1	48	36	24	24

3.1.3. Magnetic Quantities and Geometry of the Magnetic Circuit

The geometry of the stator and the rotor lamination, as well as the magnet geometry, are initially based on the estimation of the magnetic flux density in the air gap $\hat{B}_p$ at rated speed and rated power. This flux density is defined by the coupling between the stator and rotor flux and is accordingly higher than the flux density just due to the magnet, that was defined in Section 3.1.1. This effect is higher for the machines with interior magnets since their air gap is much smaller than for the machines with surface mounted magnets. The assumptions of the flux density for the four designs can be seen in Table 7. The magnetic flux per pole

$$\hat{\phi} = l \cdot \tau_p \cdot \frac{2}{\pi} \cdot \hat{B}_p \tag{11}$$

is calculated based on this assumption. By knowing the magnetic flux per pole, the tooth width and the height of the stator and the rotor yoke can be calculated by specifying the permissible magnetic flux density in the teeth $\hat{B}_t$, in the stator yoke $\hat{B}_{y,1}$ and in the rotor yoke $\hat{B}_{y,1}$ for rated operation. These values are set different for the four machine designs. The machine designs with the lower operating frequency f can be operated with higher flux densities. This is due to eddy-current losses in the iron that depend on the flux density and the frequency. For higher speed, the permissible magnetic flux density should be set lower to limit the eddy-current losses. The chosen permissible magnetic flux is shown in Table 7. To determine the yoke height

$$h_{y,i} = \frac{\hat{\phi}}{\hat{B}_{y,i} \cdot l \cdot 2}, \tag{12}$$

the magnetic flux must be divided by the length of the machine l the permitted magnetic flux density and a factor of 2, since the magnetic flux splits up in the two directions of the yoke [18]. The index i stands for either stator $i = 1$ or rotor $i = 2$.

The tooth width is calculated by multiplying the slot pitch with the ratio of the magnetic flux density in the air gap and the allowed magnetic flux density in the teeth as

$$w_{t,1} = \tau_N \frac{\hat{B}_{mp}}{\hat{B}_{t,1}}. \tag{13}$$

The results for the four machine designs are shown in Table 7.

Table 7. Magnetic quantities and geometry. Data from [18].

	PMSM-B1	PMSM-B2	PMSMS-S1	PMSM-S2
Magnetic flux density in the air gap $\hat{B}_p$	1.2 T	1.1 T	0.9 T	0.9 T
Magnetic flux density in the stator teeth $\hat{B}_{t,1}$	1.8 T	1.8 T	1.7 T	1.6 T
Magnetic flux density in the stator yoke $\hat{B}_{y,1}$	1.3 T	1.2 T	1.0 T	0.9 T
Magnetic flux density in the rotor yoke $\hat{B}_{y,2}$	1.4 T	1.3 T	1.1 T	1.0 T
Magnetic flux $\hat{\phi}$	5.1 mVs	5.1 mVs	4.3 mVs	3.5 mVs
Stator tooth width $w_{t,1}$	5.1 mm	5.6 mm	6.1 mm	5.8 mm
Stator yoke height $h_{y,1}$	13.5 mm	16.2 mm	19.9 mm	19.8 mm
Rotor yoke height $h_{y,2}$	12.5 mm	14.9 mm	18.1 mm	17.9 mm

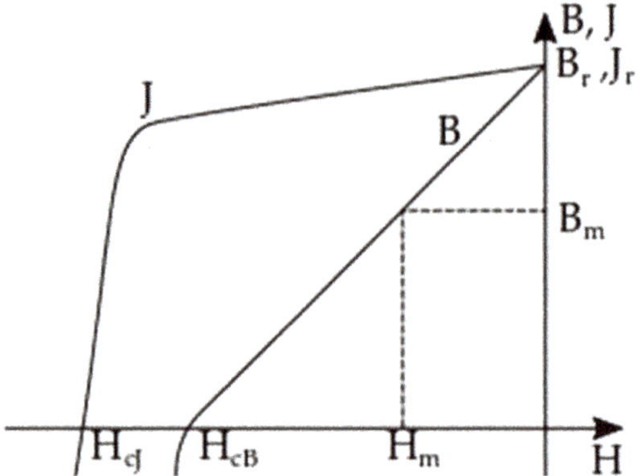

Figure 8. Schematic B-H and J-H diagram.

The size of the magnet, i.e., the width w_m and the height h_m, are set according to the flux density in the air gap and the *B-H* diagram of the magnet (cf. Figure 8) [50]. The flux density of the magnet in rated operation is determined from the estimated flux density in the air gap by

$$B_\mathrm{m} = \frac{\hat{B}_\mathrm{mp}}{\frac{4}{\pi}}.$$ (14)

To take into account the drop of magnetic voltage in the iron, in comparison to the drop of voltage over the air gap, a saturation factor of $k_\mathrm{sat} = \frac{V_\mathrm{Fe}+V_\delta}{V_\delta}$ is defined and set to $k_\mathrm{sat} = 1.6$. To take into account the slotting of the stator, the carter factor k_c is used. k_c is set to 1.1 [18]. The magnetic motive force of the magnet

$$\theta_\mathrm{m} = k_\mathrm{sat}\, k_\mathrm{c}\, \delta\, \frac{B_\mathrm{m}}{\mu_0}$$ (15)

is calculated regarding these phenomena. The magnetic field strength of the magnet H_m is determined from the *B-H* diagram of the magnet, as depicted in Figure 9. The height of the magnet is then defined by the magnetic motive force θ_m and the determined magnetic field strength H_m

$$h_\mathrm{m} = \frac{\theta_\mathrm{m}}{H_\mathrm{m}}.$$ (16)

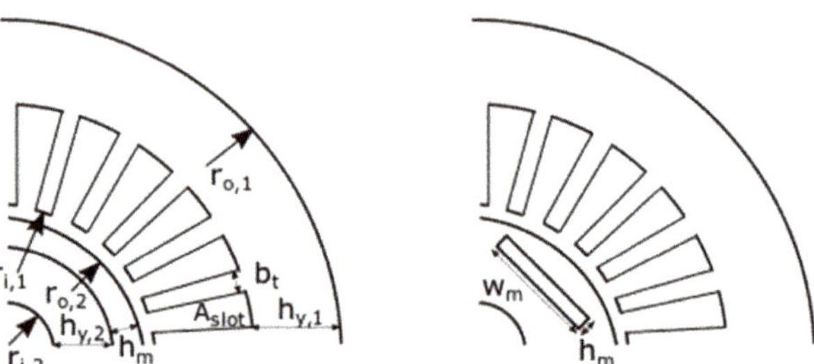

Figure 9. Machine geometry.

Since the machine shall be short-circuit-resistant, the height of the magnet needs to be validated after finalizing the winding configuration.

For a PMSM with interior magnets the width of the magnet

$$w_{\mathrm{m}} = \sin\left(\frac{\pi}{2\,p}\right) \cdot d_{\mathrm{i},1} \cdot 0.85 \tag{17}$$

is set to 85% of the side length of the equilateral triangle of one rotor pole pitch. For a PMSM with surface mounted magnets the width of the magnet is defined by the pole pitch

$$w_{\mathrm{m}} = \tau_p = \frac{\pi\,d_{\mathrm{i},1}}{2\,p}. \tag{18}$$

The bore diameter defines the magnetic flux in the machine and must be chosen accordingly.

Knowing the size of the magnet and the rotor yoke height, the rotor geometry can be completed. For a surface magnet rotor, the inner radius of the rotor is determined by

$$d_{i,2} = d_{o,2} - 2\,h_{\mathrm{m}} - 2\,h_{y,2} \tag{19}$$

If the magnets are interior in a bar shape, the depth is considered by multiplying the magnet height by a factor β to determine the inner radius:

$$d_{i,2} = d_{o,2} - 2\,\beta\,h_{\mathrm{m}} - 2\,h_{y,2}. \tag{20}$$

The factor is set to $\beta = 3$ for the machines with interior magnets. The magnet parameters are shown in Table 8.

Table 8. Magnetic quantities.

	PMSM-B1	PMSM-B2	PMSMS-S1	PMSM-S2
Magnet material		Sm2Co17		
Coercive field strength H_{cJ}		2000 kA/m		
Remanence B_r		1.13 T		
Field strength of the magnet at operating point H_{m}	280 kA/m	300 kA/m	380 kA/m	380 kA/m
Magnetic flux density of the magnet at operating point B_{m}	0.746 T	0.707 T	0.628 T	0.628 T
Magnetic motive force θ_{m}	1045 kA	990 kA	2200 kA	3960 kA
Height of the magnet h_{m}	3.7 mm	3.3 mm	5.8 mm	10.4 mm
Width of the magnet w_{m}	37.4 mm	44.3 mm	69.3 mm	62.4 mm
Inner rotor diameter $d_{\mathrm{i},2}$	67.6 mm	54.5 mm	35.5 mm	13.8 mm

3.1.4. Winding Quantities

The maximum number of series turns per phase is determined for the operating point at rated speed n_{N} and rated power P_{N}. The magnetic flux per pole $\hat{\phi}$ at this point has already been estimated with the magnetic flux density in the air gap (see Equation (9)). Assuming the winding factor $\xi_p = 0.92$ the maximum number of turns per phase

$$w_{\mathrm{calc}} = \frac{\sqrt{2}U_{\mathrm{str,max}}}{2\pi \cdot n_{\mathrm{N}} \cdot p \cdot \xi_p \cdot \hat{\phi}} \tag{21}$$

can be calculated [18]. The number of turns can be achieved with different winding configurations. The number of conductors per slot z, the number of slots per pole and phase q and the number of parallel branches a determine the winding configuration and the winding layout. The number of series turns per phase is set by these winding parameters to

$$w = \frac{z \cdot q \cdot p}{a} \tag{22}$$

and should be close to the value determined. The chosen values for the winding quantities are listed in Table 9. This winding configuration also determines the previously estimated spatial fundamental of the electric loading, calculated by

$$\hat{A}_p = \sqrt{2}\, \xi_p \frac{m\, w\, I_{N,str}}{\pi \frac{d_{i,1}}{2}}. \tag{23}$$

After setting the winding parameters, the geometry of the slot can be determined. The cross-sectional area of the conductors

$$A_{co} = \frac{I_{N,str}}{S_{1,cont}} \tag{24}$$

is calculated from the rated current $I_{N,str}$ and the current density $S_{1,cont}$. The machines are supposed to have a water jacket. The current density is accordingly assumed at $S_{1,cont} = 12\ \text{A/mm}^2$.

Depending on the type of winding and the manufacturing process, the copper fill factor k_{co} must be estimated. In this case for an automated round wire winding, it is set to $k_{co} = 0.36$ [50]. The area of the slot is then determined by

$$A_{slot} = z \frac{A_{co}}{k_{co}}. \tag{25}$$

In the case of a round wire winding, the slot can be designed as a trapezium. The width of the slot at the bore radius is set by the width of the tooth and the bore radius to

$$b_{slot,\,i} = \frac{\pi\, d_{i,1}}{N_1} - b_t. \tag{26}$$

The width of the slot at the bottom of the slot $b_{slot,u}$ and the corresponding slot height h_{slot} are calculated with the relation

$$h_{slot} = \frac{2 \cdot A_{slot}}{(b_{slot,\,l} + b_{slot,u})} \tag{27}$$

and the width of the trapezium at the bottom of the slot with

$$b_{slot,o} = \frac{2\pi \cdot r_{i,1} + 2\, h_{slot}}{N_1} - b_t. \tag{28}$$

Using the height of the slot, the outer radius of the stator is calculated as

$$d_{o,1} = d_{i,1} + 2\, h_{slot} + 2\, h_{1,y}. \tag{29}$$

Table 9. Winding configuration.

	PMSM-B1	PMSM-B2	PMSMS-S1	PMSM-S2
Calculated number of series turns per phase w_{calc}	41.37	38.56	27.8	21.57
Number of conductors per slot z	20	16	12	10
Number of slots per pole and phase q	2	2	2	2
Number of parallel branches a	4	3	2	2
Final number of series turns per phase w	40	32	24	20
Electric loading $\hat{A}_p$	108.39 kA/m	95.56 kA/m	89.71 kA/m	79.88 kA/m
Conductor cross sectional area A_{co}	2.66 mm^2	3.55 mm^2	5.32 mm^2	5.32 mm^2
Slot area A_{slot}	147.8 mm^2	157.6 mm^2	177.3 mm^2	147.8 mm^2
Inner slot width $b_{slot,i}$	2.5 mm	3.6 mm	5.4 mm	4.5 mm
Outer slot width $b_{slot,o}$	6.7 mm	8.2 mm	10.9 mm	10.1 mm
Slot height h_{slot}	31.8 mm	26.2 mm	22 mm	21.1 mm
Stator outer diameter $d_{o,1}$	208 mm	191 mm	170 mm	161 mm

Finally, the design needs to be validated to determine if it is short-circuit-resistant. This is done by calculating the magnetic motive force

$$\hat{\theta}_{\text{sc}} = 4\pi \frac{m}{2} 4 \; \sqrt{2} \; I_{\text{N,str}} \; w \frac{\xi_{\text{w}}}{2p} \tag{30}$$

in case of a sudden short circuit from rated operation occurs [51]. The calculation is based on the maximum short-circuit current. The winding attempts to keep the magnetic flux constant while the rotor keeps rotating. The short-circuit current is assumed to be 4 times the current at rated power. In the worst case, the magnetic flux is directed opposite to the magnetization direction of the magnet. The magnetic field strength of the winding, in case of the sudden short circuit, leads to a decrease of the magnetic polarization J of the magnet (cf. Figure 9). If the magnetic motive force of the winding is higher than the product of coercive field strength H_{cJ} and the height of the magnet h_{m}, the magnet will be demagnetized, leading to the requirement

$$h_{\text{m}} \geq \frac{\theta_{\text{sc}}}{H_{\text{cJ}}}. \tag{31}$$

The resulting heights of the magnets are listed in Table 10 for the four machine designs. All machine designs are short circuit resistant.

Table 10. Short circuit resistant magnet height.

	PMSM-B1	PMSM-B2	PMSMS-S1	PMSM-S2
Height of the magnet h_{m}	3.2 mm	3.4 mm	3.8 mm	3.2 mm

The active parts, i.e., rotor and stator lamination, winding and magnets, of the machine designs are shown in Figure 10. It can be noticed, that the machine size decreases with increased maximum speed, as expected.

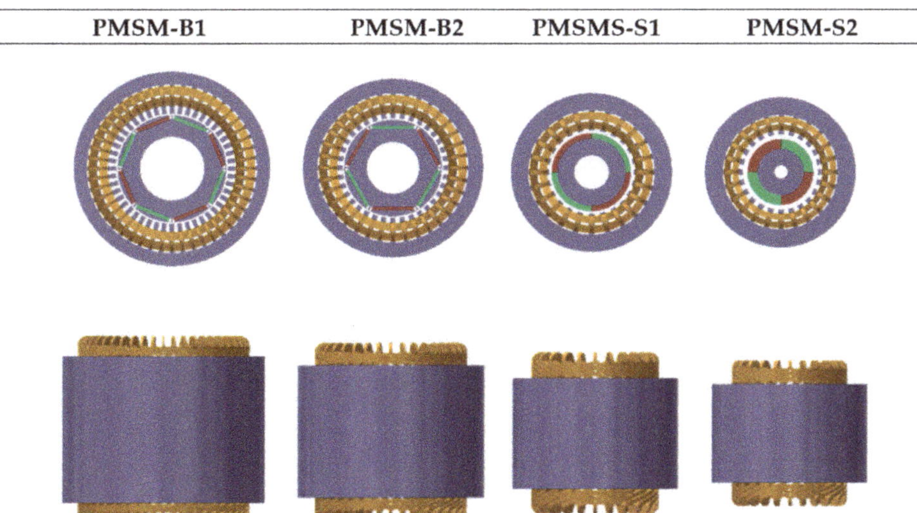

Figure 10. Final design of PMSM-B1, PMSM-B2, PMSM-S1 and PMSM-S2.

3.2. Housing Design

The housing consists of four components as depicted in Figure 11: The shaft, the water jacket and the two end shields. The components are designed as simple geometric bodies with dimensions set according to the machine designs.

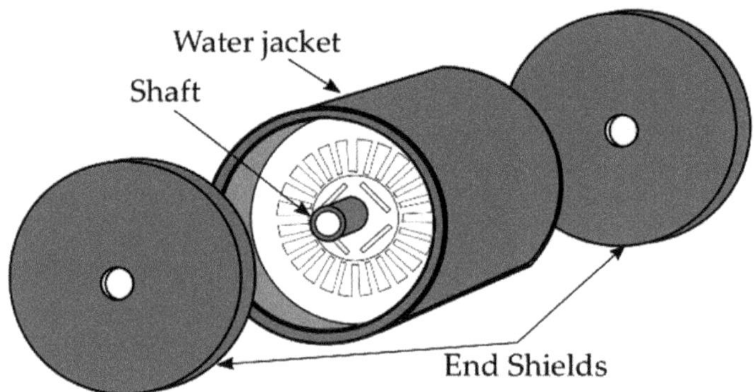

Figure 11. Housing components.

The shaft and the water jacket are approximated as hollow cylinders. The inner diameter of the shaft is set to

$$d_{i,\text{shaft}} = \frac{d_{i,2}}{1.8}. \tag{32}$$

The outer diameter is equal to the inner diameter of the rotor. The inner diameter of the water jacket is set to the outer diameter of the stator and the outer diameter is approximated by

$$d_{o,c} = d_{o,1}{\cdot}1.1. \tag{33}$$

The length of the housing

$$l_{\text{housing}} = 1.5{\cdot}l_1 \tag{34}$$

includes space for the end windings and the winding protection of the machine. The housing is closed with the end shields, that are approximated as cylinders, on both sides. The outer diameter of the end shields is equal to the outer diameter of the water jacket and the height of the end shields is set to 20 mm.

3.3. Electric Machine Design Results

Finally, the volume and the mass of the housing and the machine are calculated. The volume of the components is determined based on the derived geometric quantities. The weight is calculated based on material densities, shown in Table 11.

Table 11. Volume of the machine designs and volumetric power density.

	Rotor and Stator	Winding	Winding protection	Magnet	Housing	Shaft
Material	Electric steel	Copper	Epoxy	Sm2Co17	Aluminum	Construction steel
Density	7600 kg/m^3	8960 kg/m^3	1850 kg/m^3	8400 kg/m^3	2700 kg/m^3	7720 kg/m^3

For the housing a weight of 60% of the housing weight is added, to consider additional components, i.e., bearings, retaining rings and screws. To calculate the volume of the winding and the winding protection, the end windings need to be considered. For the PMSM-S1 and the PMSM-S2 machine designs, the volume of the end winding and winding protection is assumed to be the same size, as in the active part. For the machine design PMSM-B1 and PMSM-B2, it is assumed to be 60% of the active part. The machine designs PMSM-S1 and PMSM-S2 have a larger end winding, due to the smaller number of pole pairs p.

The weight, the volume and the volumetric and gravimetric power density, in case of maximum power, of the four designs are listed in Table 12. The results are discussed in Section 5, together with the results of the transmission.

Table 12. Volume, mass and volumetric and gravimetric power density of the machine designs.

	PMSM-B1	PMSM-B2	PMSMS-S1	PMSM-S2
Volume V	7.5 dm^3	5.8 dm^3	4.3 dm^3	3.45 dm^3
Weight m	44.1 kg	34 kg	23.6 kg	18.6 kg
Volumetric power density p_v	17.9 kW/dm^3	23.1 kW/dm^3	31.3 kW/dm^3	39 kW/dm^3
Gravimetric power density p_g	3.05 kW/kg	3.97 kW/kg	5.72 kW/kg	7.22 kW/kg

4. Conceptual Design of High-Speed Transmissions

After the introduction of the design process for the high-speed electric machines in Section 3, this section will give an overview of the design process for the associated high-speed transmissions. Like the electric machines, the transmissions are designed to match the reference vehicle's requirements for a rated power of 75 kW (cf. Table 1). In order to obtain comparable transmission designs, all systems feature the same speed and torque on the drive axle. The workflow in Figure 12 shows the design process for the transmission with the desired maximum input speed of the electric machine. After selecting the maximum speed, the necessary overall gear ratio follows directly by considering the data of the reference vehicle. The next step is the selection of the desired transmission configuration. This study considers two- and three-stage helical gear units and a combination of a planetary and helical gear set in axial parallel configurations. In the next steps, the wheel set of the transmission, including the gearings, wheel bodies, shafts and bearings is generated with the software GAP [52]. This program is able to generate transmission designs automatically and can additionally optimize the structure with regard to a large number of target variables [53]. In order to design the gearing, practical gear parameters are specified as optimization goals. To ensure comparability of the gearings, uniform target values were defined for safety factors in the load capacity calculation. The actual design of the toothing is carried out with the software STplus [54] in accordance with DIN ISO 6336 [55]. GAP also contains a structure generator for the rapid design of wheel bodies, shafts and bearings. After using the GAP structure generator, a draft of the wheel set is available that can be used to generate a model of the transmission housing. Finally, this process results in a complete concept of a transmission based on the selected configuration and overall gear ratio suitable for the reference vehicle. The following subsection will explain the individual steps of the design process in more detail.

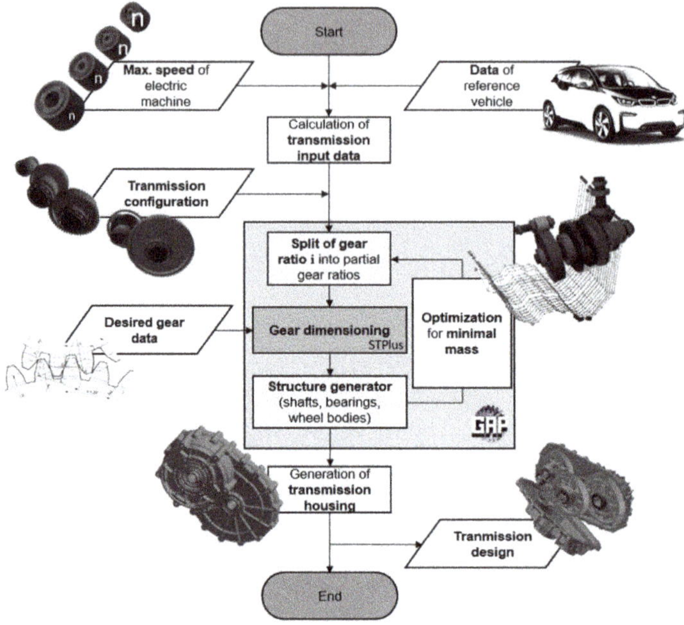

Figure 12. Transmission design process.

4.1. Calculation of Transmission Input Data

The first step of the design process is the calculation of the required input data (power, speed and torque) of the transmission. In order to obtain comparable powertrains for the reference vehicle, the transmission output parameters (see Section 1) on the drive axle are kept constant. With the maximum speed on the drive axle $n_{max,out}$ and the maximum speed of the considered electric machine n_{max}, the required overall gear ratio i follows.

$$i = \frac{n_{max}}{n_{max,out}}.$$

(35)

The overall gear ratio i of the transmission together with the known maximum torque of the reference vehicle on the drive axle $T_{max,out}$ lead to the maximum torque at the transmission input.

$$T_{max} = \frac{T_{max,out}}{i}.$$

(36)

In addition to the peak parameters at the transmission input, the rated input parameters are important for the design process. The rated P_n power of the electric machines is 75 kW and can be used to calculate the rated torque of each transmission.

$$T_n = \frac{P_N}{2 \cdot \pi \cdot n_N}.$$

(37)

Using these simple formulas, the required gear ratio and input torque can be determined for each considered input speed of the transmission.

4.2. Transmission Configurations

In the course of this study, three different transmission layouts in axial parallel configurations are considered. Figure 13 shows the transmission configurations, which are from left to right a two- (2ST) and three-stage (3ST) helical gear unit and a combination of a planetary and helical gear set (PLST). Helical gear sets with one stage are not considered, since the high gear ratios, which are considered in this study, lead to very high volumes. In case of the PLST, the first stage is the planetary gear set, since a planetary gear set on the drive axle would lead to complicated designs. Due to the small rotor diameter for higher speeds, a coaxial transmission configuration with a drive shaft guided through the rotor is not considered, as there would not be enough installation space.

2 ST	3ST	PLST

Figure 13. Considered transmission configurations.

4.3. Transmission Design with GAP

After the selection of the transmission configuration, the next step is to design the transmission system using the GAP software. The software is suitable for the rapid generation of transmission designs including the gearings, shaft dimensioning and calculation of the bearing loads. It is possible to define target functions such as minimal mass, volume or maximum efficiency and load capacity to optimize the design [56]. Since the study focuses on the power density of the powertrain, the transmissions are optimized to reach minimal mass. Therefore, the first step of the actual design process is the splitting of the overall gear ratio i into the transmissions' partial gear ratios.

4.3.1. Determination of the Transmissions Partial Gear Ratios

The overall gear ratio i of a transmission is generally made up of the transmission's stage gear ratios i_k, which are in series. Bansemir [57] developed an approach with a parameterized power function for determining the stage gear ratios of the k_{max} stages, taking into account an exemplary shaft mass.

$$i_k = a_k \cdot i^{b_k} \qquad \text{for } k < k_{max} \tag{38}$$

$$i_{k_{max}} = \frac{i}{\prod_{k=1}^{k_{max}-1} i_k} \tag{39}$$

The index k identifies the current number of the considered stage and k_{max} accordingly for the number of stages of the transmission. The parameters a_k and b_k, which control the splitting of the overall gear ratio into the stage gear ratios, are the result of an optimization for each transmission configuration and will be shown in Section 4.4.

4.3.2. Gear Dimensioning and Design

With the partial gear ratios of the transmission, the actual design of the gearings are calculated with the STplus software [54]. This software includes the load capacity calculation according to DIN ISO 6336 [55] and additional content regarding the geometrical design and machining of spur gear stages. The starting point of the design process are uniform values of the safety factors for all transmissions that lead to comparable gearing designs regarding the load carrying capacity. Table 13 shows the minimal safety factors for the load carrying capacity calculation. $S_{F,min}$ stands for the minimal safety factor against tooth root breakage, $S_{H,min}$ against pitting and $S_{B,min}$ represents the minimal required safety against scuffing. In addition to the safety factor that should be achieved, the considered load cases play an important role in gear dimensioning.

Table 13. Safety factors for the load carrying capacity calculation.

Safety Factors	
$S_{F,min}$	1.3
$S_{H,min}$	1.1
$S_{B,min}$	1.3

Table 14 gives an overview of the 5 load cases for the gear design process. Load case 1 is a load spectrum for a middle-class BEV, which has a total operating time of 2500 h and covers the map of the electric machine by scaling the torque and speed according to their rated and maximum operating points. In addition to the load collective, the overload case (2) covers an operation of the system with maximum torque and an application factor of $K_A = 1.6$. The possible operating time of this case should be at least 3 h. Load case 3 represents a static overload due to inappropriate utilization with maximum torque and an application factor of $K_A = 2$. Load cases 4 and 5 ensure the scuffing load capacity, whereby critical operating points with high-speed and load (4) and maximum load at rated torque (5) are taken into account.

Table 14. Load cases for the gearing design.

Load Case	Input Torque T	Input Speed n	Operating Time
1—Load spectrum	$0 \dots T_{max}$	$0 \dots n_{max}$	2500 h (200,000 km)
2—Overload	$T_{max} \cdot 1.6$	Rated speed n_N	>3 h
3—Static overload	$T_{max} \cdot 2$	-	Short term
4—Scuffing	T_{max} at n_{max}	n_{max}	Short term
5—Scuffing	T_{max}	n_N	Short term

To receive comparable gearing designs, uniform values for central gear parameters and other design parameters are also specified. Table 15 shows important design factors according to DIN ISO 6336 [55] and design targets for the gear design optimization. The dynamic factor K_V considers additional forces in the tooth contact due to dynamic effects. The load factors $K_{H\alpha}$ and $K_{H\beta}$ represent the influence of an uneven load distribution over the profile and width of the flank. Furthermore, in order to control the gear design process, design targets for the contact ratios ε_α and ε_β as well as the width-to-diameter ratio b/d of the wheel bodies are defined. The selected design values are suitable for the practical preliminary design of gearings, but would have to be determined more precisely with higher-quality methods for detailed designs.

Table 15. Design factors and design targets for the gear design process.

Design Factors	
Dynamic factor K_V	1.2
Transverse load factor $K_{H\alpha}$	1.1
Face load factor $K_{H\beta}$	1.2
Design targets	
Transverse contact ratio ε_α	1.6
Overlap ratio ε_β	2.1
Aspect ratio b/d	< 0.7

4.3.3. Structure Generator

To complete the transmission wheel set, the shafts, wheel bodies and bearing are generated in the next step. The shaft length follows from the numbers of bearings and wheel bodies, which are located on the shaft. Each element on the shaft is associated with an individual section. For these sections, an equivalent stress according to Niemann et al. [42] is calculated from the torsional and bending moments T and T_b to define the minimum shaft diameter $d_{V,a,min}$ (Equation (38). The allowable stress $\sigma_{b,all}$ is derived from the fatigue strength σ_w of the material, which is 20MnCrS5 in case of the shafts.

$$d_{V,a,min} = 2.17 \cdot \sqrt[3]{\frac{T \cdot 1000 \cdot \sqrt{a_b^2 + 0.4}}{\sigma_{b,all}}} \quad \text{with } a_b = \frac{T_b}{T} \quad \text{and } \sigma_{b,all} = \frac{\sigma_w}{S} \tag{40}$$

Furthermore, Parlow's [53] approach for designing a hollow shaft is used, which calculates the minimum outer diameter $d_{HW,a,min}$ and maximum inner diameter $d_{HW,i,min}$ to achieve the identical moment of resistance as the solid shaft. While the shafts of the spur gear sets are solid shafts, the design of the planetary gear set requires a hollow shaft for the planet carrier.

Figure 14 illustrates the simplified approach to determine the length of the transmissions' shafts. When calculating the shaft length, the input variables are the tooth widths b_w, the distances between the wheel bodies d_{ww}, the distances between the wheel bodies and bearings d_{wb} and bearing widths b_b. The distances are fixed at a value of $d_{wb} = d_{ww} = 10$ mm, the toothing widths b_w are a result of the gear dimensioning and the bearing widths are changed depending on the torque at the specific shaft. Arrangements with a locating and a non- locating bearing, using a cylindrical roller and a deep groove ball bearing are used for all shafts. Since the loads on the output shafts are very similar for all transmission designs, a uniform bearing concept was chosen. The bearing sizes and weights of the remaining bearings were scaled from the design with the maximum input speed according to the torque and speed.

Figure 14. Schematic structure for definition of the shaft length.

Finally, the wheel body structures are calculated in the structure generator. Figure 15 shows the geometry of the wheel bodies. The geometry depends on the root diameter d_f and the tooth width b parameters, where the web width is $b_s = 0.25 \cdot b$. The thickness of the gear rim follows the recommendations of the ISO6336 standard, in which a minimum ring gear thickness of three times the module m_n is required. Therefore, the diameter results in $d_{rim} = d_f - 6 \cdot m_n$. The hub geometry of the wheel body is defined by the inner and outer hub diameters d_{ih} and d_{oh} and the hub width, which is the tooth width. The inner hub diameter is equal to the outer shaft diameter, and the wheel hubs have a thickness of three times the module according to the procedure in case of the gear rim.

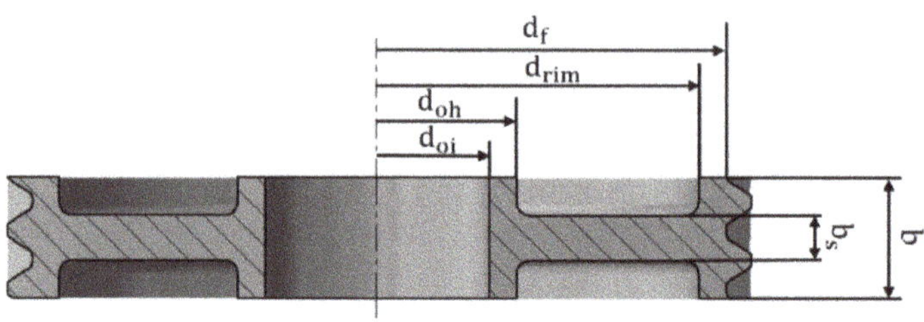

Figure 15. Schematic geometry of the wheel bodies.

4.3.4. Optimization of Transmission Design for Minimal Mass

With the conclusion of the structure generator, a first draft of a complete wheel set for the considered configuration and gear ratio is available. Since this study primarily looks at the power density of the system, the next step is to optimize the transmission to find an optimal design regarding systems minimal mass. The optimization used in the study is based on the simulated annealing algorithm [53,58], which is part of GAP and provides values for the parameters a_k and b_k; this is introduced in Section 4.4 for the splitting of the gear ratio. Table 16 shows the optimized factors in case of the three considered transmission configurations. In case of the PLST configuration, an additional requirement for a sufficient center distance was introduced to ensure the installation space for the drive axle.

Table 16. Parameters for splitting of overall gear ratio to reach minimal mass.

Transmission Configuration	a_k	b_k
2ST	$a_1 = 1.0151$	$b_1 = 0.6011$
3ST	$a_1 = 1.0151$	$b_1 = 0.6240$
	$a_2 = 1.7350$	$b_2 = 0.2236$
PLST	$a_1 = 0.4488$	$b_1 = 0.6859$

4.3.5. Generation of Simplified Housing Geometry

Analogous to the design of the electric machines, the housing structure is also considered for the transmissions. To do this, an envelope curve is placed around the wheelset, the shaft angles being adjusted to reach a minimum base area of the housing. Figure 16 shows the base area of the considered three-stage configuration 3ST, which can be described by simple geometric shapes (circular sectors and trapezoid). When calculating the housing volume, a distance of at least 10 mm from all rotating components to the housing wall and a constant wall thickness of 5 mm is assumed.

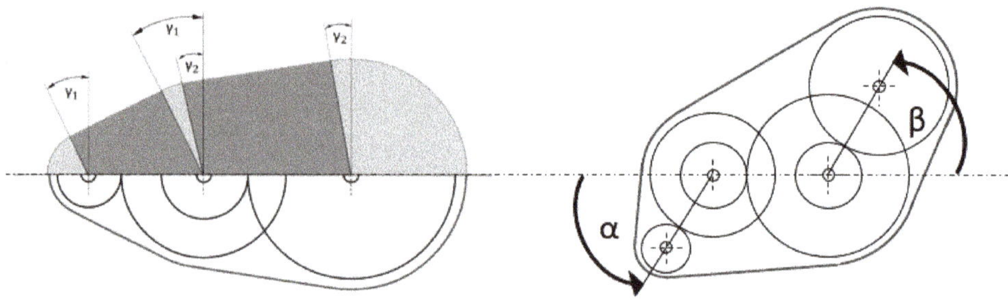

Figure 16. Exemplarily generation of a simplified housing geometry for 3ST.

Considering the maximum axis angles without collision of wheel bodies, the shaft angles α and β were optimized for each transmission design to achieve a minimum base area of the housing and thus a minimum volume and mass. A common transmission-housing alloy, AlSi9, with a density of $\rho_{AlSi9} = 2.65 \ \text{kg/dm}^3$ was assumed as material. In addition, stiffening structures and variable wall thicknesses of the housing are considered according to Linke [59] by a correction factor of $f_h = 1.5$, which is multiplied with the housing mass derived from the described optimization.

4.4. Transmission Design Results

With the designed wheelset and the simplified housing geometry, the transmission design with the desired maximum input speed and gear ratio are finalized. An overview of the results of the masses and volumes of the designed transmissions is given in this section. Figure 17 shows the transmission masses of the three configurations and total gear ratios from 5 to 50 with the associated maximum speed of the electric machines. The PLST configuration shows the lowest transmission mass over the entire range of input speeds and gear ratios, whereby the mass increases with increasing gear ratios. As expected, ST3 has the largest overall transmission mass due to the additional shaft and wheel bodies, whereby a decreasing transmission mass with increasing gear ratio can be found. The reason for the decreasing mass in case of ST3 is the decreasing torque with increasing maximum speed of the electric machine and thus a smaller possible size of the gears and shafts. The ST2 configuration is between PLST and ST3, whereby ST2 becomes heavier than ST3 for gear ratios larger than i = 40.

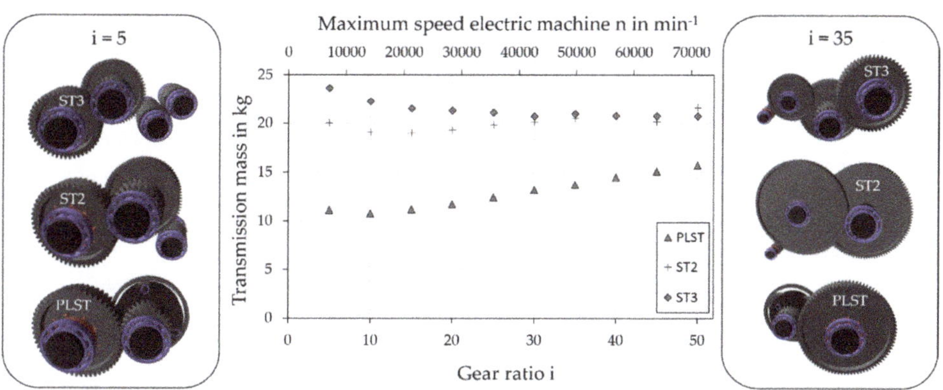

Figure 17. Transmission masses for the three considered configurations over the gear ratio.

Figure 18 shows the resulting volumes of the three transmission configurations. Again, the PLST configuration achieves the lowest values for the entire values of the volumes for the entire evaluation range. In case of the two-stage ST2 configuration, the large required wheel diameters quickly lead to a large overall system volume. The three-stage ST3 configuration shows the smallest increase in volume with increasing gear ratios and its total is in between the compact PLST and larger ST2 configuration.

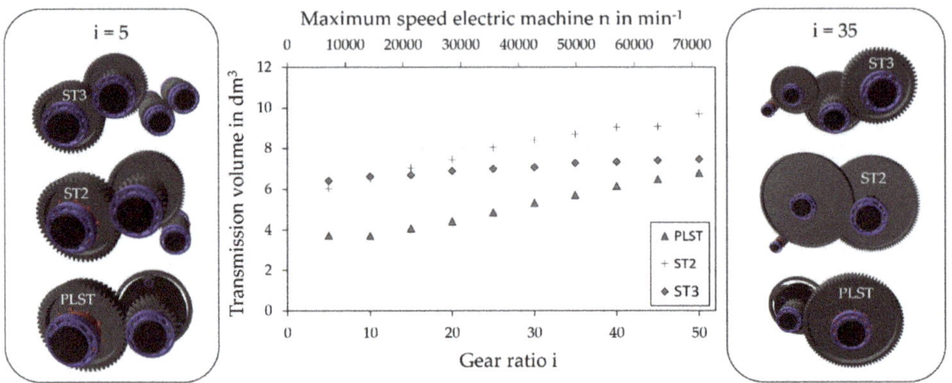

Figure 18. Transmission volume for the three considered configurations over the gear ratio.

5. Merged Design of High-Speed Powertrain

In this chapter, the results of the electric machine and transmission designs from Sections 3 and 4 are combined and presented. A total of four different electric machines with various maximum speeds but identical power values were designed, as described in Section 3. Analogously, suitable transmissions were designed in Section 4, which adapt the maximum speed of the electric machine with constant output parameters by adjusting the total gear ratio. In order to make a statement about characteristics of the whole powertrain, the transmissions that exactly match the designed electric machines were selected and combined. Through this procedure, twelve possible powertrain systems for the BMW i3 reference vehicle (cf. Chapter 1) will be presented, which differ in maximum speeds of the electric machine and transmission design. Based on these powertrain designs, the system is examined for total mass and volume and thus the power density of the electromechanical powertrain.

Figure 19 (left) shows the masses of the resulting powertrain designs over the gear ratio, respectively the maximum speed of the electric machines. The total masses show a clear decrease in powertrain mass for all transmission configurations. Since PLST achieves the lowest transmission masses, the powertrain system with this transmission configuration has the lowest overall mass, too. The increase of the maximum speed of the electric machines from 12,000 to 50,000 min^{-1} reduces the weight for the PLST configuration from 55.0 to 32.4 kg, which is 22.6 kg or a mass reduction of 41.1%. In case of the two-stage 2ST configuration, the mass reduction is 24.2 kg from 63.4 to 39.2 kg or 38.2%, whereas for the three-stage 3ST configuration this is 27.1 kg from 66.7 to 39.6 kg or 40.7%. The weight of the electric machine is reduced by 57.6% from 44.1 to 18.7 kg by increasing the maximum speed of the electric machine. The diagram on the right in Figure 19 further clarifies the composition of the total mass from the electric machine and the transmission for the PLST configuration. It shows a significantly increasing share of the transmission mass in the total mass of the powertrain due to the rapidly decreasing weight of the electric machine. However, the transmission mass is not increased by the same extent as the weight of the electric machine is decreased, which makes it possible to strongly reduce the overall weight of the powertrain.

In addition to the powertrain design of this study, the diagram in Figure 20 (left) also shows the powertrain mass of the reference vehicle including the electric machine and transmission. The transmission of the reference vehicle has a gear ratio of 9.665 and a mass of 23.2 kg. Together with the electric machine, the reference powertrain achieves a total mass of 73.1 kg [60]. This value is close to the data point of this study with a gear ratio of 8.2.

Figure 20 shows similarly to Figure 19 the volumes of the powertrain designs with the four electric machines and the three different transmission configurations. Here, the increase of maximum speed from 12,000 to 50,000 min^{-1} reduces the volume for the PLST configuration from 11.3 dm^3 to 9.2 dm^3, which is 2.09 dm^3 or a reduction of 18.5%. In case of the two-stage 2ST configuration, the volume reduction results in 1.8 dm^3 from 14.0 dm^3 to 12.2 dm^3 or 13.1%. The largest volume reduction can be found for the ST3 configuration with 3.3 dm^3 from 14.1 dm^3 to 10.8 dm^3 or 23.6%. The volume of the electric machines decreases equivalently as the mass.

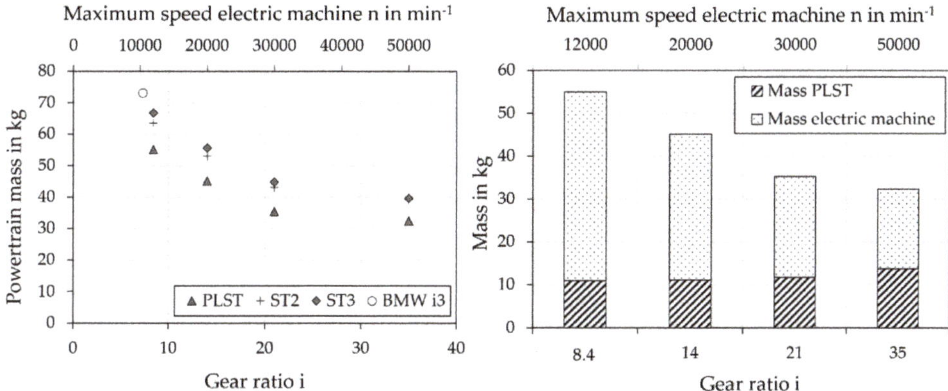

Figure 19. Powertrain masses (**left**) and its split into transmission and electric machine (**right**).

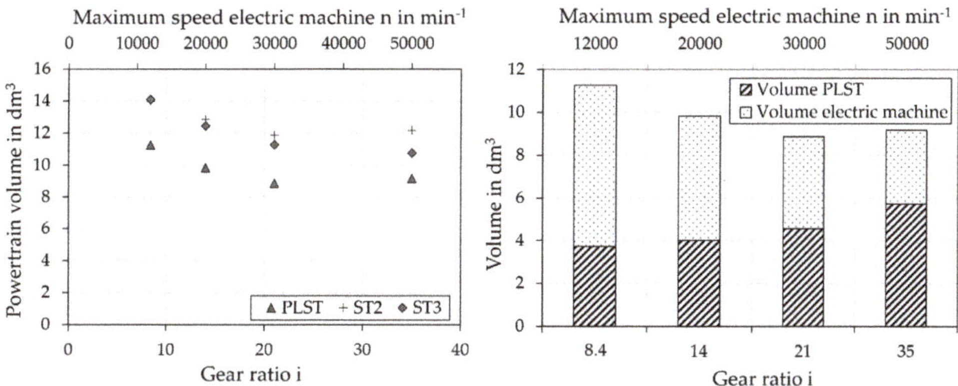

Figure 20. Powertrain volumes (**left**) and its split into transmission and electric machine (**right**).

In contrast to the total mass, the increase in the maximum speed from 30,000 to 50,000 min^{-1} in the case of the PLST and ST2 configurations already shows an increasing total volume. The volume of these transmission configurations increases to a greater extent than the electric machine decreases its volume because of the large wheel diameters. While the share of the transmission mass for PLST in the total mass at 50,000 min^{-1} is 42.3% (cf. Figure 20), the share for the volume is 62.3%, which is higher than the electric machine volume.

Figure 21 (left) shows the gravimetric power density of the powertrain designs and an increasing power density with increasing maximum speed of the electric machines in all three transmission configurations. Since PLST achieves the lowest masses, this configuration achieves the highest power density. The increasing difference between the PLST, ST2 and ST3 configurations, respectively with increasing gear ratio is due to the increasing share of the transmission mass to the total mass. In addition, the gravimetric power density of the reference vehicle's powertrain of 1.85 kW/kg is supplemented into the diagram. In comparison with the PLST configuration at maximum speed of 50,000 min^{-1}, the gravimetric power density can be increased by 125.7% to 4.17 kW/kg.

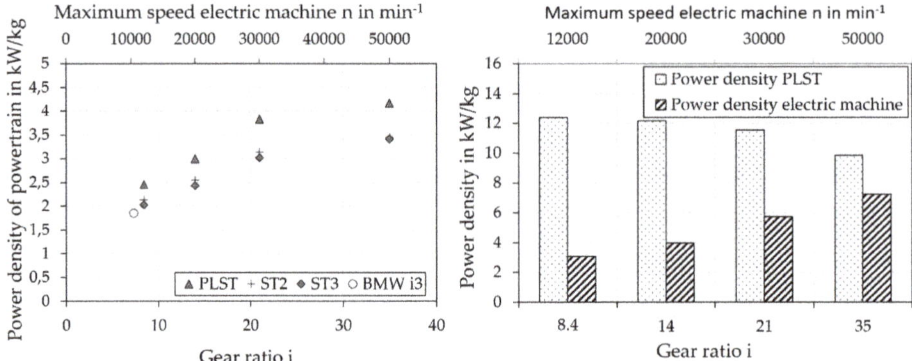

Figure 21. Gravimetric power density of powertrain (**left**), electric machine and transmission (**right**).

Figure 21 (right) shows the evolution of the gravimetric power density of the electric machine and the transmission PLST with increasing gear ratio. While there is a large gap between the very high value for the transmission and the low value of the power density of the electric machine at a low gear ratio, this gap is becoming lower and the values converge with increasing gear ratio.

When looking at the results of the volumetric power density in Figure 22, the effect of the disproportionately increasing transmission volume becomes apparent. Even at a gear ratio of 21, which equals the maximum speed of 30,000 min^{-1}, the electric machine has a higher volumetric power density than the transmission and expands the distance to the transmission with further increase of the maximum speed. Like in the case of the powertrain mass, the transmission shows higher volumetric power densities for lower gear ratios. A balanced ratio of the volumetric power density can be expected in the range of the overall gear ratio between 14 and 21.

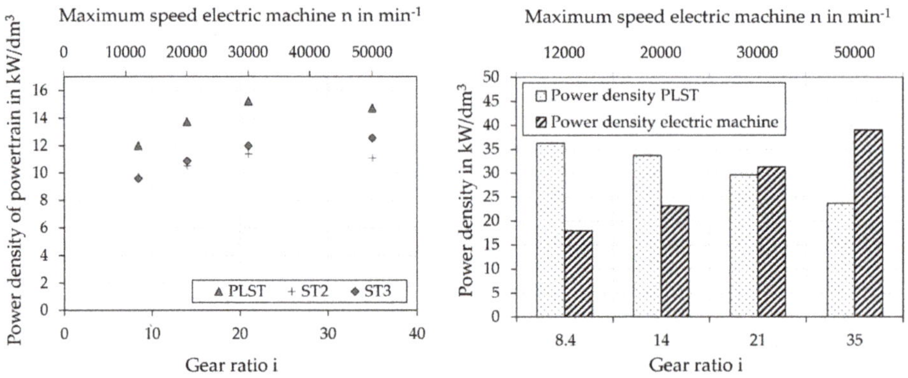

Figure 22. Volumetric power density of powertrain (**left**), electric machine and transmission (**right**).

6. Conclusions

This study investigated the influence of maximum speed of the electric machine on the power density of electromechanical powertrains. For this purpose, powertrains with an electric machine and suitable transmission were designed for different maximum speeds of the electric machines. An increase in the speed of the electric machine results in a higher required gear ratio of the transmission to reach the same transmission output conditions. Four different electric machines with maximum speeds from 12,000 min^{-1} to 50,000 min^{-1} were designed. This defined four powertrain designs, which are completed by suitable transmissions with three different configurations. The machines and transmissions were calculated based on conceptual design. Further investigations and adaptions may be necessary to meet all requirements of a detailed design.

The results of the powertrain designs show that the mass of the electric machines decrease significantly with increasing maximum speed. Contrarily, the mass of the transmissions increases, whereby the increase of mass is less than the reduction in mass for the electric machines. As a result, the gravimetric power density can be increased significantly by increasing the maximum speed of the electric machines. Compared to the reference vehicle, which has a maximum speed of 11,400 min^{-1}, the gravimetric power density could be increased by 125.7% from 1.85 to 4.17 kW/kg by increasing the maximum speed up to 50,000 min^{-1}. In addition to the mass, the volume of the powertrains can also be reduced, whereby the reduction is smaller due to a disproportionately increase of the transmission volume.

For both parameters, the gravimetric and volumetric power density, a convergent growth could be found. In case of the volumetric power density, a peak can be determined between 20,000 min^{-1} and 30,000 min^{-1}, while the peak of the gravimetric power density lies above the maximum speed of 50,000 min^{-1} of the fastest electric machine considered in this study. The results of this study clearly show the potential of increasing both the gravimetric and volumetric power density of electromechanical powertrains significantly by increasing the maximum speed of the electric machine.

7. Outlook

In addition to the power density of electromechanical powertrains, there are other important parameters whose behavior for increasing maximum speed should be investigated in future studies.

The increasing speed limits the bore volume, due to mechanical stress in the rotor and bend critical speeds. Consequently, the output power of the machine is limited [5]. These limits have not been exceeded in the presented machines. Thus, the increase of the output power at high maximum speed will lead to a limitation in the power of the electric machine. Further investigation needs to be carried out to distinguish these limits. Furthermore, effects such as field weakening behavior in case of the electric machines or the manufacturing need to be discussed further on.

One crucial aspect in the design process of electromechanical powertrains is the efficiency behavior of the overall powertrain with increasing maximum speed that determines the size of the battery or the range of the car. In the electric machine, the hysteresis and eddy-current losses rise due to the higher frequency. Furthermore, the frequency dependent ohm's losses need to be investigated in depth to make a statement about the efficiency of the electric machine. For the transmission, the impact of the higher speeds on the losses has to be investigated. Concerning the no-load losses, higher speeds of the bearings and gearings are expected to have an impact towards higher losses. Optimizing the oil supply and oil flow as well as low viscosity fluids in the transmission can reduce the increase of these losses. In case of load-dependent losses, the increased speed causes reduced normal force on the tooth flanks due to the reduced torque and therefore reduced bearing loads, reducing the load-dependent losses. Also, high circumferential speeds can support fluid film lubrication regime of tribological contacts. Further investigations on the efficiency are to be carried out to describe the interdependency of higher circumferential speeds and the reduced forces acting in transmissions and powertrains with increasing maximum speed. In addition, the potential of low-loss gears and innovative base oils like water-containing gear fluids have to be evaluated, also in the context of holistic thermal managements.

Another aspect is the NVH-behavior of the transmission and the electric machine. Even though the eigenfrequencies of the electric machine are increasing, due to the more compact design, the frequency range of the exiting forces' waves increases as well with the maximum speed. For the transmission, it becomes more difficult to operate the first stages subcritically over the entire operating range. The interaction of the electric machine with its larger frequency range can also lead to a more complex excitation behaviour for the transmission and powertrain system, which should be examined in detail.

Furthermore, the increase of the maximum speed result in the use of new technologies, which currently may be not present or lead to increased costs of the powertrain system. For the electric machine, one technology leap is the use of surface mounted magnets and the fixation with the bandage onto the rotor. The bandage material is quite expensive and the shrinkage on the rotor a challenging process. Furthermore, the amount of magnet material increases to keep the magnetic flux density in the air gap high with increasing size of the air gap. To achieve the high rotor speeds, bearings with higher accuracy and special materials to limit the centrifugal forces are needed. The lower torque leads to smaller sizes of the gearings, which means that influences from manufacturing errors are more important. Therefore, the quality of the gearings may be increased in order to obtain a sufficient load capacity and acoustical behaviour of the meshes under load.

Author Contributions: Conceptualization: D.S. and M.E.G.; methodology, D.S. and M.E.G.; software, D.S., A.T. and M.E.G.; validation, D.S., A.T., and M.E.G.; investigation, B.M., A.H.; resources, Institute of Machine Elements (TUM, Munich), Institute for Drive Systems and Power Electronics (LUH, Hannover); data curation, D.S. and M.E.G.; writing—original draft preparation, D.S., B.M., M.E.G. and A.H.; writing—review and editing, T.L., M.O., K.S. and B.P.; visualization, D.S. and M.E.G.; supervision, T.L., M.O., K.S. and B.P.; project administration, K.S.; All authors have read and agreed to the published version of the manuscript.

Funding: Supported by: Federal Ministry of Economic Affairs and Energy on thebasis of a decision by the German Bundestag.

Conflicts of Interest: The authors declare no conflict of interest.

References

1. *Green Paper—A 2030 Framework for Climate and Energy Ploicies*; European Commission: Brussels, Belgium, 2013.
2. Deiml, M.; Schneck, M.; Eriksson, T.; Tan-Kim, A.; Peyman, J.; Anthony, S. High Speed Electric Drive-AVL solution for next generation EDU. In Proceedings of the 13th CTI SYMPOSIUM "Automotive Drivetrain, Intelligent and Electrified", Novi, MI, USA, 13–14 May 2019.
3. Sedlmair, M.; Fischer, P.D.; Lohner, T.; Pflaum, H.; Stahl, K. Holistic Investigations on the innivative High-Speed Powertrain Speed2E for Electric Vehicles. In Proceedings of the 15th International CTI Symposium Automotive Transmissions, HEV and EV Drives, Berlin, Germany, 4 December 2017.
4. Merwerth, J. The Hybrid-Synchronous Machine of the New BMW i3&i8: Challenges with Electric Traction Drives for Vehicles. In Proceedings of the Workshop University Lund, Lund, Sweden, 1 April 2014.
5. Binder, A.; Schneider, T. High-Speed Inverter-Fed AC Drives. In Proceedings of the International Aegean Conference on Electrical Machines and Power Electronics, Bodrum, Turkey, 10–12 September 2007; pp. 9–16.
6. Binder, A.; Schneider, T.; Klohr, M. Fixation of buried and surface-mounted magnets in high-speed permanent-magnet synchronous machines. *IEEE Trans. Ind. Appl.* **2006**, *42*, 1031–1037. [CrossRef]
7. Sarlioglu, B.; Morris, C.T.; Han, D.; Li, S. Benchmarking of Electric and Hybrid Vehicle Electric Machines, Power Electronics, and Batteries. In *ACEMP-OPTIM-ELECTROMOTION 2015 Joint Conference, Intl Aegean Conference on Electrical Machines & Power Electronics (ACEMP), Side, Turkey, 2–4 September 2015*; Ertan, H.B., Ed.; IEEE: Piscataway, NJ, USA, 2015; pp. 519–526.
8. Gerada, D.; Mebarki, A.; Brown, N.L.; Gerada, C.; Cavagnino, A.; Boglietti, A. High-Speed Electrical Machines: Technologies, Trends, and Developments. *IEEE Trans. Ind. Electron.* **2014**, *61*, 2946–2959. [CrossRef]
9. Henke, M.; Narjes, G.; Hoffmann, J.; Wohlers, C.; Urbanek, S.; Heister, C.; Steinbrink, J.; Canders, W.-R.; Ponick, B. Challenges and Opportunities of Very Light High-Performance Electric Drives for Aviation. *Energies* **2018**, *11*, 344. [CrossRef]
10. *Soft MAgnetic Cobalt-Iron Alloys Vacoflux and Vacodur*; Vacuumschnelze GmbH & Co. KG.: Hanau, Germany, 2016.
11. Huang, W.-Y.; Bettayeb, A.; Kaczmarek, R.; Vannier, J.-C. Optimization of Magnet Segmentation for Reduction of Eddy-Current Losses in Permanent Magnet Synchronous Machine. *IEEE Trans. Energy Convers.* **2010**, *25*, 381–387. [CrossRef]
12. L TYPE LAMINATED MAGNETS. *Reduce Eddy Current Loss in High-Efficiency Motors*; Arnold Magnetic Technologies: New York, NY, USA, 2020.
13. Mirzaei, M.; Binder, A.; Funieru, B.; Susic, M. Analytical Calculations of Induced Eddy Currents Losses in the Magnets of Surface Mounted PM Machines with Consideration of Circumferential and Axial Segmentation Effects. *IEEE Trans. Magn.* **2012**, *48*, 4831–4841. [CrossRef]
14. M. Woite GmbH. Material Datasheet for 42CrMo4, 1.7225, Chrome-Molybdenum Alloyed Heat Treatable Steel. Available online: http://www.woite-edelstahl.com/17225en.html (accessed on 11 April 2020).

15. M. Woite GmbH. Material Datasheet for NiMo16Cr16Ti, 2.4610, Hastelloy®C4, Nicrofer®6616 hMo, Nickel Chrome Molybdenum Alloy. Available online: http://www.woite-edelstahl.com/24610en.html (accessed on 11 April 2020).

16. Niederstadt, G. *Ökonomischer und Ökologischer Leichtbau mit Faserverstärkten Polymeren. Gestaltung, Berechnung und Qualifizierung*; mit 31 Tabellen; 2., völlig neubearb. Aufl; expert-Verl.: Renningen, Germany, 1997; ISBN 9783816914167.

17. Glaessel, T.; Pinhal, D.B.; Masuch, M.; Gerling, D.; Franke, J. Manufacturing Influences on the Motor Performance of Traction Drives with Hairpin Winding. In Proceedings of the 2019 9th International Electric Drives Production Conference (EDPC), Esslingen, Germany, 3–4 December 2019; pp. 1–8, ISBN 978-1-7281-4319-4.

18. Müller, G.; Vogt, K.; Ponick, B. *Berechnung Elektrischer Maschinen*, 6th ed.; Wiley-VCH: Weinheim, Germany, 2008.

19. Sedlmair, M.; Fischer, P.; Stahl, K. *FVA Nr. 716—Schlussbericht—Speed2E—Innovatives Super-Hochdrehzahl-Mehrgang-Konzept für den Elektrifizierten Automobilen Antriebsstrang für Höchste Effizienz und Höchsten Komfort*; Forschungsvereinigung Antriebstechnik e.V.: Frankfurt am Main, Germany, 2017.

20. Mileti, M.; Schweigert, D.; Pflaum, H.; Stahl, K. Speed4E: Hyper-High-Speed Driveline and Gearbox for BEVs. In Proceedings of the CTI Symposium, Novi, MI, USA, 13–16 May 2019.

21. Schweigert, D.; Mileti, M.; Morhard, B.; Sedlmair, M.; Fromberger, M.; Otto, M.; Lohner, T.; Stahl, K. Innovative Transmission Concept for Hyper-High-Speed Electromechanical Powertrains. In Proceedings of the International VDI-Congress E-Drive—Drivetrain for Vehicles, Bonn, Germany, 10–11 July 2019; pp. 541–554.

22. Morhard, B.; Schweigert, D.; Mileti, M.; Sedlmair, M.; Fromberger, M.; Otto, M.; Lohner, T.; Stahl, K. A First Efficiency and Dynamics Analysis of the Innovative Hyper-High-Speed Electromechanical Powertrain Speed4E. In *E-MOTIVE Expert Forum Electric Vehicle Drives*; MediaTUM: München, Germany, 2019.

23. Lang, C.; Wilson, B. Simulation and Optimization of Noise for a Highly Integrated EV Drivetrain. In Proceedings of the 8th International CTI Symposium North America, Rochester, NY, USA, 14–15 May 2014.

24. Epskamp, T.; Butz, B.; Doppelbauer, M. Design and Analysis of a High-Speed Induction Machine as Electric Vehicle Traction Drive. In Proceedings of the 18th European Conference on Power Electronics and Applications EPE, Karlsruhe, Germany, 5–9 September 2016.

25. Yilmaz, M.; Mirza, M.; Lohner, T.; Stahl, K. Superlubricity in EHL Contacts with Water-Containing Gear Fluids. *Lubricants* **2019**. [CrossRef]

26. Hirano, M.; Shinjo, K. Atomistic locking and friction. *Pyhsical Rev.* **1990**, *41*, 11837–11851. [CrossRef] [PubMed]

27. Wimmer, A.J. Lastverluste von Stirnradverzahnungen—Konstruktive Einflüsse, Wirkungsgradmaximierung, Trobologie. Ph.D. Thesis, Technische Universität München, Munich, Germany, 2006.

28. Hinterstoißer, M.; Sedlmair, M.; Lobner, T.; Stahl, K. Minimizing Load-dependent Gear Losses. *Tribol. Schmier.* **2019**, *66*, 15–25.

29. Yilmaz, M.; Lohner, T.; Michaelis, K.; Stahl, K. Minimizing gear friction with water-containing gear fluids. *Forsch Ing.* **2019**, *83*, 327–337. [CrossRef]

30. Yilmaz, M.; Lohner, T.; Michaelis, K.; Stahl, K. Bearing Power Losses with Water-Containing Gear Fluids. *Lubricants* **2020**. [CrossRef]

31. Ruellan, A.; Ville, F.; Kleber, X.; Arnaudon, A.; Girodin, D. Understanding white etching cracks in rolling element bearings: The effect of hydrogen charging on the formation mechanisms. *Proc. Inst. Mech. Eng. Part J. Eng. Tribol.* **2014**, *228*, 1252–1265. [CrossRef]

32. Evans, M.-H. White structure flaking (WSF) in wind turbine gearbox bearings: Effects of 'butterflies' and white etching cracks (WECs). *Mater. Sci. Technol.* **2012**, *28*, 3–22. [CrossRef]

33. Errichello, R.; Budny, R.; Eckert, R. Investigations of Bearing Failures Associated with White Etching Areas (WEAs) in Wind Turbine Gearboxes. *Tribol. Trans.* **2013**, *56*, 1069–1076. [CrossRef]

34. Gwinner, P.; Otto, M.; Stahl, K. Vibration Behaviour of High-Speed Gearings for Electrically Driven Powertrains. In Proceedings of the Drivetrain for Vehicles, Bonn, Germany, 21–22 June 2016.

35. Gwinner, P. Schwingungsarme Achsgetriebe Elektromechanischer Antriebsstränge-Auslegung Schwingungsarmer Stirnradverzahnungen für den Automobilen Einsatz in Hochdrehenden, Elektrisch Angetriebenen Achsgetrieben. Ph.D. Thesis, Technische Universität München, Munich, Germany, 2017.

36. Höhn, B.-R. Modern Gear Calculation. In Proceedings of the International Conference, Munich, Germany, 13–15 March 2002.

37. Deiml, M.; Eriksson, T.; Schneck, M.; Tan-Kim, A. Hochdrehende E-Antriebseinheit für die nächste Fahrzeuggeneration. *ATZ Automob.* **2019**, *121*, 42–47. [CrossRef]

38. DIN 628-1:2008-01: Rolling Bearings—Angular Contact Radial Ball Bearings—Part 1: Single Row Self-Locking. Available online: https://www.beuth.de/de/norm/din-628-1/102569644 (accessed on 24 June 2020).

39. Super Precision Bearings, Spindle Bearings, Super Precision Cylindrical Roller Bearings, Axial Angular Contact Ball Bearings: Data Sheet 2019. Available online: https://www.schaeffler.co.uk/content.scha effler.co.uk/en/news_media/media_library/index.jsp?tab=mediathek-pub&uid=114232&subfilter=app: dc;language-vid:167;language-pub:167;mediatyp-pub:all;referencetyp-pub:0 (accessed on 24 June 2020).

40. Super-Precision Angular Contact Ball Bearings: High-Speed, B Design, Sealed as Standard: Data Sheet 2012. Available online: https://www.skf.com/binary/57-75457/0901d196801b304f-Super-precision-angular-contact -ball-bearings-High-speed-B-design-sealed---06939_5-EN.pdf (accessed on 24 June 2020).

41. ISO 6194-1:2007-09: Rotary Shaft Lip-Type Seals Incorporating Elastomeric Sealing Elements—Part 1: Nominal Dimensions and Tolerances 2007. Available online: https://www.beuth.de/de/norm/iso-6194-1/103111254 (accessed on 24 June 2020).

42. Niemann, G.; Winter, H.; Höhn, B.-R.; Stahl, K. *Maschinenelemente 1*; Springer: Berlin/Heidelberg, Germany, 2019.

43. Matthias, A. Das Durchflussverhalten von Labyrinthdichtungen bei unterschiedlichen Betriebsbedingungen. Ph.D. Thesis, Technische Universität Wien, Vienna, Austria, 2007.

44. Neuberger, S.; Bock, E.; Haas, W.; Lang, K. Gas-lubricated mechanical face seals reduce CO_2 emissions. *Seal. Technol.* **2014**, *2014*, 8–12. [CrossRef]

45. Mileti, M.; Strobl, P.; Pflaum, H.; Stahl, K. Design of a Hyper-High-Speed Powertrain for EV to Achieve Maximum Ranges. In Proceedings of the CTI SYMPOSIUM, Berlin, Germany, 3–6 December 2018; pp. 265–273.

46. Continental Automotive. High Voltage Axle Drive for Hybrid and Electric Vehicles. Available online: https://www.continental-automotive.com/en-gl/Passenger-Cars/Technology-Trends/Electrification/High-Voltage-Tec hnologies (accessed on 24 June 2020).

47. Bosch Mobility Solutions. eAxle—The Modular eAxle Drive System—Compact, Cost Attractive and Efficient. Available online: https://www.bosch-mobility-solutions.com/en/products-and-services/passenger-cars-and -light-commercial-vehicles/powertrain-systems/electric-drive/eaxle/ (accessed on 24 June 2020).

48. *E-Axle: Nidec Has Developed This System Targeting EV and PHEV*; Nidec Corporation: Kyoto, Japan, 2018.

49. BorgWarner's iDM eAxle Transforming EV and Hybrid Propulsion; BorgWarner Knowledge Library. Available online: https://cdn.borgwarner.com/docs/default-source/default-document-library/borgwarner_i dm_en.pdf?sfvrsn=89a3853c_8 (accessed on 24 June 2020).

50. Binder, A. *Elektrische Maschinen und Antriebe. Grundlagen, Betriebsverhalten*; Springer: Berlin/Heidelberg, Germany, 2012.

51. Helmer, R. Kopplung Numerischer und Analytischer Verfahren zur Berechnung des Betriebsverhaltens von Synchronmaschinen. Ph.D. Thesis, Leibniz Universität Hannover, Hannover, Germany, 2011.

52. Höhn, B.-R.; Bansemir, G.; Dyla, A. Application of a Functional Product Model for Product Development (Example: Gear Units). In Proceedings of the SIMVEC—Numerical Analysis and Simulation in Vehicle Engineering, Baden, Germany, 26–27 November 2008.

53. Parlow, J.; Otto, M.; Stahl, K. Anwendungsflexible Getriebeauslegung nach Wunsch. In Proceedings of the 4. Kongress zu Einsatz und Validierung von Simulationsmethoden für die Antriebstechnik, Koblenz/Lahnstein, Germany, 17–18 September 2014.

54. Fromberger, M.; Otto, M.; Stahl, K. Programmanleitung FVA-Stirnradprogramm STPlus 10F. Frankfurt am Main, Germany, 11 February 2020.

55. ISO. *ISO 6336: Calculation of Load Capacity of Spur and Helical Gears, Parts 1–6*; ISO: Geneva, Switzerland, 2019.

56. Parlow, J. Entwicklung einer Methode zum Anforderungsgerechten Entwurf von Stirnradgetrieben. Ph.D. Thesis, Technische Universität München, Munich, Germany, 2016.

57. Bansemir, G. Konstruktionsleitsystem für den Durchgängig Rechnerbasierten Zahnradgetriebeentwurf. Ph.D. Thesis, Technische Universität München, Munich, Germany, 2012.

58. McCall, J. Genetic algorithms for modelling and optimization. *J. Comput. Appl. Math.* **2005**. [CrossRef]

59. Linke, H. *Stirnradverzahnung. Berechnung—Werkstoffe—Fertigung, 2., vollständig überarb. Aufl.*; Hanser: München, Germany, 2010.
60. Ramesh, P.; Lenin, N.C. High Power Density Electrical Machines for Electric Vehicles—Comprehensive Review Based on Material Technology. *IEEE Trans. Magn.* **2019**, *55*. [CrossRef]

 *vehicles*

Article

Benchmarking of Dedicated Hybrid Transmissions

Christian Sieg * and Ferit Küçükay

Institute of Automotive Engineering, Technische Universität Braunschweig, 38106 Brunswick, Germany;
f.kuecuekay@tu-braunschweig.de
* Correspondence: c.sieg@tu-braunschweig.de; Tel.: +49-531-391-2641

Received: 20 December 2019; Accepted: 10 February 2020; Published: 13 February 2020

Abstract: For many manufacturers, hybridization represents an attractive solution for reducing the energy consumption of their vehicles. However, electrification offers a wide range of possibilities for implementing powertrain concepts. The concepts can differ regarding their mechanical complexity and the required power of the electrical machines. In this article, drive concepts that differ in their functionality and drive train topology are compared. Based on requirements for the C, D, and E segment, the mechanical and electrical effort of the concepts is analyzed. The results show that the mechanical effort in the C segment can be reduced as long as the electrical effort is increased. In case of higher vehicle segments, the electrical effort can increase considerably, making concepts with increased mechanical complexity more suitable. The driving performance and efficiency in hybrid operation are evaluated via simulation. The results show that the difference of acceleration times in hybrid operation between a charged and discharged battery is lower for mechanically complex concepts. At the same time, they achieve lower CO_2 emissions. Therefore, these concepts represent a better compromise regarding performance and efficiency. Despite lower transmission efficiencies in hybrid operation, they achieve conversion qualities similar to simpler concepts and lower emissions with lower electrical effort.

Keywords: dedicated hybrid transmission; benchmarking; hybrid electric vehicle; efficiency; topology optimization; drive train optimization; powertrain concepts

---

1. Introduction

To reduce the energy consumption, electrification of the drive train represents a suitable solution. Since one or more electric machines (EM) can be integrated into the drive train at different positions, a large number of possible drive concepts for hybrid electric vehicles (HEV) can be realized. The possible drive trains differ in their characteristics and operating modes. On the one hand, powertrains of conventional vehicles can be electrified, resulting in comparatively complex mechanical concepts. On the other hand, it is possible to implement transmissions that are dedicated for use in HEV and have a much simpler design.

Due to this diversity, the question which kinds of drive train concepts represent suitable solutions in which vehicle segment needs to be answered. It is of interest to be able to make a statement whether the mechanical effort of a transmission can be reduced in low vehicle segments without having to accept a loss of driving performance or efficiency. At the same time, the question should be answered whether higher vehicle segments require an increase in the mechanical complexity of the transmission. The aim of this article is to answer these questions. A detailed analysis regarding the influence of multi-speed transmissions on various basic dedicated hybrid transmission (DHT) concepts is carried out by [1]. However, coaxial multi-mode DHT (MM-DHT) with several planetary gear sets (PGS), which enable parallel as well as power-split operating modes, are not included in the considerations.

For these reasons, this article compares drive train concepts with different mechanical and electrical complexity. Several add-on concepts as well as DHT are considered. Furthermore, similar concepts with different mechanical complexity are analyzed. Based on equal driving performance requirements for all concepts, it is shown which EM power is necessary to meet the requirements for plug-in hybrid electric vehicles (PHEV) in the C, D, and E segments. The results confirm the correlation between the concept's mechanical and electrical complexity identified by [1]. They show that a reduction of the mechanical complexity is only possible in connection with a simultaneous increase of the electrical power. Based on the results, it can be derived which concepts can be used in which segment. Furthermore, characteristic properties of the considered concepts are derived.

Within the scope of a concept comparison for vehicles in the C segment, the concepts are evaluated regarding their efficiency in hybrid operation and driving performance. Apart from these factors, especially costs for mechanical and electrical components as well as the necessary installation space have a major influence on the concept decision. However, the costs depend on the quantities of produced transmissions. In addition, they may be influenced by using existing components from either previous or other vehicle models so that the costs of a powertrain concept are manufacturer-specific. Furthermore, not every component is produced by vehicle manufacturers so that the transmission costs also depend on the suppliers. To assess the required installation space, a detailed design of the transmission is necessary. Therefore, the mechanical components such as gears, bearings, shafts, and the transmission housing need to be dimensioned regarding worst-case scenarios, which are unknown for new concepts in early development phases and can only be estimated. In addition, the available installation space depends on a specific vehicle model. In early concept phases, it is unknown in which segment or vehicle a powertrain concept will be used. Considering the legislation, it can be expected that the energy consumption of vehicles needs to be reduced even further in the future. Thus, it is a major influencing factor when comparing powertrain concepts. For these reasons, this paper focuses on efficiency and driving performance.

The results show that concepts with high mechanical effort offer a good compromise between driving performance and efficiency. The advantage of these concepts is that in hybrid operation there are minor differences in acceleration times when the battery can be discharged or needs to be charged. If the concepts are simpler, a higher share of the power of the internal combustion engine (ICE) must be transmitted via the EM, resulting in higher losses and thus lower driving performance.

2. Drive Train Concepts

In this article, different powertrain concepts for hybrid vehicles are compared. The topologies differ considerably regarding their structure, the number of EM, the power flows occurring in the drive train and the operating modes. The mechanical and electrical complexity of the concepts is another important distinguishing feature.

Hybrid powertrains can be divided into two main categories. In addition to so-called add-on hybrids based on conventional powertrains, this article examines DHT especially designed for use in hybrid vehicles. Furthermore, hybrid drive trains can be divided into parallel hybrids, series-parallel hybrids, power-split hybrids, and multi-mode DHT (MM-DHT).

2.1. Parallel Hybrid Concepts

Due to their high market share, two parallel hybrids are analyzed in this article. They are based on a conventional drive train and differ with respect to the positioning and the number of EM. In addition to the P2 topology, explained in Section 2.1.1, a P1P4 topology, shown in Section 2.1.2, with two EM is investigated. In both concepts, an 8-speed transmission with a wet dual-clutch is used.

2.1.1. P2 HEV

The drive train topology of a P2 hybrid is shown in Figure 1. In this configuration, an EM is positioned between the ICE and the transmission. Furthermore, there is a separating clutch between EM and ICE, so that the ICE can be disconnected from the drive train. Thus, the vehicle can be driven either electrically without ICE drag losses or by the combustion engine solely or in hybrid mode.

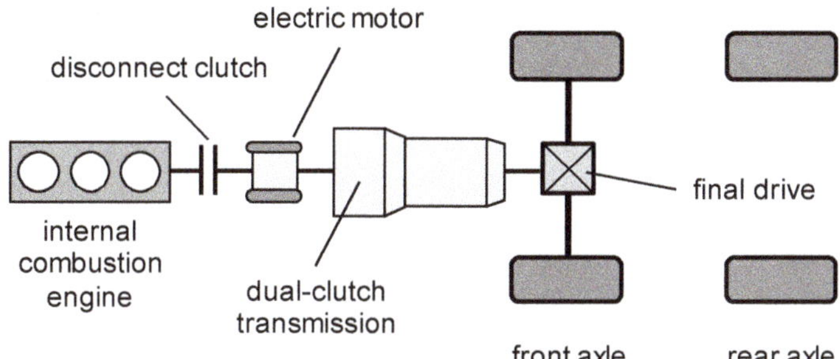

Figure 1. P2 drive train topology.

The transmission of this add-on topology is a wet 8-speed dual-clutch transmission (DCT). Consequently, it offers eight hybrid and eight electric driving modes.

Since P2 HEV are based on conventional drive trains, they are characterized by a high number of mechanical components compared to other concepts considered in this paper. To estimate and compare the mechanical effort of different concepts, the number of relevant mechanical components can be totaled. In the P2 HEV from Figure 1, the relevant components include the double clutch at the transmission input, the separating clutch between ICE and EM, the final drive (FD) at the front axle and the eight gears. This results in a total mechanical effort of 11.

2.1.2. P1P4 HEV

Apart from P2 HEV, vehicles with an electrified rear axle, so-called P4 hybrids, show a high market share. Thus, a parallel hybrid with an electrified rear axle is investigated in this paper, see Figure 2. To enable battery charge at standstill and to start the ICE quickly and comfortably while driving, the drive train additionally comprises an electric motor between ICE and transmission. Since there is no disconnect clutch, electric driving is only possible by using the rear axle.

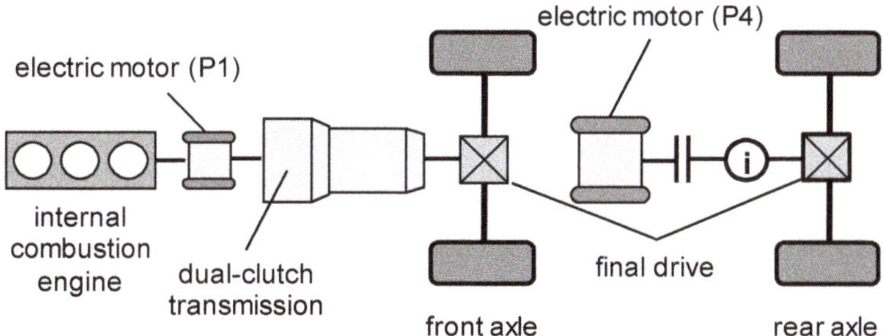

Figure 2. P1P4 drive train topology.

Because of the positioning of the electric motor on the rear axle, the P1P4 topology offers temporary all-wheel drive. To avoid over-speeding of the rear electric motor, a disconnect device is installed between the EM and the reduction gear on the rear axle. The concept comprises the same 8-speed DCT as the P2 HEV concept. It offers an electric driving mode and 16 hybrid modes since the vehicle can be driven in eight gears with or without the rear axle.

Similar to the P2 hybrid, the P1P4 hybrid has a high mechanical complexity. Both the eight gears of the transmission and the FD on the front axle are identical to the P2 concept. Differences result from the fact that there is no separating clutch between the ICE and the P1 EM and that the EM is connected to a second FD at the rear axle via a reduction gear with a disconnect device. This results in a higher mechanical effort of 13 compared to the P2 HEV.

2.2. Series-Parallel DHT

Another topology suitable for HEV is the powertrain shown in Figure 3. On the one hand, it enables a serial operating mode in which there is no mechanical connection between the ICE and the wheel, and the EM can transmit the power electrically. On the other hand, the ICE and the wheel can be connected by closing a separating clutch so that the ICE drives the vehicle at higher wheel powers. When the clutch is open, an electrical mode is possible. The shift matrix of the concept is shown in Table 1.

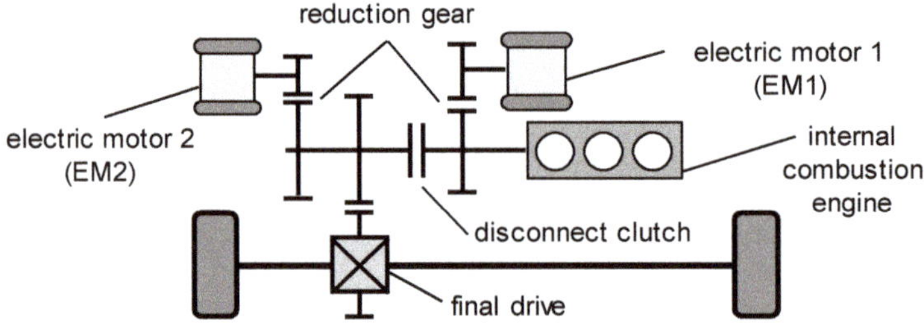

Figure 3. Drive train topology of the series-parallel DHT.

Table 1. Shift matrix of series-parallel DHT.

Operating Mode	C
Electric	
Parallel	•
Serial hybrid	

The series-parallel concept differs greatly from the other concepts in this article in terms of its design, operating modes, and mechanical complexity.

The mechanically relevant components include the two reduction gears, the separating clutch and the FD, so that the mechanical effort of 4 is lower compared with most other concepts.

2.3. Power-Split DHT Concepts

In addition to add-on hybrids, the Toyota Prius' power-split hybrid drive system has established itself on the market. In addition to a variant known from the Prius, a drive system based on the same concept and supplemented by a 4-speed transmission is considered. This allows determination of the influence of increased mechanical complexity on Power-Split-DHT (PS-DHT).

2.3.1. Power-Split DHT

An important DHT concept with a high market share is the power-split concept introduced by Toyota in 1997. In this paper, the transmission structure of the 4th generation of Toyota's power-split concept, introduced in [2], is investigated. The drive train structure is shown in Figure 4.

The most important component of the transmission is a PGS. The ICE is connected to the planet carrier C. One EM is connected to the sun gear S and one to the ring gear R. The EM connected to the ring gear is connected to the driven axle via a reduction gear. Furthermore, a one-way clutch (OWC) is connected to the PGS so that the vehicle can be driven by both EM in electric driving. Without the OWC, EM1 would need to be controlled so that the planet carrier would be stationary to turn off the ICE.

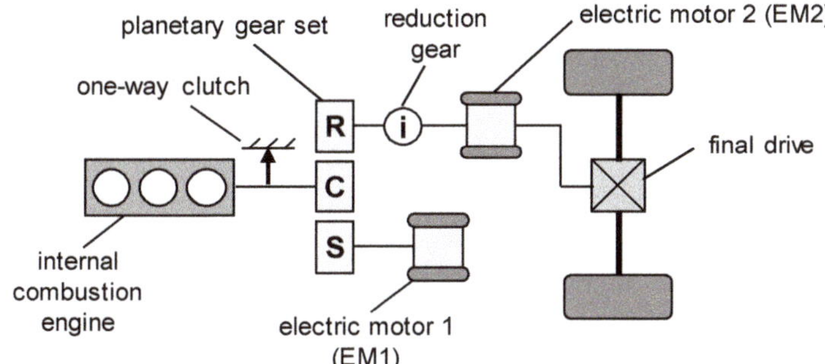

Figure 4. PS-DHT drive train topology with planetary gear set—R: ring gear, C: planet carrier, S: sun gear.

Apart from electric driving, the PS-DHT offers a hybrid mode with a continuously variable gear ratio. This can be realized by controlling the speed of EM1 so that the mode can also be called electronic continuously variable transmission (eCVT) mode. In this operating mode, the speed and torque of the ICE can be chosen freely.

The concept is comparatively simple regarding the mechanical effort. In addition to the planetary gear set, the relevant components include the OWC, the reduction gear for EM2 and the FD. This results in a mechanical effort of 4.

2.3.2. Power-Split DHT with 4-Speed Transmission

The PS-DHT structure can be combined with a 4-speed transmission. A similar concept is introduced in [3,4]. Figure 5 shows a modified structure with an OWC between ICE and the planet carrier so that both EM can operate during electric driving. The 4-speed transmission is positioned between EM2 and the FD so that the speed of EM2 and the ring gear can be changed. Thus, it influences the power flow between the energy converters but does not add an output-split or compound-split operating mode. Instead, four input-split eCVT modes and four electric driving modes can be selected.

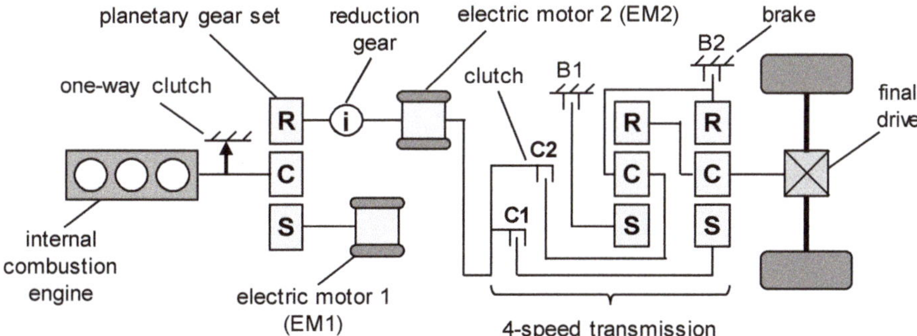

Figure 5. Drive train topology of PS-DHT enhanced by 4-speed transmission—R: ring gear, C: planet carrier, S: sun gear.

The 4-speed transmission consists of two PGS and four shifting elements. By controlling two clutches C1 and C2 and two brakes B1 and B2, four gear ratios can be selected. The shift matrix is shown in Table 2.

Table 2. Shift matrix of PS-DHT with 4-speed transmission.

Operating Mode	OWC	B1	B2	C1	C2
eCVT 1st			•	•	
eCVT 2nd		•		•	
eCVT 3rd				•	•
eCVT 4th		•			•
electric 1st	•		•	•	
electric 2nd	•	•		•	
electric 3rd	•			•	•
electric 4th	•	•			•

Compared to the PS-DHT without the 4-speed transmission, the powertrain is more complex. The increased mechanical effort directly results from the 4-speed transmission. For the quantification of the mechanical effort, three PGS, 5 shifting elements and the FD must be taken into account. Thus, the mechanical complexity can be estimated with 9.

2.4. Multi-Mode DHT Concepts

MM-DHT represent another category of hybrid transmission. They differ from PS-DHT since they do not only have an input-split eCVT mode in addition to electrical operation, but also offer additional functions. Depending on the structure of the concepts, several parallel or additional power-split modes, such as compound-split modes, can be implemented. This paper analyzes two concepts that are used in production vehicles. These include the concept from the second-generation Chevrolet Volt, introduced in [5] and the concept from the Cadillac CT6 PHEV, introduced in [6]. The influence of increased mechanical complexity compared to the basic concept can also be determined for the MM-DHT.

2.4.1. Multi-Mode-DHT with two PGS

Figure 6 shows the structure of the MM-DHT with two PGS and two active shifting elements. The concept shown, corresponds to the drive system of the Chevrolet Volt 2, see [5]. The ICE is connected to the ring gear of the first PGS and can drive the vehicle. The sun gears of the PGS are each connected to EM, while the planet carriers form a common output. The sun gear of the first PGS can be connected to the second one via a clutch C. In addition, the ring gear of the second PGS can be stationary, when brake B is actuated. With the PGS and three shifting elements, the concept enables two electric, two power-split, and one parallel operating mode. The corresponding shift matrix is shown in Table 3.

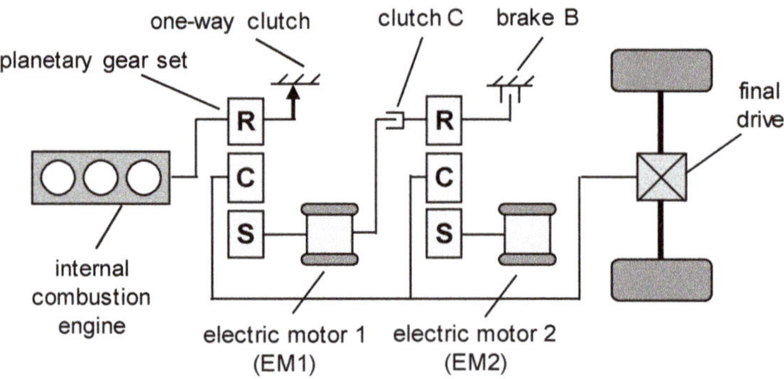

Figure 6. Drive train topology of MM-DHT with two planetary gear sets—R: ring gear, C: planet carrier, S: sun gear.

Table 3. Shift matrix of MM-DHT with two PGS.

Operating Mode	OWC	B	C
eCVT 1st		•	
parallel mode		•	•
eCVT 2nd			•
electric 1st	•	•	
electric 2nd		•	

In mode eCVT 1st there is an input-split power flow while in mode eCVT 2nd there is a compound-split power flow. The shift matrix in Table 3 shows that in mode electric 1st both EM can drive the vehicle, while in mode electric 2nd only EM2 can operate.

With the two PGS, three shifting elements and the axle drive, the mechanical effort for the concept can be quantified to 6. This results in a complexity between the PS-DHT and the PS-DHT with 4-speed transmission.

2.4.2. Multi-Mode-DHT with three PGS

An extension of the MM-DHT with two PGS is the MM-DHT with three PGS, see Figure 7. Up to the third PGS, the design is identical. The drive system is derived from the Cadillac CT6 PHEV, see [6]. The planet carrier of the first two PGS no longer forms the transmission output but is connected to the sun gear of the third PGS. The EM2 is connected to the ring gear of the third PGS via a clutch C2. The ring gear R is stationary when brake B2 is actuated. The output is the planet carrier of the third PGS.

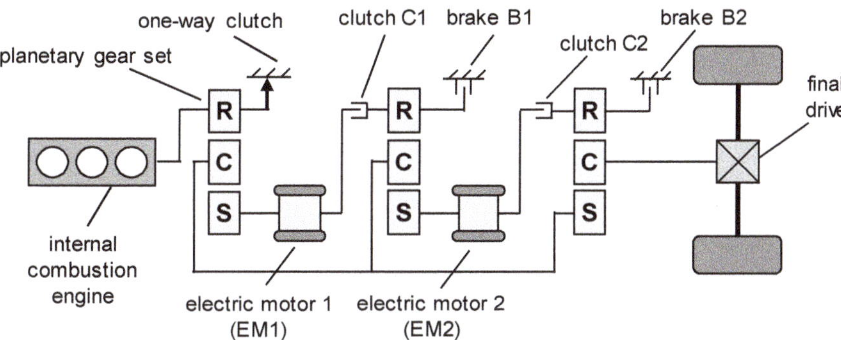

Figure 7. Drive train topology of MM-DHT with three planetary gear sets—R: ring gear, C: planet carrier, S: sun gear.

With the three PGS and five shifting elements, 11 operating modes can be implemented. These include four electrical operating modes in which either one or both EM can operate. In addition, four eCVT modes are available in which either an input-split or compound-split power transmission occurs. In addition, three parallel modes with a constant gear ratio between ICE and wheel can be selected. The shift matrix of the concept is shown in Table 4.

Table 4. Shift matrix of MM-DHT with three PGS.

Operating Mode	OWC	B1	B2	C1	C2
electric 1st	•	•	•		
electric 2nd	•		•	•	
electric 3rd	•			•	•
electric 4th	•	•			•
eCVT 1st		•	•		
parallel 1st		•	•	•	
eCVT 2nd			•	•	
parallel 2nd			•	•	•
eCVT 3rd				•	•
parallel 3rd		•		•	•
eCVT 4th		•			•

Compared to the MM-DHT with two PGS, the MM-DHT with three PGS is more complex. An additional PGS and two additional shifting elements increase the mechanical complexity from 6 to 9.

3. Vehicle Parameters

One aim of this paper is to show that there is a connection between the total electrical power required to meet the driving performance requirements and the mechanical effort of a concept. For this purpose, the three high-volume vehicle segments C, D, and E are considered. For all concepts, equal driving performance requirements are defined within a segment. Table 5 shows the vehicle parameters of the three segments.

Table 5. Vehicle parameters for C, D, and E segment vehicle.

Parameter	Unit	C Segment	D Segment	E Segment
mass	kg	1600	1700	1900
max. additional load	kg	500	550	650
frontal area	m^2	2.2	2.2	2.35
drag coefficient	-	0.27	0.26	0.25
rolling resistance coefficient	-		8×10^{-3}	
residual braking force	N	40	40	45
wheelbase	m	2.65	2.8	2.95
weight distribution front to rear, empty vehicle	%	60:40	60:40	50:50
weight distribution front to rear, loaded vehicle	%	50:50	50:50	45:55
center of gravity height	m		0.5	
wheel radius	m	0.31	0.32	0.32
max. power of ICE	kW	100	150	225
driven axle	-	front	front	rear
battery capacity	kWh	10	10	10

This paper examines concepts suitable for use in PHEV. For this reason, typical vehicle parameters are defined. The vehicle mass of the C segment is assumed 1600 kg. It increases to 1700 kg for the D segment and 1900 kg for the E segment. The maximum payload in the C segment is 500 kg, 550 kg in the D segment and 650 kg in the E segment. The wheelbase increases from 2.65 m for the C segment over 2.8 m in the D segment to 2.95 m in the E segment. Another important parameter is the center of gravity of the vehicles. In the unloaded case, the vehicles in the C and D segments are front-loaded with a weight distribution of 60:40, while the vehicle in the E segment has a balanced weight distribution. If the vehicles are loaded by their maximum payload, the center of gravity of the vehicle changes. In case of the C and D segments, a balanced weight distribution results. The vehicle of the E segment has a weight distribution of 45:55 when fully loaded.

Another important difference between the segments is the maximum power of the ICE. For the C segment vehicle, a naturally aspirated engine with a maximum output of 100 kW is considered. The D segment is based on a supercharged engine with 150 kW, while the E segment is based on a supercharged engine with a maximum power of 225 kW. For all concepts within a segment, the equal combustion engine is used to determine the effects of a certain drive train structure on important properties such as performance and efficiency. It should be noted that it is neglected that the ICE can be optimized for some concepts.

As the reference vehicle is a PHEV and there are no requirements for the electrical range, the battery capacity is 10 kWh.

The driven axle also influences the driving performance requirements. In the C and D segments, the vehicles are front-wheel-drive (FWD) and in the E segment rear-wheel-drive (RWD).

The residual brake force describes an additional driving resistance resulting from residual braking torques of the wheel brake and wheel bearing friction.

4. Driving Performance Requirements

In hybrid drive trains, multiple energy storages and energy converters are present. Depending on their interaction, the power available on the wheel may vary. Therefore, several drive train states should be taken into account when defining driving performance requirements. According to [1], there can be several ways in which the energy can be provided by either the battery or the ICE in case of HEV. In this article, three corresponding states are taken into account when defining driving performance requirements. In hybrid mode, the battery can either be charged by the ICE or provide additional power to drive the vehicle. In addition to hybrid operation, it is also possible to define driving performance requirements for electrical operation. The power available at the wheel varies as a result of the power provided by the energy converters. This is shown in Figure 8 for different drive states.

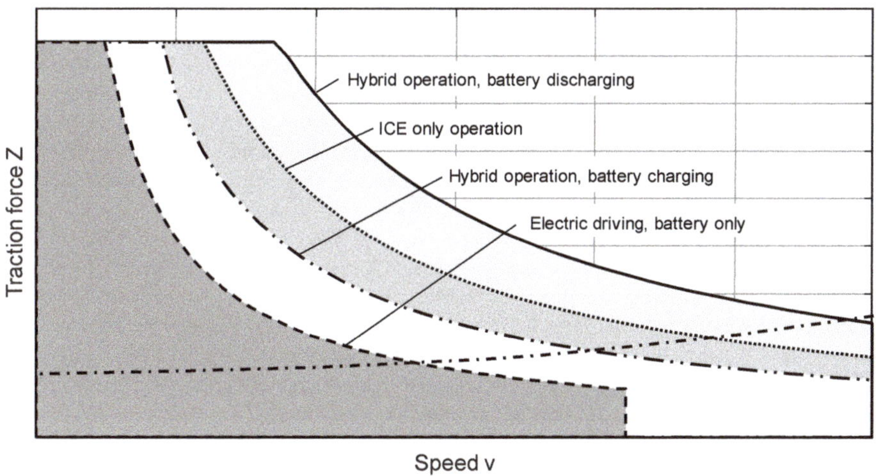

Figure 8. Traction force at wheel level for different drive train states.

The number of defined driving performance requirements should be as low as possible to be able to consider many possible variants within a parameter variation. At the same time, it needs to be ensured that a sufficient number of driving situations, i.e., tractive forces at a certain speed, are taken into account. Therefore, there is a conflict between the number of requirements and the required computation time. In this paper, driving performance requirements for different drive states are defined at a small number of vehicle speeds, so that there are several requirements per drive state in order to keep the computing time as short as possible. The requirements are distributed over the entire speed range of the vehicle to ensure that the resulting delivery maps contain realistic driving situations. It should be noted that the specified driving performance requirements represent minimum requirements, which all variants of each concept must meet. The requirements within one segment are the same for all concepts to enable a fair comparison. In addition, the designed concepts may exceed the minimum requirements. Table 6 summarizes the requirements.

Table 6. Drive train state specific driving performance requirements for C, D, and E segment vehicle.

Drive Train State	Parameter	Unit	C Segment	D Segment	E Segment
Hybrid operation, battery discharging	traction force limit loaded vehicle $v \leq 10$ km/h	-	yes	yes	Yes
	power at ICE level due to boosting at $v = 80$ km/h	kW	100 + 40	150 + 50	225 + 60
	max. speed	km/h	200	225	250
Hybrid operation, battery charging	traction force limit loaded vehicle $v \leq 10$ km/h	-	yes	yes	yes
	power at ICE level due to charging at $v = 80$ km/h	kW	100–10	150–10	225–20
	max. speed	km/h	200	225	250
Electric driving	traction force limit loaded vehicle $v \leq 10$ km/h	-	yes	yes	yes
	$t_{60\text{-}100}/a(v = 80$ km/h)	s	6	5	4
	max. speed	km/h	135	145	160

It should be noted that the requirements in hybrid operation for charging or discharging the battery refer to the transmission input. The requirement at wheel level is calculated for a powertrain efficiency of 90%.

In hybrid operation, when the battery is discharged, all variants, irrespective of the segment, must reach the traction force limit of the fully loaded vehicle in a speed range up to at least 10 km/h to ensure sufficient gradeability. In all DHT concepts considered, one of the EM is responsible for providing sufficient torque on wheel level until the traction force limit is reached. Since EM can deliver their maximum torque over a wider speed range, there are delivery maps in which the traction force limit is reached at speeds higher than 10 km/h. It is, therefore, sufficient to require the concepts to reach the traction force limit at least in a speed range between 0 km/h and 10 km/h.

For the medium speed range at 80 km/h, it is required that the variants must achieve a certain traction force. To be able to quantify this traction force, ICE operation without discharging the battery is the starting point. For hybrid operation with battery discharge, a boost power is defined that must be applied in addition to the maximum ICE power. For vehicles in the C segment, 140 kW of power must be available at the transmission input. This power results from a nominal ICE power of 100 kW in combination with a boost power of 40 kW. Taking into account a drive train efficiency of 90%, the necessary power or traction force can thus be calculated at wheel level. For the other segments, there is a differentiation due to a higher ICE maximum power as well as a higher boost power.

To define a driving performance requirement in the higher speed range and to limit the demand map of the vehicles, the variants in the C segment are required to achieve a maximum speed of 200 km/h in hybrid operation when the battery is discharged. In the D and E segment, the maximum speed is 225 km/h and 250 km/h, respectively.

In addition to discharging the battery, hybrid operation while charging the battery is a relevant drive train state since less power is available on the wheel. The approach to determine the corresponding tractive force is identical to hybrid operation while discharging the battery. The only difference is that no additional power is provided, but is reserved for charging the battery. In the C and D segment, a reserve of 10 kW is defined, while in the E segment the maximum ICE power is reduced by 20 kW.

This does not mean that the battery is charged with the specified power, but takes into account the fact that some concepts in eCVT mode or serial mode require the battery to be charged in a certain speed range, resulting in increased losses. In the driving performance calculation of the simulation model, see Section 5, no charging power of the battery is specified. The model only ensures a negative battery power. In hybrid operation while charging the battery, the requirements are identical to the drive train state in which the battery is discharged.

It is characteristic for PHEV that they can achieve higher distances in electrical operation due to their battery capacity. Compared to mild or full hybrids, the driving performance requirements for electrical operation are therefore becoming more important. For the vehicles of all segments, it is specified that they must reach the traction force limit when the vehicle is fully loaded. The only exception is the P1P4 topology with an electric all-wheel drive. In electric operation, this vehicle would have RWD and would therefore offer a considerably higher traction potential than vehicles with FWD in the C and D segment. This would lead to significantly higher tractive forces required for electrical operation and would contradict a fair comparison of concepts. Therefore, the variants of the P1P4 topology are required to reach the traction force limit of a FWD vehicle in electrical operation.

To define a driving performance requirement in the medium speed range, an acceleration time from 60 km/h to 100 km/h is used as a distinguishing feature. It is assumed that below 60 km/h there are minor differences regarding the acceleration time between different vehicle segments. Therefore, the acceleration time between 60 km/h and 100 km/h can be used to determine an average longitudinal acceleration at 80 km/h. This acceleration requirement directly corresponds to a traction force requirement. It should be noted that the acceleration time only servers as an orientation for deriving an average acceleration or traction force and does not represent a requirement to be achieved.

For the variants in the C segment, the acceleration time is 6 s, in the D segment it declines to 5 s and in the E segment to 4 s.

In electrical operation, the requirement for the maximum vehicle speed in the C segment is based on the speed profile of the Worldwide harmonized Light Duty Test Cycle (WLTC). To ensure that the cycle can be fully completed in electrical operation, a maximum speed of 135 km/h must be reached. So that vehicles in higher segments differ from those of the C segment, the maximum speed in the D segment is 145 km/h and 160 km/h in the E segment.

5. Simulation Model

A modular simulation model developed by [1] makes a significant contribution to the results in this paper. The structure and the essential functionality are explained in this section. Explanations that are more detailed can be found in [1].

An essential feature of the simulation model is that the calculation of operating points and the efficiency of drive concepts is coupled with a driving performance calculation. Based on the possible driving performance, i.e., the delivery maps of a drive train concept, an efficiency calculation can be carried out in any speed profile. Therefore, the model can be used on the one hand for estimating driving performance and thus for determining appropriate drive train variants. On the other hand, it enables the simulation of almost any drive topology in cycle or customer operation.

A schematic structure of the model is shown in Figure 9. Based on driving resistances due to the vehicle parameters, operating points of the drive train components and energy converters are calculated backwards through the drive train. Based on efficiency maps of the energy converters, the vehicle's energy consumption can be determined.

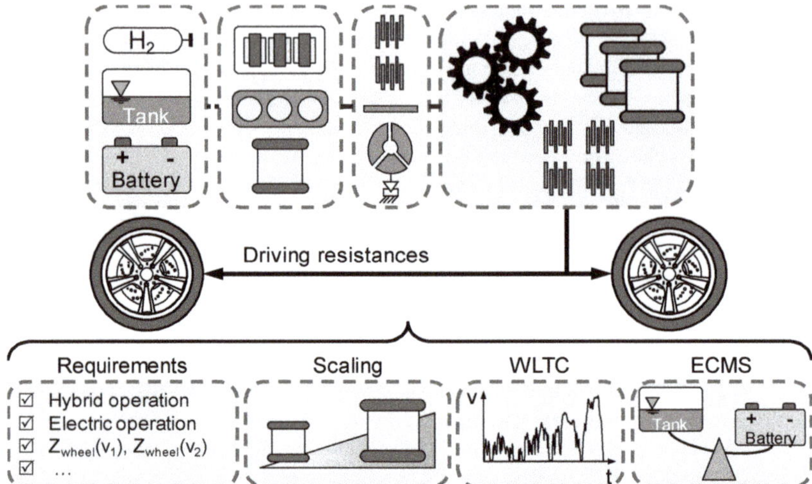

Figure 9. Schematic structure of the modular simulation model developed and inspired by [1].

A central component of the model is a modular transmission calculation, which calculates the associated speeds and torques within the transmission based on all possible operating points of the energy converters. The calculation has a modular structure so that a wide variety of transmission concepts can be calculated within a short time with little parameterization effort. In addition to the calculation of conventional transmissions, it is possible to simulate concepts in which several EM are integrated into the transmission. This feature is particularly necessary for the simulation of complex DHT with several EM.

To calculate the transmission losses, two approaches can be selected. In addition to a simplified approach, which describes the efficiency depending on the output power of the transmission, mode-specific efficiency maps can be implemented. In this paper, however, the simplified approach is used to reduce the calculation time.

Two primary energy converters can be positioned outside the transmission. They can be connected to the transmission module via a launch element. The primary energy converters include ICE, fuel cells and EM.

In addition to the modules for calculating driving performance and efficiency, driving performance requirements can be defined. With the help of the performance calculation, it is then possible to determine which variants of a drive train concept can meet the requirements. Within the scope of parameter variations, fractional or full factorial scaling plans can be considered.

To select the most efficient operating points, two operation strategy approaches are implemented. In addition to a globally optimal control strategy, presented in [7,8], there is a locally optimal operating strategy based on the equivalent consumption minimization strategy (ECMS) [9,10]. This local optimal approach is used in the following considerations.

6. Dimensioning of Electric Motors

The simulation tool described in Section 5 is used to determine powertrain designs that meet the requirements described in Section 4. For each concept considered in this paper, a range of design parameters is defined. These include upper limits for the maximum power of the EM. Furthermore, intervals are defined in which ratios of PGS, the FD or the reduction gears are varied.

The driving performance calculation allows determination of which variants meet the requirements. The variants whose total EM power is minimal are of particular importance. Designs with a significantly higher total power are of less importance as they are oversized regarding the requirements. Based on the driving performance requirements and the drive train concepts, the above-mentioned relationship between mechanical and electrical effort can be confirmed. Furthermore, the investigations allow drawing of conclusions about the suitability of concepts in different vehicle segments.

Figure 10 shows the minimum required power of the EM which is necessary to meet the requirements. Furthermore, the mechanical effort quantified by a figure, see Section 2, is compared.

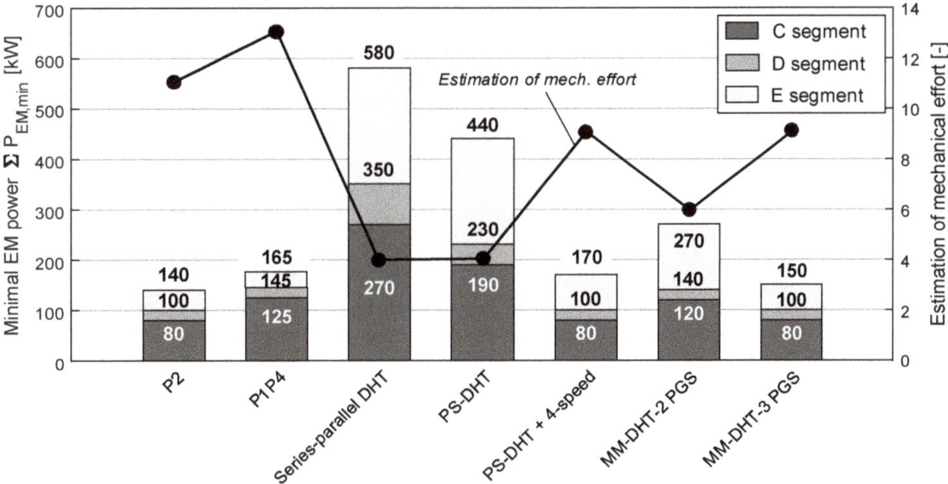

Figure 10. Concept-specific minimal required EM power to meet the driving performance requirements in the C, D, and E segment in comparison with the estimated mechanical effort of the concepts.

6.1. Analysis of Total Necessary EM Power for C Segment Requirements

In the C segment, the minimum required EM power for the P2 HEV is 80 kW. Due to the eight gears of the DCT, the necessary EM power does not result from the requirements in hybrid operation, but is due to the requirement in electric operation at 80 km/h. Reaching the traction force limit at full load can also be ruled out as a cause since the first gear already enables a high gear ratio. A power of 80 kW is therefore not required to reach the traction force limit.

The total necessary electrical power of the P1P4 HEV increases to 125 kW. This consists of 15 kW for the P1-EM and 110 kW for the P4-EM. It should be noted that a power of 15 kW has been defined for the P1-EM for all variants of the P1P4 and in all segments. Compared to the P2 HEV, the higher power of the EM results from the fact that there is only a constant gear ratio between EM and wheel and that the EM can only be separated from the drive train when its speed limit is reached. To reach the maximum speed in electrical operation, the reduction gear ratio for the EM must not exceed a certain value. The P4-EM must therefore provide a sufficiently high torque to allow the vehicle to reach the traction force limit when launching with full load.

The total necessary electrical power in case of the series-parallel DHT increases considerably compared to the add-on concepts. In total, 270 kW of electrical power must be installed in the drive train to meet the requirements in the C segment. This high power demand results from the requirements in hybrid operation when charging the battery. In this case, the concept offers two operating modes. In parallel mode, the ICE can only provide the required wheel power in a certain speed range since there is a constant gear ratio between ICE and wheel. In case the ICE cannot provide enough power to meet the requirement, the serial mode needs to be selected. In this mode, the ICE provides power to drive the vehicle and EM1, see Figure 10, needs to have a similarly high power to be able to transmit the power electrically so that EM2 can drive the vehicle. EM2 therefore at least needs to have a power which is similar to the power of EM1.

To reach the traction force limit, EM2 must provide a high torque during charging of the battery in hybrid operation and during electric driving. There is a similar correlation as with the P4-EM, so that a high torque requirement results in a correspondingly high power for EM2.

The PS-DHT requires a lower total EM power than the series-parallel concept. In the C segment, a total power of 190 kW is required. In this case, EM1 needs to have a power of 60 kW and EM2 is required to provide 130 kW. The significantly higher power of EM2 is needed since the traction force limit must be reached in electric and hybrid operation. Compared to the PS-DHT concept from [4], the minimum required power of EM1 is higher, since high requirements have been defined for hybrid operation with simultaneous charging of the battery.

If a 4-speed transmission is added to the PS-DHT, the minimum power required is reduced to a total amount of 80 kW. This results in an electrical effort comparable to the P2 HEV. The 4-speed transmission, see Figure 5, allows adjustment of the operating point of the EM2 and the speed of the ring gear. This reduces the torque required for EM2 to reach the traction force limit. Furthermore, the speed of the ring gear influences the power flow in power-split mode, so that the power of EM1 can also be reduced.

The required power of the EM of the MM-DHT with two PGS increases to 120 kW. Compared to the PS-DHT, the power is 70 kW lower. Due to a higher number of possible operating modes, including two power-split eCVT modes and a parallel mode, more degrees of freedom are available in hybrid operation when the battery is discharged.

The MM-DHT with three PGS requires as much electrical power as the P2 HEV and the PS-DHT with 4-speed transmission. By increasing the number of available operating modes, including four eCVT modes and three parallel modes, it is possible to install a lower total electrical power than in the similar concept with two PGS.

It can be concluded that with the EM maps used and the requirements applied, the minimum electrical power required in the C segment is 80 kW.

6.2. Analysis of Total Necessary EM Power for D and E Segment Requirements

Due to the higher requirements in the D segment, the total electrical power required for all concepts increases. However, a comparison between the concepts shows that the increase differs among the concepts. For example, the P2 HEV's and the P1P4 HEV's output increases by 20 kW. It should be noted that the increase for the P1P4 HEV only affects the P4-EM, as it was specified that the P1-EM should have an output of 15 kW in all segments.

A higher increase results in case of the series-parallel DHT. For this concept, the electrical effort increases by 80 kW to 350 kW. This corresponds to an increase of approx. 29.6%. On the one hand, the increase in power can be explained by the higher traction force limit due to the vehicle parameters. On the other hand, the increased electrical power requirement results from hybrid operation when the battery is charged. In this case, only the ICE is available as a power source. The power provided by the ICE is received by EM1, see Figure 3. As a result of this dependency, the power of EM1 is directly dependent on the power of the ICE.

The power of EM1 is also dependent on the power of the ICE in case of the PS-DHT. When charging the battery with low power, however, a portion of the power provided by the ICE can be transferred mechanically to the wheel. Therefore, the EM1 of the PS-DHT does not have to receive the maximum power of the ICE. As a result, the total electrical power required for the PS-DHT increases by approx. 21% compared to the C segment. The total required power of the EM is 230 kW.

The 4-speed transmission allows adjustment of the operating point of EM2 and the speed of the ring gear. This reduces the total electrical power required compared to the PS-DHT without a multi-speed transmission. Therefore, in hybrid operation with discharged battery less power must be transmitted via the EM in the case of split power mode, which directly reduces the power requirement of the EM. As a result of this, an electrical effort of 100 kW is required in the D segment.

A total electrical power demand of 140 kW results for the MM-DHT with two PGS in the D segment. Compared to the C segment the increase of approx. 16.7% is lower. In comparison to the PS-DHT, less electrical power is required because two eCVT modes and one parallel mode are available.

If the MM-DHT with two PGS is supplemented by an additional PGS, 100 kW of electrical power are required in the D segment. As in the C segment, this concept has a similar electrical effort as the P2 HEV and the PS-DHT with 4-speed transmission.

In the E segment, the total electrical power required for all concepts increases as a result of increased requirements. The P2 HEV, PS-DHT with 4-speed transmission, and MM-DHT with three PGS continue to have the lowest electrical effort. However, there are slight differences between these concepts in the E segment. The total necessary EM power for the series-parallel DHT increases to 580 kW. The relative increase is even higher for the PS-DHT, but with 440 kW less electrical power is required in total. Compared to the PS-DHT, the required EM power of the MM-DHT with two PGS in the E segment increases to a comparatively lower value of 170 kW.

6.3. Relationship between Mechanical and Electrical Effort

In addition to the total electrical power required, Figure 10 shows an estimate of the mechanical effort for the considered concepts. If this is compared with the required electrical power, a correlation between electrical and mechanical effort is shown.

The results show that the concepts with high mechanical effort require a comparatively low electrical effort in all vehicle segments. The P2 hybrid has a mechanical effort of 11 and is the concept with the second highest mechanical complexity according to the estimation used in this paper. In all segments, the P2 HEV requires the lowest total electrical power.

Due to the additionally driven rear axle, the mechanical effort of the P1P4 HEV rises to 13. Compared to the P2 HEV, a higher overall EM power is required in all segments. Therefore, no linear relationship between the mechanical effort and the electric effort can be shown. However, a clear trend can be identified based on the results. For future considerations, however, a different parameter should be used to estimate the mechanical effort to be able to describe a clearer relationship.

The PS-DHT with 4-speed transmission and the MM-DHT with three PGS also are concepts with a high mechanical effort. Each of these concepts is a mechanically more complex variant of a basic concept. In the C and D segment, they have a similarly low necessary EM power as the P2 HEV, but with lower mechanical effort. Although there are differences in the E segment, the overall EM power is significantly lower compared to the other concepts with lower mechanical complexity.

If the mechanical effort of the concepts is significantly reduced, the required power of the EM increases significantly. Both the series-parallel DHT and the PS-DHT have a comparatively simple design with a mechanical effort of 4. At the same time, these concepts require the highest EM power compared to the other concepts. The difference is particularly high in the E segment.

6.4. Application of Powertrain Concepts in Different Vehicle Segments

Based on the results in Figure 10, it can be concluded which concepts should be used in which vehicle segment. For this purpose, the ratio of electrical and mechanical effort can provide a helpful orientation. It should be noted that the assessment of the suitability of drive train concepts in different segments depends on the underlying requirements to a considerable extent. For example, a reduction of the requirements in hybrid operation while charging the battery would lead to significantly reduced EM power. Especially concepts with low mechanical effort would benefit from this. Nevertheless, it can be expected that mechanically complex concepts would still have the lowest power demand for electrical power.

However, based on the driving performance requirements in this paper, the use of the concepts for segments C, D, and E can be assessed.

The E segment shows considerable differences between the total electrical power required by the concepts. The largest difference with 440 kW is between the P2 HEV and the series-parallel DHT. The difference between P2 HEV and PS-DHT is 300 kW. In comparison to the MM-DHT with two PGS, the difference decreases to 130 kW. This leads to the conclusion that the series-parallel DHT as well as the PS-DHT in the E segment do not represent the most favorable solutions. For the MM-DHT with two PGS it is shown that the electrical effort can be reduced considerably compared to concepts with higher complexity. Thus, the required power output decreases by 120 kW in case of the MM-DHT with three PGS and by 100 kW in case of the PS-DHT with 4-speed transmission if the mechanical effort is increased by 3. Therefore, it can be concluded that the MM-DHT with two PGS should not be used in the E segment either. Instead, concepts with a mechanical complexity of 9 or higher represent promising solutions.

In the D segment, the series-parallel DHT as well as the PS-DHT require significantly lower EM power compared to the E segment. However, regarding a good compromise between electrical and mechanical effort, other concepts lead to a better solution. Compared to the PS-DHT, the electrical effort of the MM-DHT with two PGS is reduced by almost 100 kW with a moderate increase of the mechanical complexity. This leads to the conclusion that the MM-DHT with two PGS is a better concept than the PS-DHT and the series-parallel DHT in the D segment. Concepts that are mechanically more complex have a lower electrical power requirement. To be able to assess the applicability with sufficient quality in these cases, further criteria such as costs or required installation space need to be taken into consideration.

Compared to the D segment, the difference between the series-parallel DHT and the PS-DHT is similarly large in the C segment. The difference in the D segment between the series-parallel DHT compared to the P2 hybrid is about 338%, while in the C segment it is 350%. For PS-DHT, there are relative differences of 230% in the D segment and about 238% in the C segment. Although the difference between the concepts is similar, the absolute required power decreases, so that the series-parallel DHT should not be excluded from application in the C segment. However, it becomes apparent that the PS-DHT as well as the MM-DHT with two PGS represent a better compromise between mechanical and electrical effort. For the mechanically more complex concepts, less clear statements can be made, since although a low total electrical power is required, criteria such as cost, scalability, or modularity play an important role and are not considered in this paper.

7. Benchmark Analysis

The approach described in Section 6 allows identification of the minimal required EM power which needs to be installed in the considered concepts to meet the performance requirements. Furthermore, the simulation model provides requirement-compliant transmission ratios, which are varied within a parameter variation. The gear ratios include the FD ratio for the P2 and P1P4 hybrids. In the case of the series-parallel DHT, the gear ratios of the reduction gear for EM2 and the FD are varied. In case of the PS-DHT and MM-DHT concepts, the PGS ratios, the gear ratio of the FD and, if available, the gear ratio of the reduction gear for EM2 are varied.

For the concepts considered in this paper, the parameter variations result in a high number of different drive train variants. These can differ from each other in terms of driving performance and efficiency. To be able to make a statement which concepts and which dimensioning represent promising solutions, the concepts are compared with each other with regard to their driving performance in hybrid operation and in electric operation as well as their efficiency in charge sustaining operation.

The analysis of the minimum required EM power shows that all considered concepts require a low electrical effort in the C segment compared to higher segments. Additionally, the differences between the concepts are smaller than in the D or E segment. Therefore, the question arises whether a high mechanical effort should be realized in the C segment to reduce the required electrical power. By comparing the driving performance and efficiency, it can be analyzed whether a high mechanical complexity results in advantages regarding driving performance and efficiency. Furthermore, the results can indicate whether concepts with high electrical effort should not be used in the C segment. For these reasons, the following considerations are limited to the results of the C segment.

To compare the efficiency of the concepts, the CO_2 emissions in the WLTC for charge sustaining operation are calculated. During charge sustaining operation, it is ensured that the state of charge (SOC) of the battery is almost identical at the beginning and end of the cycle to allow a fair comparison of the concepts. To evaluate the driving performance, the acceleration times from 0 km/h to 100 km/h in hybrid as well as in electric operation are calculated. For these calculations, the simulation model presented in Section 5 is used. A locally optimal operating strategy based on the ECMS is used.

Since it can be assumed that realistic drive trains are not over-dimensioned with regard to their driving performance requirements, variants that are in the range of the minimum required total electrical power are selected for the concept comparison. Table 7 shows the selected EM power for all concepts.

Table 7. EM-specific and total EM power selected for C segment drive train concepts.

Drive Train Concept	P_{EM1} (kW)	P_{EM2} (kW)	Total EM Power (kW)
P2 HEV	80	-	80
P1P4 HEV	15	110–120	125–135
Series-parallel DHT	110	160–170	270–280
PS-DHT	60–70	130–140	190–210
PS-DHT + 4-speed transmission	40	40	80
MM-DHT – 2 PGS	50–60	80–90	120
MM-DHT – 3 PGS	40	40–50	80–90

In addition to the driving performance and the efficiency of the concepts, this article also evaluates the conversion quality defined by [11]. The concepts are compared regarding this benchmark parameter. Based on these results, important concept-specific properties are identified.

7.1. Analysis of Efficiency in WLTC and Driving Performance

In this section, the concepts in the C segment are compared regarding their efficiency in charge sustaining operation in the WLTC as well as their hybrid and electric driving performance. In hybrid operation, a distinction is made between driving performance with fully charged and discharged battery. In this case, the SOC-neutral CO_2 emissions are not affected by the battery's state of charge, since the simulation ensures a SOC of 50% at the beginning and end of the cycle.

7.1.1. Evaluation of Efficiency and Driving Performance in Hybrid Driving while Discharging the Battery

The simulation results in case the battery can be discharged are shown in Figure 11. It should be noted that the emissions in charge sustaining mode do not correspond to the combined CO_2 emissions, since the electrical energy consumption must also be analyzed and weighted to determine the combined CO_2 emissions.

The P2 HEV variants achieve acceleration times of approx. 6.4 s from 0 km/h to 100 km/h in hybrid operation while the battery is discharged. Furthermore, CO_2 emissions of about 111 g/km to 111.6 g/km are achieved. Compared to the other concepts, the P2 HEV variants achieve both good acceleration times and low CO_2 emissions.

As a result of electrifying the rear axle with a P4-EM with an output of 110 kW or 120 kW, considerably lower acceleration times can be achieved due to the electric all-wheel drive. Depending on the power of the EM on the rear axle, the variants achieve acceleration times in a range between 4.1 s and 4.4 s. Thus, the difference compared to the P2 HEV is about 2 s or 2.3 s. It should be noted that the simulation model allows a serial power flow in case of full-load acceleration for the P1P4 hybrid. If the front axle of the vehicle reaches the traction force limit, the P1-EM can transmit excess torque to the P4-EM positioned on the rear axle.

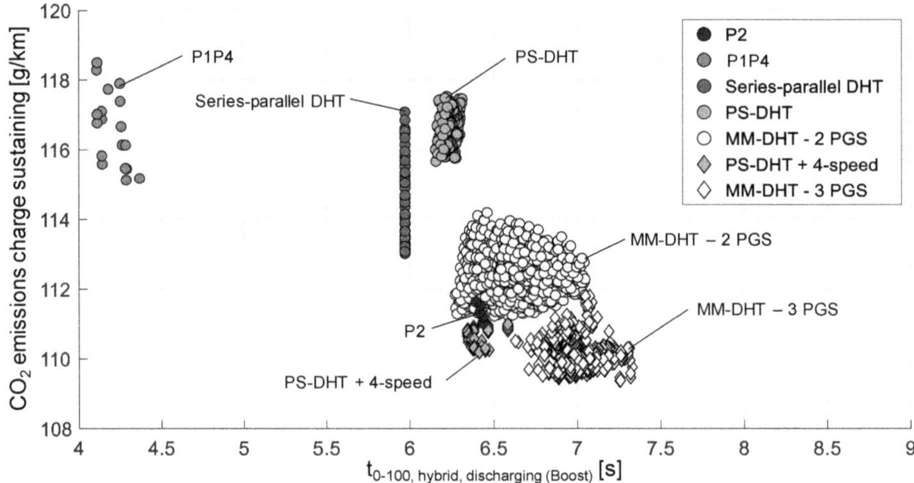

Figure 11. Comparison of charge sustaining efficiency in WLTC and hybrid driving performance for drive train concepts compliant with C segment requirements.

In contrast to the P2 HEV, however, the CO_2 emissions of the variants of the P1P4 concept are significantly higher. Depending on the dimensioning, the emissions range between 115 g/km and 118.5 g/km. The higher CO_2 emissions result from the fact that the P4-EM, which can be used during electric operation, is connected to the wheel via a constant gear ratio. Therefore, the operating point of the EM cannot be changed by a gearshift during electric driving or recuperation. In addition, due to its lower power output than the P2 HEV, the P1-EM allows the load point of the ICE to be increased to a lesser extent. This can have a less positive effect on the efficiency of the ICE. Furthermore, operating states may occur in which the load point of the ICE is increased via the P4-EM on the rear axle, which may result in high transmission losses.

With the series-parallel DHT, hybrid operation shows a comparatively good acceleration time of approx. 5.9 s due to the powerful EM. This means that all variants of this concept reach the traction force limit up to at least 100 km/h.

However, the variants of the concept show high differences in CO_2 emissions. They range between 113 g/km and 117.1 g/km. The transmission ratio between the ICE and the wheel has a major influence, since with increasing the gear ratio, the ICE is operated at higher speeds and lower torques. As a result, a higher amount of load point increase is carried out to improve ICE efficiency. The energy obtained by load point increase must later be used for electric driving or load point decrease to ensure SOC neutrality. The energy available for electrical operation is therefore generated under high losses compared to energy gained from recuperation. Thus, the CO_2 emissions are increased. Furthermore, the high power of the EM has a negative effect on CO_2 emissions as well. Due to the high power, the EM are operated at a low average power and thus at low efficiencies. As a result, their efficiency during electric driving and recuperation is lower compared to other concepts. This means that fewer driving situations can be covered in electric operation.

Compared to the series-parallel concept, the acceleration times of the PS-DHT variants increase. Depending on the variant, acceleration times between approx. 6.1 s and 6.3 s are possible. In contrast to the series-parallel DHT, the driving performance is slightly reduced, because EM2 has a lower maximum power and torque. Therefore, the variants of the PS-DHT do not reach the traction force limit up to speeds of 100 km/h.

The CO_2 emissions of the PS-DHT are between approx. 115.7 g/km and 117.5 g/km. Therefore, the PS-DHT has a lower efficiency in WTLC compared to the series-parallel DHT. Although the ICE can operate at high efficiencies due to the eCVT mode, the powertrain efficiency during electric driving and during recuperation is lower compared to the series-parallel DHT. This means that less electrical energy is available from recuperation in the traction phase and a higher amount of energy is required for electric driving. This means that the ICE operates more frequently and must provide additional energy by increasing the load point. In addition, powerful EM are necessary for the PS-DHT to meet the driving performance requirements. Thus, they operate at low average loads and have a low average efficiency.

If the PS-DHT is supplemented by a 4-speed transmission, the driving performance and CO_2 emissions are reduced. The variants achieve acceleration times between 6.3 s and 6.6 s and CO_2 emissions between 110.2 g/km and 111 g/km. Due to the lower total electrical power of the EM, the acceleration time increases slightly compared to the PS-DHT. However, four electrical and four power-split operating modes allow adaptation of the operating points of the energy converters to the driving situation. Especially in hybrid operation, this leads to the fact that less power of the ICE must be transmitted via the electrical branch by adjusting the speed of the ring gear. This increases the transmission efficiency and reduces CO_2 emissions. Furthermore, the efficiency of the EM in both traction and thrust phase is positively influenced, which also has a positive effect on energy consumption.

The variants of the MM-DHT with two PGS achieve acceleration times between 6.3 s and 7.1 s in hybrid operation. This results in similar driving performance as the PS-DHT with a 4-speed transmission or P2 HEV.

Due to an additional eCVT mode and a parallel mode, the MM-DHT concept with two PGS achieves lower CO_2 emissions than the PS-DHT. Although the transmission losses are higher because the concept is mechanically more complex, the additional degrees of freedom regarding the choice of operating points have a positive effect on efficiency. This results in CO_2 emissions between 111.2g/km and 114.2g/km. Compared to the P2 HEV, the CO_2 emissions are similarly good.

If the MM-DHT concept is enhanced by an additional PGS, the driving performance and efficiency are reduced compared to the concept with two PGS. The acceleration times are in a range between 6.6s and 7.3s and the CO_2 emissions are between 109.4 g/km and 111.8 g/km. Compared to the concept with two PGS, the number of operating modes increases, so that more efficient operating points of the energy converters can be selected in WLTC. On the one hand, the concept offers four modes in electrical operation in contrast to the concept with two PGS with two modes. On the other hand, more degrees of freedom are also available in hybrid operation with three parallel and four eCVT modes.

The results in Figure 11 show that mechanically complex concepts offer a higher efficiency in the C segment due to additional operating modes. At the same time, they show only slightly increased acceleration times compared to concepts with more powerful EM. Therefore, the concepts P2, MM-DHT with two PGS, PS-DHT with 4-speed transmission and MM-DHT with three PGS offer a better compromise in terms of performance and efficiency in hybrid operation. A higher mechanical complexity leads to a better efficiency.

However, it should be noted that the differences in efficiency and driving performance are small. For example, the best variants of the series-parallel DHT have good CO_2 emissions, which are about 1.7 g/km to 2 g/km higher than the best variants of P2 HEV or MM-DHT with two PGS. In addition, the concepts have to meet high performance requirements in hybrid operation. For example, the concepts must reach the traction force limit between 0 km/h and 10 km/h even when loaded, while the battery is being charged or the vehicle is driven purely electrically. This leads to a comparatively high total EM power for the PS-DHT and the series-parallel DHT, see Section 6. Due to the high requirements, these two concepts show a lower efficiency and driving performance. It can be assumed that different results are obtained if the driving performance requirements are reduced.

Furthermore, other important factors such as the required installation space and the costs influence the evaluation of the drive concepts. As these two important variables are not considered in this paper, the concepts are only assessed in terms of efficiency and driving performance.

7.1.2. Comparison of Efficiency in Hybrid Driving and Driving Performance in Electric Operation

In addition to the driving performance in hybrid operation, this paper also examines the driving performance in electric operation. Figure 12 shows the CO_2 emissions in hybrid operation versus the acceleration time from 0 km/h to 100 km/h in electric operation.

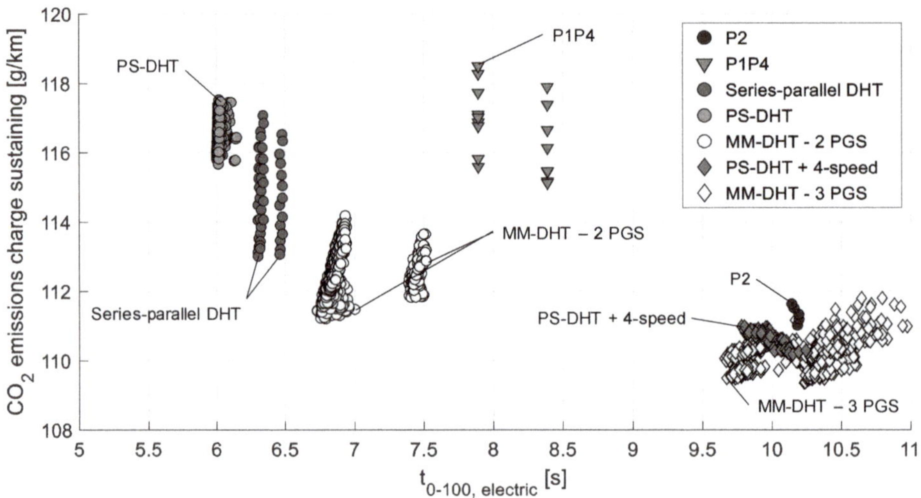

Figure 12. Comparison of charge sustaining efficiency in WLTC and electric driving performance for drive train concepts compliant with C segment requirements.

It can be seen that concepts with a high total electrical power achieve lower acceleration times in electrical operation. The variants of the PS-DHT and series-parallel DHT have acceleration times of approx. 6 s to 6.5 s, as the EM have total outputs of 170 kW–190 kW and 270 kW–290 kW, respectively. It should be noted that the variants of the PS-DHT have lower acceleration times, although less EM power is installed. The reason for this is that with the PS-DHT, both EM can drive the vehicle in electric mode because of the OWC between the ICE and PGS. In contrast, in series-parallel DHT only the EM2 can drive the vehicle.

The lower the power available in electrical operation, the higher the acceleration time. Therefore, the acceleration time of the MM-DHT with two PGS is between 6.6 s and 7.5 s. In addition, the results show that two different total EM powers are considered. In case of the P1P4 concept, the acceleration time increases to 7.9 s or 8.4 s. For mechanically more complex concepts, EM powers of 80 kW to 90 kW are required. For the P2 hybrid, this results in an acceleration time of about 10.1 s. For the PS-DHT with 4-speed transmission, the acceleration time increases to 9.8 s to 10.3 s. The largest differences are found in the MM-DHT with three PGS. The variants of this concept achieve acceleration times between 9.7 s and 11 s.

In comparison to hybrid operation with a charged battery, these acceleration times therefore differ by approx. 3.1 s in case of the MM-DHT with three PGS and 3.8 s for the P2 HEV. For the P1P4 the difference is also 3.8 s. Furthermore, the difference is considerably smaller for the mechanically simple concepts with high total electrical power. The MM-DHT with two PGS offers a good compromise between the acceleration time in hybrid operation and in electrical operation. For this concept, the acceleration time in electrical operation is 0.5 s higher.

The results in Figure 12 therefore show that the mechanically complex concepts have lower driving performance in electrical operation than the concepts with a simpler mechanical design because of their lower total electrical power. However, it should be noted that no requirement regarding the acceleration time was specified for electrical operation. In this case, it can be assumed that the mechanically more complex concepts would achieve this acceleration time with a lower total electrical power.

7.1.3. Comparison of Hybrid Driving Performance with Charged and Discharged Battery

In addition to electrical operation, the consistency of the acceleration behavior also plays a role in the evaluation of the concepts. Therefore, the concepts in this section are examined regarding the difference between charged and discharged battery in hybrid operation. For this purpose, Figure 13 shows the corresponding acceleration times from 0 km/h to 100 km/h. A solid black line shows the range of equal driving performance with fully charged and discharged battery in hybrid mode. To the right of this line, the driving performance is better with a discharged battery than with a charged battery. Furthermore, isolines indicate a constant time difference between the operation with charged and discharged battery.

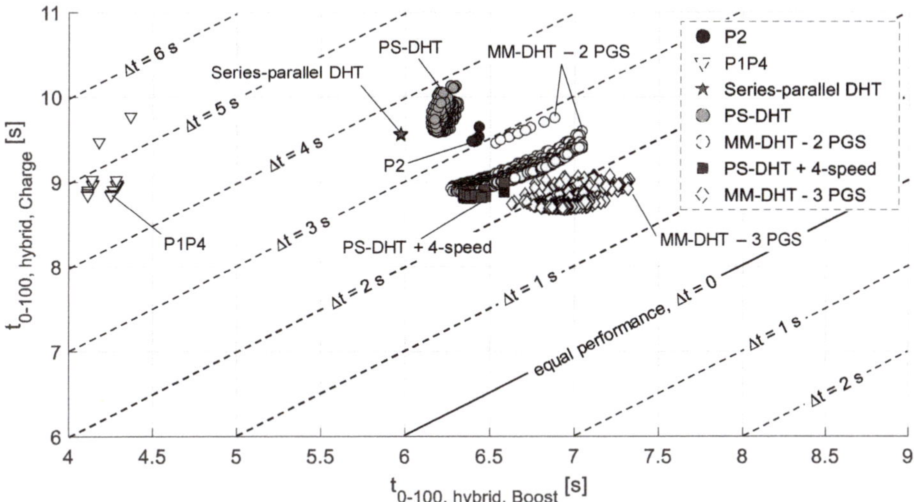

Figure 13. Comparison of acceleration time from 0 km/h to 100 km/h in hybrid operation in case of charging or discharging the battery.

The highest difference between hybrid operation with charged and discharged battery occurs in case of the P1P4 HEV. At high SOC, additional drive power is available at the rear axle of the vehicle resulting in a good traction potential. However, in case of a low SOC, drive power is only available at the front axle. Thus, the difference in acceleration time is more than 4 s. Therefore, the concept has disadvantages regarding the similarity of driving performance. Furthermore, Figure 13 shows that some variants of the P1P4 HEV have a shorter acceleration time at low SOC than the variants of the P2 HEV. This is due to the possibility to transfer a part of the drive power in series mode to the rear axle of the vehicle. The P1-EM works as a generator, while the P4-EM can provide its power at the rear axle with losses. In the low speed range, when the traction force limit of the front axle is reached, the P1P4 concept can thus offer advantages over the P2 HEV.

A smaller difference results in case of the series-parallel DHT and the PS-DHT. Their acceleration times differ by more than 3 s and less than 4 s. The P2 hybrid can be rated slightly better. A disadvantage of the series-parallel DHT is that the drive power of the ICE must be transmitted in series mode in a certain speed range, depending on the gear ratio between ICE and wheel. This can result in high conversion losses, so that the transmission efficiency declines and less drive power is available at the wheel.

A slightly better compromise than P2 HEV is offered by the MM-DHT variants with two PGS. With this concept the difference in acceleration time of the hybrid operation at high and low SOC is less than 3 s. Compared to the PS-DHT, the two eCVT modes as well as the parallel mode have a positive effect on the acceleration time, since less drive power has to be transmitted via the EM. This means that a higher proportion of the drive power is transmitted to the wheel mechanically, resulting in lower power losses.

If additional mechanical components are added to the PS-DHT, the difference between high and low SOC is reduced. The 4-speed transmission enables transmission of a higher share of the drive power mechanically in eCVT mode, resulting in shorter acceleration times compared to the PS-DHT. For the MM-DHT with two PGS there are similar acceleration times and differences between the two states of charge.

The lowest difference in acceleration time is shown by the MM-DHT variants with three PGS. Depending on the dimensioning, the difference is slightly higher than 2 s and in the best case only about 1.5 s. Since this concept offers the highest number of operating modes, the difference in driving performance is the lowest.

The results in Figure 13 therefore lead to the conclusion that the increase in mechanical effort in the transmission leads to less difference in driving performance in hybrid operation at high and low SOC. The additional operating modes provided by additional mechanical components can lead to a higher share of the drive power provided by the ICE being transferred mechanically to the wheel, resulting in better driving performance at low SOC. As a result, the difference in hybrid operation is lower and the driving performance is better. It should be noted that the MM-DHT with three PGS and the PS-DHT with 4-speed transmission offer better performance in comparison to the P2 HEV in case of a low SOC. This is due to the four eCVT modes offered by both concepts. Compared to the P2 HEV with 8-speed DCT, they provide better adaptation to the traction force hyperbola, resulting in lower acceleration times. In addition, there are variants of the MM-DHT with two PGS, which offer advantages compared to the P2 HEV.

An increase in performance in hybrid operation at low SOC would only be possible in concepts with low mechanical effort if the EM could achieve better efficiencies or if the ICE offered a higher maximum power.

It should be noted that no explicit requirements were made for a small difference in acceleration times. In this case, the required total electrical power of the concepts would differ from the results shown in this paper.

7.2. Evaluation of Conversion Quality and Transmission Efficiency

For further evaluation of the concepts in hybrid operation, the average transmission efficiency and the average conversion quality in WLTC are calculated. The conversion quality is introduced in [11] and describes the ratio of the theoretical optimum efficiency of the ICE to the actual efficiency of the ICE. The parameter thus describes the ability of a transmission to operate an energy converter at its power-specific optimum efficiency.

Figure 14 shows the average conversion quality and the average transmission efficiency in hybrid operation in the WLTC. There are differences between add-on hybrid concepts and DHT on the one hand and between mechanically simple and complex concepts on the other hand.

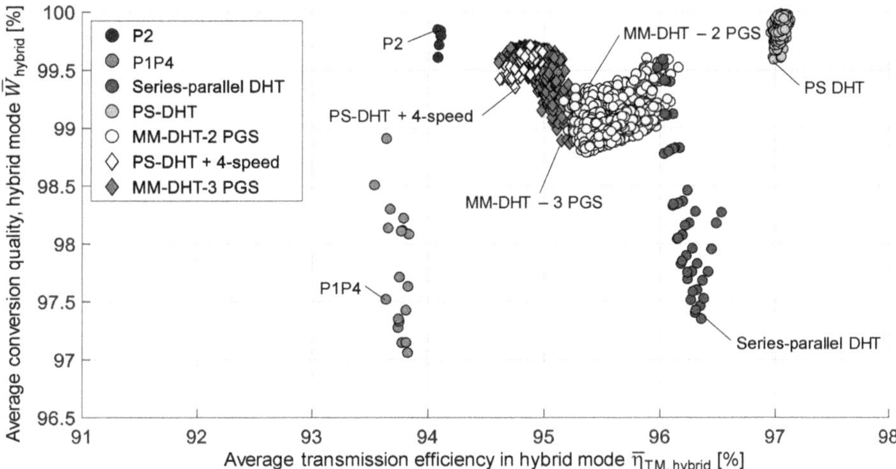

Figure 14. Comparison of average conversion quality of the ICE and average transmission efficiency in hybrid operation in WLTC.

The variants of the PS-DHT achieve very good transmission efficiencies of around 97% in hybrid operation. A positive effect is that the transmission is mechanically simple, so that mechanical losses are low. In addition to a good transmission efficiency, all variants of the PS-DHT achieve high conversion qualities of over 99.5%. This is due to the eCVT operating mode, which allows infinite variation of the transmission ratio, so that the ICE can operate at its power-specific optimum efficiency operating point.

Compared to the PS-DHT, the variants of the series-parallel DHT have lower transmission efficiencies in hybrid operation. Although this concept also has a simple mechanical design and thus low mechanical losses, the EM are much more powerful. Since the conversion losses in the EM are taken into account in the transmission efficiency in case of serial operation, the efficiency is lower compared to the variants of the PS-DHT. In addition, the transmission efficiency in hybrid operation depends on the wheel power in which the hybrid operation occurs. It is thus possible that hybrid operation in case of PS-DHT takes place at higher wheel power, whereby the transmission efficiency tends to be better, since load-independent losses cause a smaller share of the losses.

In contrast to the PS-DHT, the variants of the series-parallel DHT have lower conversion qualities in hybrid operation. Compared to the other concepts, there are greater differences. In the best case, the conversion quality is over 99.5% and in the worst case less than 97.5%. This variation results from the scaling of the FD ratios. An increase in the FD ratio leads to a lower conversion quality, since the ICE is operated at higher speeds and lower torques due to the constant gear ratio to the wheel. Thus, load point increase of the ICE is necessary to increase the efficiency of the ICE. Since the load point increase does not lead to an increased efficiency in all areas, the average conversion quality decreases. Furthermore, an increase in the FD ratio means that lower speeds can be achieved in parallel mode. Therefore, there may be variants in which the serial mode must be selected at high speeds. Since in serial mode both the speed and the torque of the ICE can be chosen freely, the conversion quality is positively influenced in serial mode.

Compared to the mechanically simple DHT, the P2 HEV and P1P4 HEV add-on hybrids achieve lower average transmission efficiency. In case of the P2 HEV it is over 94% and in case of the P1P4 it is around 93.5%. The reason for this is that mechanical losses are higher due to the more complex design.

Regarding the average conversion quality, there are greater differences. The P2 HEV variants achieve similarly good conversion qualities as PS-DHT of over 99.5%. On the one hand, the combination of gear ratios and ICE efficiency map has a positive effect on the conversion quality. On the other hand, with the P2 HEV it is possible to operate the ICE in more favorable efficiency ranges by increasing the load point.

In contrast, the extent of load point increase is limited for the P1P4 HEV. On the one hand, the P1-EM only has a power of 15 kW. On the other hand, the load point increase via the P4-EM is associated with high losses. For this reason, the FD ratio of the P1P4 HEV variants also has a strong influence on the average conversion quality. As with the series-parallel DHT, an increase of the FD ratio leads to a reduction of the conversion quality of the ICE.

A good compromise between conversion quality and transmission efficiency in hybrid operation is offered by the mechanically complex DHT. The variants of the MM-DHT with two PGS achieve transmission efficiencies between approx. 95% and 96.1% in hybrid operation with conversion qualities of 98.7% to 99.6%. With a higher mechanical complexity as with the MM-DHT with three PGS, the transmission efficiency declines to a range between 94.6% and 95.3%. However, the number of operating modes gained through the higher mechanical complexity leads to an increase in the conversion quality compared to the MM-DHT with two PGS. It lies between approx. 98.7% and 99.7%.

The extension of the PS-DHT by a 4-speed transmission has a negative effect on the average transmission efficiency in hybrid operation. It lies between 94.6% and 95% and is significantly lower than in case of the PS-DHT. The conversion quality is also reduced compared to the PS-DHT. However, it is still high with about 99.3% to 99.7%. It should be noted that the local optimal operation strategy approach always achieves an optimal compromise between high conversion quality of the ICE and low losses in the EM. Therefore, a lower conversion quality of the ICE does not directly lead to higher CO_2 emissions, see Figure 11.

The results in Figure 14 show characteristic differences between different types of hybrid transmission concepts. For example, mechanically simple DHT such as PS-DHT or series-parallel DHT in hybrid operation show very high transmission efficiencies. The concepts have a low mechanical complexity, so that mechanical losses are reduced. In addition to a high transmission efficiency, these concepts can achieve high conversion qualities of over 99%. However, as the minimum required total EM power in Figure 10 shows, powerful EM are necessary. DHT with increased mechanical complexity have lower transmission efficiencies in hybrid operation, but these concepts can achieve comparably high conversion qualities as mechanically simpler concepts, so that with reduced total EM power, lower CO_2 emissions are achieved in SOC-neutral operation. Therefore, especially MM-DHT are a good solution. Although add-on-HEV concepts can also achieve high conversion qualities, their transmission efficiency is lower due to the higher mechanical effort.

8. Summary

The electrification of the drive train creates considerable degrees of freedom in the design of drive trains. Depending on the number and positioning of EM, there may be significant differences regarding the topology and functionality of the concepts. On the one hand, additional EM can supplement conventional drive trains. Common solutions to add-on concepts are P2 HEV as well as electrified all-wheel drive systems such as P1P4 HEV. In addition, it is also possible to implement DHT. These transmissions are dedicated to HEV and allow different operating modes than parallel hybrids. Depending on the structure, DHT can be of varying complexity.

In this paper, various powertrain concepts, which differ in their design and operating modes, are examined. These include parallel hybrids, series-parallel DHT, PS-DHT, and MM-DHT with different mechanical complexity. In this paper, the concepts are briefly described regarding their structure and operating modes. Furthermore, the mechanical complexity of the concepts is estimated and quantified by considering mechanically relevant components.

To compare powertrain concepts driving performance requirements in different speed ranges are defined for a C, D, and E segment vehicle. These include reaching the traction force limit in the low speed range, requirements in the medium speed range, which are quantified by boost or charge power, and maximum speed requirements.

A comparatively small number of requirements is defined to allow for short computing times while at the same time allowing for a high number of powertrain variants. Therefore, a full factorial scaling plan can be used so that the system behavior can be identified.

The consideration of different speed ranges also ensures that the delivery maps of the concepts match realistic demand maps of the drive concepts.

The requirements are defined for different drive train states to take into account the interaction of ICE and battery. In hybrid operation, a distinction is made as to whether the battery can be discharged or needs to be charged. In addition, requirements for electrical operation are defined.

Using a modular simulation environment described in this paper, variants that meet the requirements can be identified based on a driving performance simulation. Differences arise especially regarding the required EM power. This differs between the segments, since the tractive forces resulting from the requirements depend on vehicle and drive train parameters. Furthermore, concept-specific differences regarding the required EM power are analyzed. This electrical effort is compared to the quantified mechanical effort.

The results show that a reduction in the number of mechanical components requires comparatively high EM power. Responsible for this is a constant gear ratio between the EM, which in most situations operates as a motor. This EM must, on the one hand, reach the maximum speed of the vehicle in hybrid operation and, on the other hand, enable reaching the traction force limit both in hybrid operation and electric driving. Therefore, this EM must have a comparatively high torque and thus a high power output. In addition, due to their structure, the concepts with low mechanical effort must transmit a high share of the drive power electrically. In case of the series-parallel DHT, the serial mode must be selected in a certain speed range when the battery is discharged. In this case, the drive power is only provided by the ICE and transmitted via both EM. Depending on the gear ratio, the PS-DHT also needs to transmit a certain share of the drive power via the EM. The fact that an EM must receive the power of the ICE and therefore operate as a generator results in a dependency on the maximum power of the ICE.

For these reasons, the required EM power of the mechanical simple concepts increases distinctly with higher vehicle segments or higher requirements. In these cases, an increase in the mechanical complexity of the transmission is advisable. Additional components allow transmission of higher shares of the drive power mechanically, which can reduce the required EM power. In addition, several gears may be available for the EM, which is usually operating as a motor. Mechanically complex concepts are therefore better suited for use in higher vehicle segments. Depending on the segment and the EM used, they do not require significantly more EM power than concepts with one EM, such as P2 HEV.

The comparison of the required EM power shows that in the C segment there are the least differences between the drive concepts. For this reason, hybrid operation is examined within a benchmark analysis. This paper focuses on efficiency and performance since both costs and the required installation space can only be evaluated with comparably high uncertainty in early concept phases. To assess the efficiency, the SOC-neutral CO_2 emissions in the WTLC are considered. To compare the driving performance, the acceleration time from 0 km/h to 100 km/h for hybrid operation and electric driving is analyzed. The results show that mechanically complex concepts offer a good compromise between driving performance and efficiency in hybrid operation. On the one hand, they achieve low CO_2 emissions. On the other hand, they show small differences between the acceleration times of hybrid operation with charged and discharged battery.

In contrast, the CO_2 emissions of the mechanically simpler concepts increase. Regarding hybrid operation, they therefore do not represent more attractive solutions in the context of these investigations. However, it should be noted that due to the higher electrical power, the driving performance in electric operation increases. In addition, there are only minor differences in the acceleration time during hybrid operation with a charged battery and electric operation.

To be able to derive concept-specific properties, the average transmission efficiency and the conversion quality in the WLTC are also investigated in hybrid operation. The results show that mechanically simple concepts such as PS-DHT or series-parallel DHT can provide both good transmission efficiency and high conversion quality. However, the PS-DHT offers advantages due to the eCVT mode and can achieve higher conversion qualities.

With mechanically more complex DHT such as MM-DHT with two or three PGS or a PS-DHT with a 4-speed transmission, the average transmission efficiency and conversion quality decline slightly. Compared to the mechanically simpler concepts, however, considerably less electrical power must be installed in the drive train. Therefore, these concepts offer the best compromise between transmission efficiency and conversion quality.

The results also show that the considered parallel hybrids have the lowest average transmission efficiencies in hybrid operation. In case of the P2 HEV, however, a very high conversion quality can be achieved with lower required EM power.

Author Contributions: C.S. and F.K.; methodology, C.S.; software, C.S.; validation, C.S., F.K.; formal analysis, C.S.; investigation, C.S. and F.K.; resources, C.S.; data curation, C.S.; writing—original draft preparation, C.S.; writing—review and editing, C.S.; visualization, C.S.; supervision, F.K. All authors have read and agreed to the published version of the manuscript.

Funding: This research received no external funding

Conflicts of Interest: The authors declare no conflict of interest.

References

1. Lange, A. Optimierung modularer Elektro- und Hybridantriebe. Ph.D. Dissertation, Technische Universität Braunschweig, Braunschweig, Germany, 6 February 2018.
2. Suzuki, Y.; Nishimine, A. *Development of New Plug-In Hybrid Transaxle for Compact-Class Vehicles*; SAE International: Warrendale, PA, USA, 2017.
3. Imamura, T.; Tabata, A. *Concept and Approach of Multi Stage Hybrid Transmission*; SAE International: Warrendale, PA, USA, 2017.
4. Kato, S.; Ando, I. Development of Multi Stage Hybrid System for New Lexus Coupe. *SAE Int. J. Alt. Power* **2017**, *6*, 136–144. [CrossRef]
5. Conlon, B.; Blohm, T. *The Next Generation "Voltec" Extended Range EV Propulsion System*; SAE International: Warrendale, PA, USA, 2015.
6. Holmes, A.; Liu, J. *General Motors Electric Variable Transmission for Cadillac CT6 Sedan*; SAE International: Warrendale, PA, USA, 2016.
7. Lange, A.; Küçükay, F. A new, systematic approach to determine the global energy optimum of a hybrid vehicle. *Automot. Engine Tech.* **2016**, *1*, 35–46. [CrossRef]
8. Lange, A.; Küçükay, F. Ermittlung des globalen energetischen Optimums von Hybridantrieben. *ATZ* **2017**, *119*, 106–111. [CrossRef]
9. Sciaretta, A.; Guzzella, L. Control of Hybrid Electric Vehicles. *IEEE Control Syst. Mag.* **2007**, *27*, 60–70.
10. Sciaretta, A.; Back, M.; Guzzella, L. Optimal Control of Parallel Hybrid Electric Vehicles. *IEEE Trans. Control Syst. Technol.* **2004**, *12*, 352–363. [CrossRef]
11. Lange, A.; Küçükay, F. Benchmark parameters of transmissions. *Automot. Engine Tech.* **2018**, *3*, 1–10. [CrossRef]

 *vehicles* 

Article

Development of a PHEV Hybrid Transmission for Low-End MPVs Based on AMT

Yongcheng Zhen [1], Yong Bao [1], Zaimin Zhong [1,*], Stephan Rinderknecht [2,*] and Song Zhou [1]

[1] School of Automotive Studies, Tongji University, No. 4800 Cao'an Road, Jiading District, Shanghai 201804, China; 1831658@tongji.edu.cn (Y.Z.); baoyong@tongji.edu.cn (Y.B.); 1933561@tongji.edu.cn (S.Z.)

[2] Institute for Mechatronic Systems in Mechanical Engineering, Technische Universität Darmstadt, Otto-Berndt-Straße 2, L101/208, 64287 Darmstadt, Germany

* Correspondence: zm_zhong@tongji.edu.cn (Z.Z.); rinderknecht@ims.tu-darmstadt.de (S.R.)

Received: 27 December 2019; Accepted: 19 March 2020; Published: 25 March 2020

Abstract: In order to improve the fuel economy of vehicles, based on the automated mechanical transmission (AMT), a plug-in hybrid electric vehicle (PHEV) hybrid transmission for low-end multi-purpose vehicles (MPVs) is developed. To obtain the statistics of the best-selling models, we took several best-selling models in the Chinese market as the research object to study the relationship between power demand, energy demand, weight, and cost. The power requirements and energy requirements of PHEVs are decoupled. According to the decoupled theory, a single-motor parallel scheme based on the AMT is adopted to develop a PHEV hybrid transmission. In the distribution of engine and motor power, the engine just needs to meet the vehicle's constant driving power, and the backup power can be provided by the motor, which means we can use an engine with a smaller power rating. The energy of short-distance travel is mainly provided by the motor, which can make full use of the battery, reducing the fuel consumption. The energy of long-distance travel is mainly provided by the engine, which can reduce the need for battery capacity. The working modes of the electrified mechanical transmission (EMT) are proposed, using P3 as the basic working mode and setting the P2 mode at the same time, and the gear ratios are designed. Based on the above basic scheme, two rounds of prototype development and assembling prototype vehicles for testing are carried out for the front-engine-front-drive (FF) layout. The test results show that the vehicle's economy has been improved compared to the unmodified vehicle, and the fuel-saving rate of 100 km has been achieved at 35.18%. The prototype development and the vehicle matching verify the effectiveness of the new configuration based on AMT.

Keywords: hybrid electric vehicle; plug-in hybrid electric vehicle; electromechanical coupling; dedicated hybrid transmission; electrified mechanical transmission; multi-purpose vehicle

---

1. Introduction

Environmental pollution and energy shortage in the world urgently require the development of new energy vehicles, which has become the world consensus [1,2]. As two important types of new energy vehicles, electric vehicles (EVs) and plug-in hybrid electric vehicles (PHEVs) have become the main research directions of new energy vehicles in the world [3,4]. PHEVs can completely solve the user's mileage anxiety problem, and become one of the main choices of users [5,6].

The dedicated hybrid transmission (DHT) is the most important component of the PHEV's powertrain, which couples the engine and the motor [7]. According to the different power coupling forms, DHTs can be divided into series, parallel, and power-split configurations [8]. The series DHT can only be used on extended-range electric vehicles, such as Chevrolet Volt, BMW i3, Nissan Note e-power, etc. However, it cannot drive the vehicle directly by engine. The engine power must be converted into electric energy to drive the vehicle [9,10]. The parallel DHT couples the engine power and the motor power in parallel, enabling two powers to drive the vehicle individually or together. Typical parallel DHTs include Volkswagen's single-motor dual-clutch transmission and BYD's DM-II transmission [11,12]. The power-split DHT enables more flexible coupling between the engine and the motor, allowing the vehicle to combine the advantages of both series and parallel [13,14], with better fuel economy, such as Toyota's Prius, the GM's VOLTEC II, and so on [15–17]. However, the structure of the power-split DHT is complex, with two sets of motors and motor controllers, which greatly increases the cost of the transmission. Another scheme is on the basis of the automated mechanical transmission (AMT), such as BYD's DM, the Eaton's hybrid system, and so on [18,19]. Compared with the power-split DHT, the AMT-based solution has a simpler structure and a lower cost, meaning that it is more suitable for developing a PHEV low-cost hybrid transmission.

Based on the above design ideas, a new hybrid transmission based on AMT is proposed in this paper. We have developed prototypes, assembled prototype vehicles, and conducted tests to verify the effectiveness of the scheme.

2. Statistical Law of New Energy Vehicles

2.1. PHEV Power Demand

The best-selling 58 internal combustion engine vehicles (ICEVs) and 30 PHEVs (Statistics for the first 6 months of 2019) are statistically analyzed. All statistics are from two statistics websites, Autohome (http://www.autohome.com.cn/) and Yiche.com (http://bitauto.com/). The statistical results of engine specific power are shown in Figure 1. In the figure, SUV is short for sport utility vehicle, and MPV is short for multi-purpose vehicle.

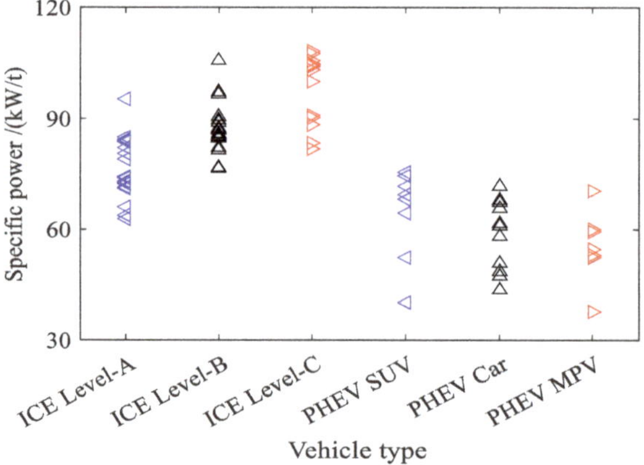

Figure 1. Engine specific power of the best-selling vehicles.

According to Figure 1, the engine specific power of the PHEVs is significantly lower than that of the ICEVs. The maximum engine specific power of the PHEV MPVs is 70.4 kW/t, the minimum is 37.6 kW/t, and the median is 54 kW/t.

The required power of ICEVs under all conditions is only provided by the engine. This includes the constant driving power at constant speed required to overcome rolling resistance and air resistance, as well as the backup power required for acceleration and climbing. For acceleration and climbing conditions, the vehicle's power requirements are shown in Figure 2.

In Figure 2a, the three curves represent three different acceleration situations. In Figure 2b, the four curves in the figure represent the power requirements of the four different slopes (0%, 5%, 10%, 15%).

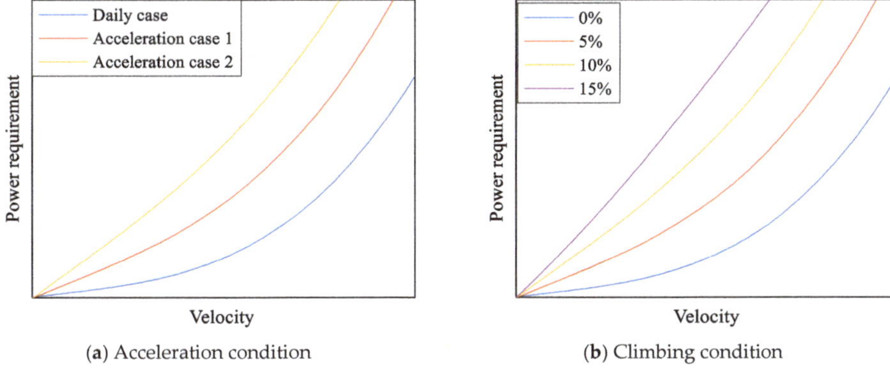

(**a**) Acceleration condition   (**b**) Climbing condition

Figure 2. Vehicle's power requirements.

According to Figures 1 and 2, the required power of the vehicle can be divided into two parts: the constant driving power and the backup power. The constant driving power and backup power of ICEVs can only be provided by the engine. But the PHEV engine only needs to meet the constant driving power demand. Hence, the engine specific power of PHEVs is significantly lower than that of ICEVs. When the vehicle requires a large backup power, the engine and the motor drive the vehicle together. Therefore, it can be considered that the backup power of PHEVs can be provided by the motor.

2.2. PHEV Energy Demand and Cost Analysis

China currently has no official data on the daily mileage statistics. According to the National Household Travel Survey (NHTS) of 2009 in America, approximately 90% of car commuters in the U.S. travel less than 30 miles to work. That is, if the pure electric driving range of PHEVs is set to 50 km, it can meet the commuting needs of about 90% of users.

The best-selling 54 electric vehicles (EVs) are statistically analyzed. The statistical results of vehicle mass and price per kilometer are shown in Figure 3.

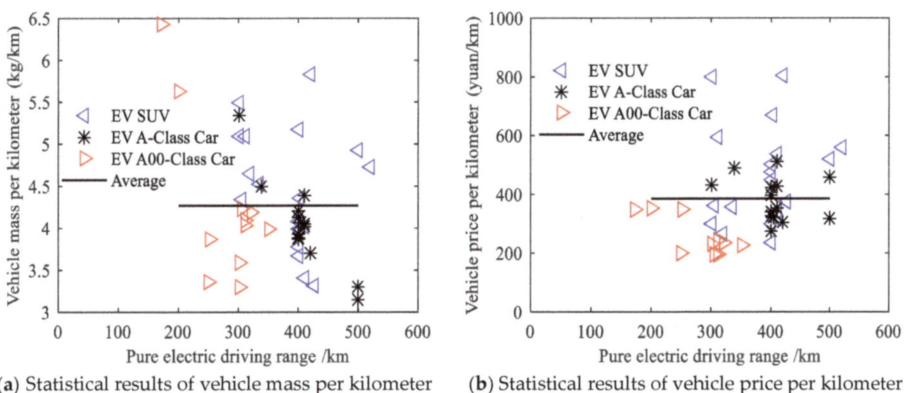

(**a**) Statistical results of vehicle mass per kilometer   (**b**) Statistical results of vehicle price per kilometer

Figure 3. Statistical results of vehicle mass and price per kilometer.

According to Figure 3, under the current technology and market conditions, the pure electric driving range of the best-selling EVs is increased by 1 km, the price increases by about 400 yuan, and the mass increases by about 4.27 kg. The price of 20 best-selling A-class PHEVs in China with a pure electric driving range of 50–60 km is statistically analyzed, and the average price is about 160,000 yuan. The average price of the A class PHEV50s (short for, the PHEVs with a pure electric driving range of

50 km) is 160,000 yuan, which is increased by 400 yuan per kilometer. The average mass of the A-Class PHEV50s is 1600 kg, which is increased by 4.27 kg per kilometer. The price of PHEV50, PHEV100 (short for, the PHEVs with a pure electric driving range of 100 km) and EV are compared and analyzed. The results are shown in Figure 4.

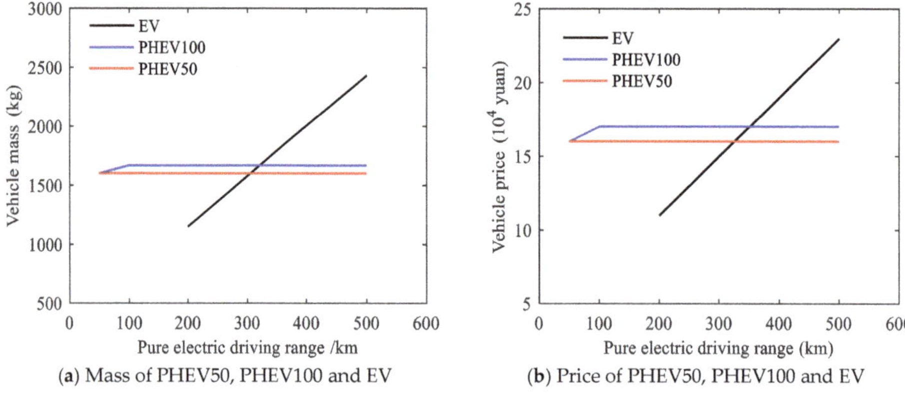

(**a**) Mass of PHEV50, PHEV100 and EV   (**b**) Price of PHEV50, PHEV100 and EV

Figure 4. Mass and price of PHEV50, PHEV100, and electric vehicles (EVs).

As can be seen from Figure 4, the mass intersection point of EV and PHEV50 is 315 km, and the price intersection point of EV and PHEV50 is 325 km. That is, if the pure electric driving range of EVs is more than 325 km, the mass and price of EVs will exceed PHEV50s. This means that EVs that pursue long driving ranges have no advantage in terms of cost.

In summary, with the increase in driving mileage, the weight of the vehicle will also increase. Increasing the quality of the car will increase the energy consumption of the car and increase the useless power of the car, which will lead to the deterioration of the vehicle's economy. So, it is difficult to find a suitable demarcation point to meet both short-distance travel and long-distance travel for EV, but PHEVs solve this problem very well. If the pure electric driving range of PHEVs is set to 50 km, it can meet the commuting needs of about 90% of users. Therefore, this paper proposes a solution based on AMT for PHEVs. We use a decoupled approach to consider power requirements and energy requirements. In the distribution of engine and motor power, the engine just needs to meet the vehicle constant driving power, and the backup power can be provided by the motor, which means we can use an engine with a smaller power rating. When driving long distances, the engine provides backup energy, which greatly reduces the need for battery capacity.

3. Configuration and Parameter Design of PHEV Hybrid Transmission Based on AMT

3.1. Transmission Configuration

Based on the AMT, an electrified mechanical transmission (EMT) is developed for PHEVs, and its configuration is shown in Figure 5. In the figure, S1/2, S3/4, and S5 are the synchronizers of first/second, third/fourth, and fifth gears of mechanical gears. S_M is the synchronizer of the mode gears. P2 and P3 are the coupling paths of the electric power flow to the input shaft and the output shaft of the transmission, respectively. P2 indicates that the drive motor is located behind the engine and before the transmission and is connected to the transmission input. That is, the power output of the motor and the engine is coupled at the transmission input. P3 indicates that the drive motor is connected to the transmission output in some way. That is, the motor output power and the engine is coupled at the output of the transmission.

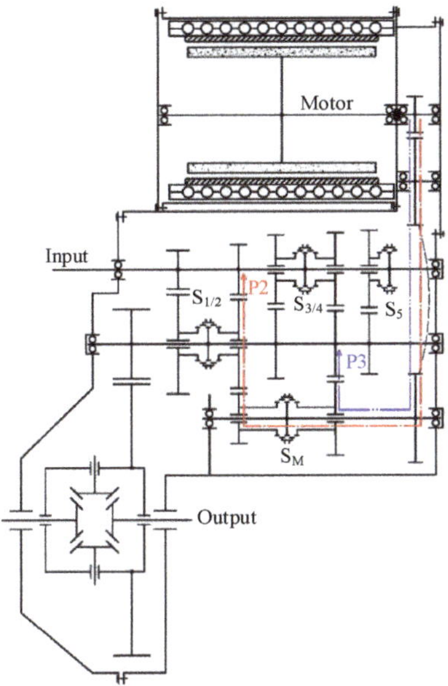

Figure 5. Configuration of electrified mechanical transmission (EMT).

The main technical features of EMT are as follows:

1. Based on fixed axis gears and AMT technology, the industrialization foundation is good.
2. Both P2 and P3 modes are considered, and the functions are complete.
3. There is no power interruption during shifting in P3 mode.
4. Multiple gears can make the engine more efficient.
5. Multiple gears can improve the power density and torque density of the motor.

3.2. Working Modes

The main working modes are shown in Table 1.

Table 1. Main working modes of EMT.

Conditions	Modes/gears	Descriptions
Launch	P3 mode	Normally, the vehicle power is provided only by the motor.
EV mode	P3 mode	No shift
HEV mode	P3 mode second/third/fourth /fifth gear	EMT works in P3 mode, and the mechanical gear is second/third/fourth/ fifth gear depending on speed. When the mechanical gear shifts, the motor provides power through P3 mode, without power interruption.
High-speed mode	P2 mode fifth gear	EMT works in P2 mode, and the mechanical gear is fifth gear. When the P2/P3 mode shifts, the engine drives the vehicle in fifth gear, without power interruption.
Parking charge mode	P2 mode neutral gear	Parking charging (charging the power battery or supplying power to the outside)
Limp mode	P2 mode first gear	Dealing with power system failures and extreme conditions

3.3. Gear Ratio Design

On the basis of an MPV, a PHEV equipped with the EMT is developed. According to the results of Section 2, the engine specific power is selected as 54 kW/t. The vehicle mass is set to 1500 kg. The calculated engine power is 81 kW, and the engine power actually selected is 77 kW. The rated power of the motor is 35 kW, and the peak power is 70 kW.

The design criteria of the gear ratio, i.e., the constraints of the gear ratio, are as follows:

1. The mechanical gear ratio design takes into account both pure mechanical driving and hybrid driving.
2. The center distances of the coupling shaft and the output shaft in P2 and P3 modes are the same.
3. The motor is as close as possible to the transmission.
4. The second/third/fourth/fifth gears in the P3 mode meet daily driving needs.
5. The first gear in the P2 and P3 modes meet extreme conditions, such as maximum climbing.

According to the above criteria, the design results of the gear ratio are shown in Table 2.

Table 2. Design results of the gear ratio.

Modes/Gears	Gear Ratio
First gear	3.462
Second gear	2.048
Third gear	1.464
Fourth gear	1.03
Fifth gear	0.838
P3 mode	2.0889
P2 mode	1.7594
Final drive ratio	4.529

Under the urban conditions, when the PHEV works in EV mode with the P3 mode, the traction force and resistance curves are shown in Figure 6.

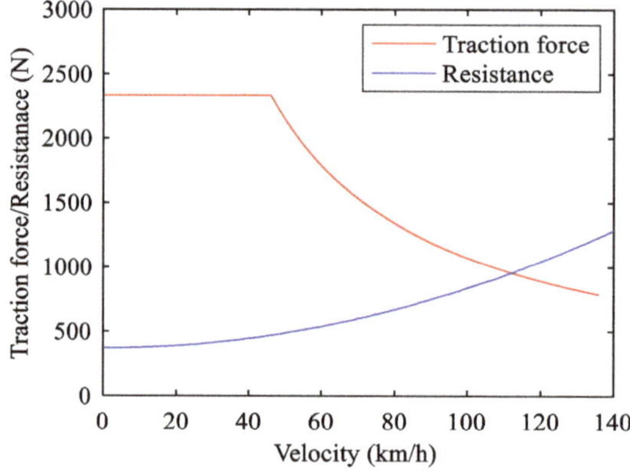

Figure 6. The traction force and resistance curves in EV mode with the P3 mode.

Under the urban conditions, when the PHEV works in HEV mode with the P3 mode and second/third/fourth/fifth gear. Under the extra-urban conditions, the PHEV works in HEV mode with the P3/P2 mode, and the traction force and resistance curves are shown in Figure 7. As can be seen from Figure 7, the maximum velocity can be 180 km/h.

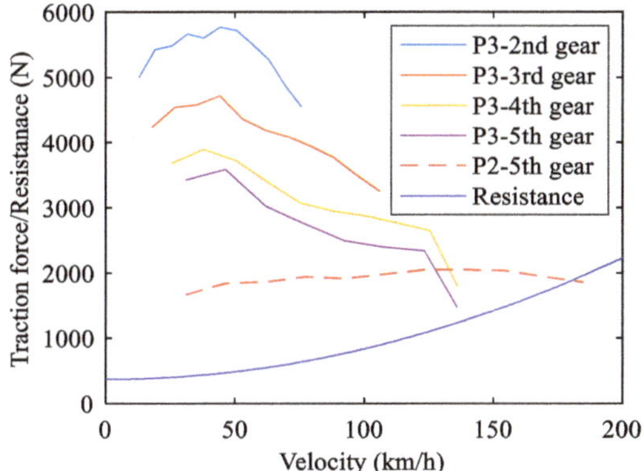

Figure 7. The traction force and resistance curves in HEV mode with the P3/P2 mode.

Under the limp mode, the PHEV works in HEV mode with the P2 mode and first gear, and the traction force and resistance curves are shown in Figure 8.

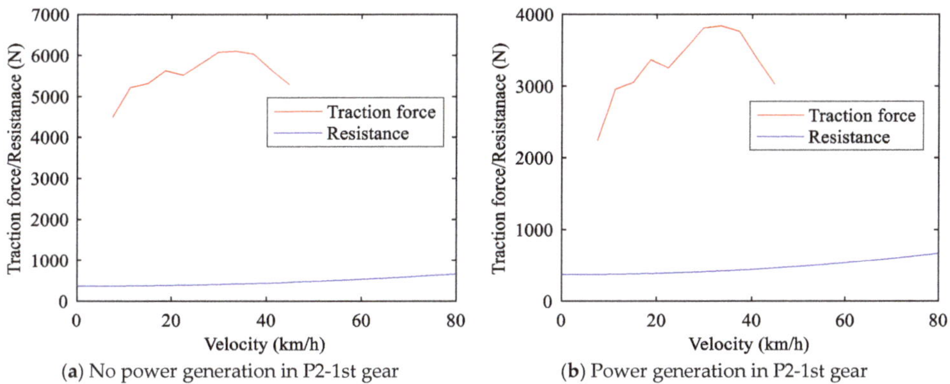

(**a**) No power generation in P2-1st gear

(**b**) Power generation in P2-1st gear

Figure 8. The traction force and resistance curves under the limp mode.

In the climbing condition, under different slopes, the traction force and resistance curves with HEV mode are shown in Figure 8. As can be seen from Figure 9, the vehicle can easily drive up a 30% slope.

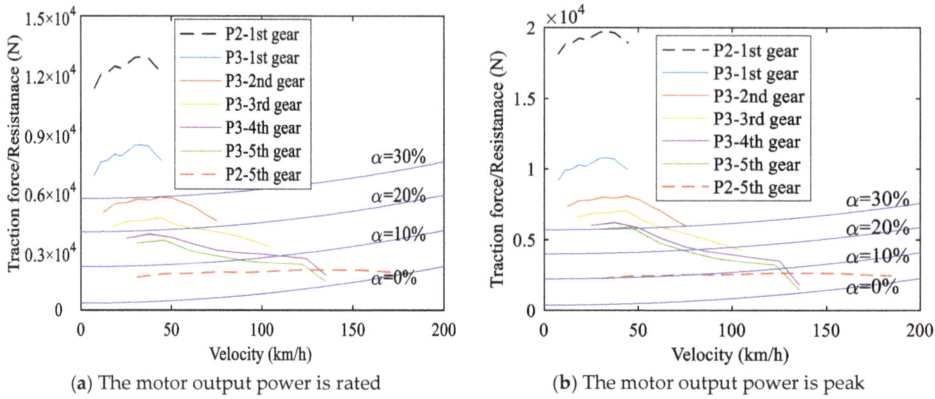

(**a**) The motor output power is rated

(**b**) The motor output power is peak

Figure 9. The traction force and resistance curves under the climbing condition.

The dynamic performance comparison between the PHEV and the benchmark vehicles is shown in Figure 10. As can be seen from Figure 10, the PHEV is powerful.

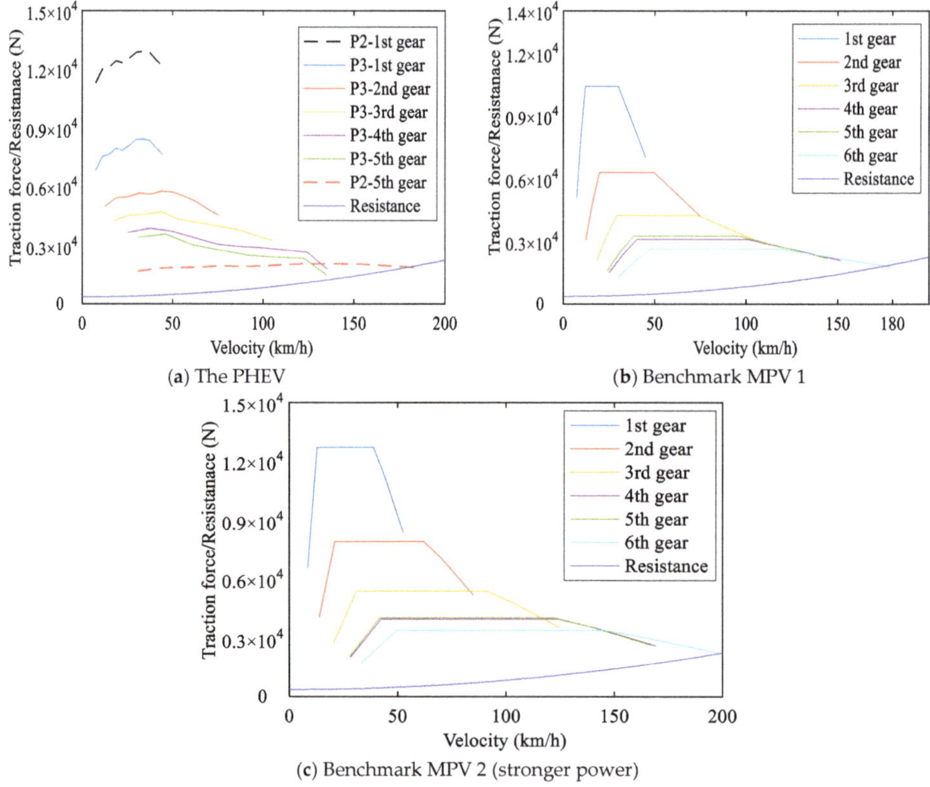

(**a**) The PHEV

(**b**) Benchmark MPV 1

(**c**) Benchmark MPV 2 (stronger power)

Figure 10. Dynamic performance comparison.

4. Prototyping and Verification

Based on the above basic scheme, two rounds of prototype development and assembling prototype vehicles for testing are carried out for the front-engine-front-drive (FF) layout.

4.1. EMT prototyping

According to the above scheme and parameter design results, the EMT prototype is developed as shown in Figure 11.

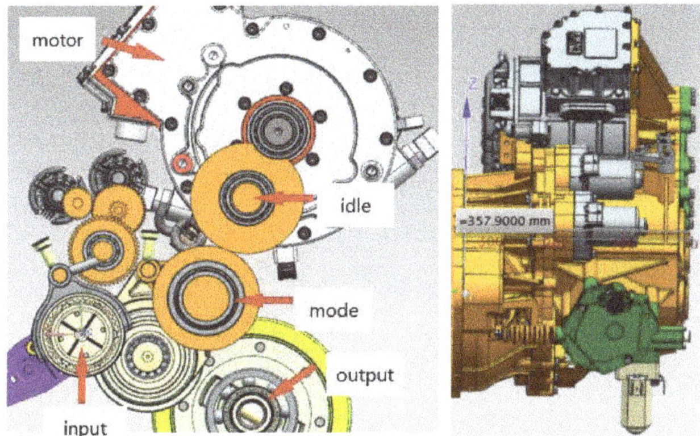

Figure 11. EMT prototype.

The shift actuator is shown in Figure 12. Cam 2 is responsible for the P2 and P3 mode switching, and cam 1 is responsible for gear shifting.

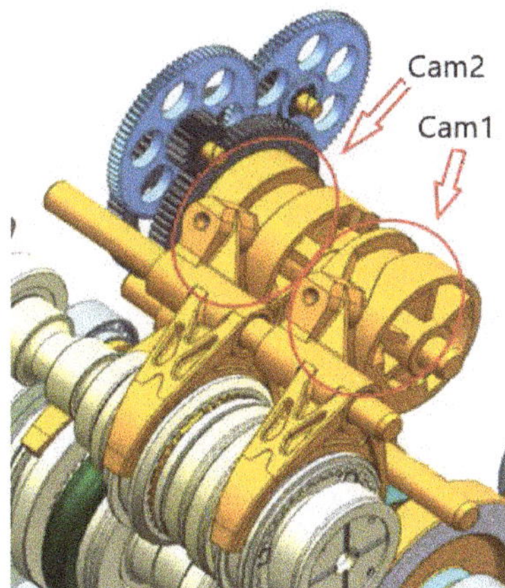

Figure 12. The shift actuator.

We designed the EMT transmission based on AMT. The addition of the motor power coupling shaft and the motor power switching synchronizer made EMT have a variety of different working modes (P2 and P3). Subsequently, the EMT assembly was parameterized, including the selection of power system components and the optimal speed ratio configuration of the transmission system. Finally, the vehicle's dynamics and economy are analyzed using the parameters of the project as the standard.

4.2. EMT and Vehicle Test

In order to verify the actual performance of EMT, we conducted a bench test in accordance with national standards.

As can be seen from Figure 13, we individually test the overall efficiency, lubrication, shift performance, and durability of the EMT. To observe the lubrication of the gearbox, we made a transparent front cover especially.

(a) Test bench

(b) Lubrication experiment

Figure 13. Test photograph.

As can be seen from Figure 14, we have selected a representative MPV for modification. We retained the engine of the original car, replaced the transmission with EMT, and added a three-electric system. This vehicle is an FF layout, so the EMT assembly is arranged in the front cabin. The PCU assembly and other accessories are installed in the rear cabin.

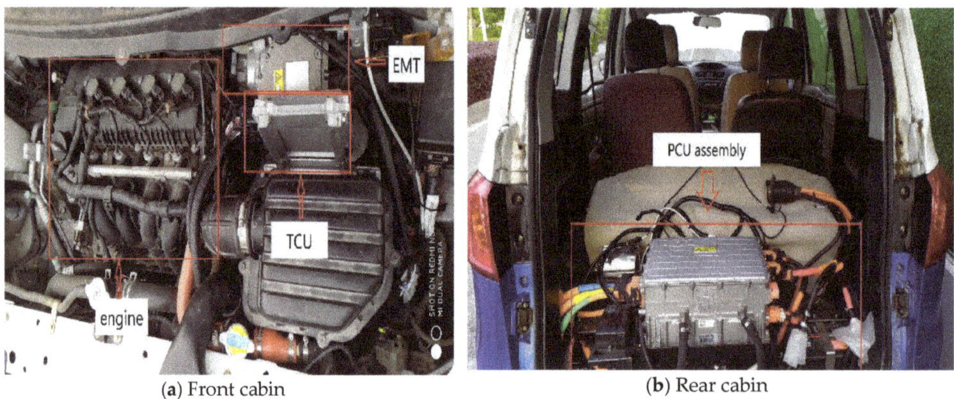

(a) Front cabin

(b) Rear cabin

Figure 14. Vehicle modification photograph.

As can been seen in Figure 15, we tested the power and economy of the prototype on the chassis dynamometer. For vehicle dynamics, we tested the vehicle's 100-km acceleration time, maximum speed, etc. For the economy of the vehicle, we mainly tested the New European Driving Cycle (NEDC) fuel consumption of the vehicle to calculate the equivalent fuel consumption of 100 km.

Figure 15. Test photograph.

The test results are shown in Table 3. The results show that the EMT can simultaneously meet the dynamic and economic requirements of PHEVs. The prototype development and the vehicle matching verify the effectiveness and practicability of the new configuration based on AMT.

Table 3. The test results on the chassis dynamometer.

Mode	Index	Test results
HEV mode	Maximum velocity	≥160 km/h
	Maximum slope	≥30%
	0~100 km/h acceleration time	≤12 s
	Fuel consumption per hundred kilometers	5.38 L/100 km
EV mode	Maximum velocity	≥130 km/h
	Maximum slope	≥30%
	0~100km/h acceleration time	≤17 s
	Pure electric driving range	≥50 km

Because we did not modify the engine, the comparison of vehicle dynamics is not meaningful. So, we only conducted economic tests on unmodified cars. Compared with the experimental results of unmodified vehicles, it is found that the vehicle's power and economy have been improved, and the fuel-saving rate of 100 km has been achieved at 35.18% ((8.30 − 5.38)/8.30 ≈ 0.3518) (Table 4). The prototype's dynamic and economical tests validate the design's effectiveness.

Table 4. The economical test results before modification.

Index	Test Results
Fuel consumption per hundred kilometers	8.30 L/100 km

5. Conclusions

Based on the AMT, a PHEV hybrid transmission for low-end MPVs is developed. The parameter design and prototype development are carried out, and the following conclusions are obtained:

1. The power demand and energy demand of PHEVs can be decoupled. In the distribution of engine and motor power, the engine just needs to meet the vehicle's constant driving power, and the backup power can be provided by the motor, which means we can use an engine with a smaller power rating. Compared with EV, PHEV can easily meet the needs of short-distance travel and long-distance travel with a small battery capacity.
2. The scheme is based on the AMT, with a single motor parallel design, using P3 as the basic working mode and setting the P2 mode at the same time. The energy of short-distance travel is

mainly provided by the motor, which can make full use of the battery, reducing fuel consumption. The energy of long-distance travel is mainly provided by the engine, which can reduce the need for battery capacity. So, the energy of daily driving is provided by the battery, while the backup energy for long-distance driving is provided by the engine.

3. Compared with the experimental results of unmodified vehicles, it is found that the vehicle's economy has been improved, and the fuel-saving rate of 100 km has been achieved at 35.18% ((8.30 − 5.38)/8.30 ≈ 0.3518). The prototype's dynamic and economical tests validate the design's effectiveness.

Author Contributions: Conceptualization, Z.Z. and S.R.; methodology, Y.Z. and Y.B.; validation, Y.Z. and Y.B.; formal analysis, Y.Z. and Y.B.; writing—original draft preparation, Y.B.; writing—review and editing, S.Z.; funding acquisition, Z.Z. All authors have read and agreed to the published version of the manuscript.

Funding: This research was funded by the National Key Research and Development Project of China (Grant No. 2017YFB0103200).

Conflicts of Interest: The authors declare no conflict of interest.

References

1. Liu, Z.; Guan, D.; Crawford-Brown, D.; Zhang, Q.; He, K.; Liu, J. A low-carbon road map for China. *Nature* **2013**, *500*, 143. [CrossRef] [PubMed]
2. Hao, H.; Ou, X.; Du, J.; Wang, H.; Ouyang, M. China's electric vehicle subsidy scheme: Rationale and impacts. *Energy Policy* **2014**, *73*, 722–732. [CrossRef]
3. Du, J.; Ouyang, M.; Chen, J. Prospects for Chinese electric vehicle technologies in 2016 e 2020: Ambition and rationality. *Energy* **2020**, *120*, 584–596. [CrossRef]
4. Wu, G.; Zhang, X.; Dong, Z. Powertrain architectures of electrified vehicles: Review, classification and comparison. *J. Frankl. Inst.* **2015**, *352*, 425–448.
5. Markel, T.; Simpson, A. Cost-Benefit Analysis of Plug-In Hybrid Electric Vehicle Technology. *World Electr. Veh. J.* **2007**, *1*, 294–301. [CrossRef]
6. Amjad, S.; Neelakrishnan, S.; Rudramoorthy, R. Review of design considerations and technological challenges for successful development and deployment of plug-in hybrid electric vehicles. *Renew. Sustain. Energy Rev.* **2010**, *14*, 1104–1110. [CrossRef]
7. Ehsani, M.; Gao, Y.; Miller, J.M. Hybrid Electric Vehicles: Architecture and Motor Drives. *Proc. IEEE* **2007**, *95*, 719–728. [CrossRef]
8. Chan, C.C.; Bouscayrol, A.; Chen, K. Electric, Hybrid, and Fuel-Cell Vehicles: Architectures and Modeling. *IEEE Trans. Veh. Technol.* **2010**, *59*, 589–598. [CrossRef]
9. Conlon, B.M.; Blohm, T.; Harpster, M.; Holmes, A.; Palardy, M.; Tarnowsky, S.; Zhou, L. The Next Generation "Voltec" Extended Range EV Propulsion System. *SAE Int. J. Alt. Power.* **2015**, *4*, 248–259. [CrossRef]
10. Rahman, K.M.; Jurkovic, S.; Stancu, C.; Morgante, J.; Savagian, P.J. Design and Performance of Electrical Propulsion System of Extended Range Electric Vehicle (EREV) Chevrolet Volt. *IEEE Trans. Ind. Appl.* **2015**, *51*, 2479–2488. [CrossRef]
11. Jelden, H.; Pelz, N.; Haußmann, H.; Kloft, M. The Plug-in Hybrid Drive of the VW Passat GTE. *MTZ Worldw.* **2015**, *76*, 16–23. [CrossRef]
12. Neusser, H.-J.; Jelden, H.; Bühring, K.; Philipp, K. The Powertrain of the Jetta Hybrid from Volkswagen. *MTZ Worldw.* **2013**, *74*, 4–11. [CrossRef]
13. Meisel, J. An Analytic Foundation for the Toyota Prius THS-II Powertrain with a Comparison to a Strong Parallel Hybrid-Electric Powertrain. In Proceedings of the SAE 2006 World Congress & Exhibition, Detroit, MI, USA, 3–6 April 2006.
14. Yang, Y.; Hu, X.; Pei, H.; Peng, Z. Comparison of power-split and parallel hybrid powertrain architectures with a single electric machine: Dynamic programming approach. *Appl. Energy* **2016**, *168*, 683–690. [CrossRef]
15. Kim, J.; Kim, N.; Hwang, S.; Hori, Y.; Kim, H. Motor control of input-split hybrid electric vehicles. *Int. J. Automot. Technol.* **2009**, *10*, 733. [CrossRef]
16. Zhang, X.; Li, S.E.; Peng, H.; Sun, J. Design of Multimode Power-Split Hybrid Vehicles—A Case Study on the Voltec Powertrain System. *IEEE Trans. Veh. Technol.* **2016**, *65*, 4790–4801. [CrossRef]

17. Zhang, X.; Li, C.; Kum, D.; Peng, H. Prius$^+$ and Volt$^-$: Configuration Analysis of Power-Split Hybrid Vehicles With a Single Planetary Gear. *IEEE Trans. Veh. Technol.* **2012**, *61*, 3544–3552. [CrossRef]
18. Cornils, H. Medium-Duty Plug-in Hybrid Electric Vehicle for Utility Fleets. *SAE Int. J. Commer. Veh.* **2010**, *3*, 90–100. [CrossRef]
19. Baseley, S.; Smaling, R.M.; Hu, H. *Advanced Hybrid Powertrains for Commercial Vehicles*; SAE International: Warrendale, PA, USA, 2012.

Article

Environmental and Economic Benefits of a Battery Electric Vehicle Powertrain with a Zinc–Air Range Extender in the Transition to Electric Vehicles

Manh-Kien Tran [1], Steven Sherman [1], Ehsan Samadani [2], Reid Vrolyk [1], Derek Wong [1], Mitchell Lowery [3] and Michael Fowler [1,*]

[1] Department of Chemical Engineering, University of Waterloo, Waterloo, ON N2L 3G1, Canada; kmtran@uwaterloo.ca (M.-K.T.); sbsherman@uwaterloo.ca (S.S.); rvrolyk@uwaterloo.ca (R.V.); d74wong@uwaterloo.ca (D.W.)

[2] Department of Mechanical and Mechatronics Engineering, University of Waterloo, Waterloo, ON N2L 3G1, Canada; sesamada@uwaterloo.ca

[3] Department of Economics, University of Waterloo, Waterloo, ON N2L 3G1, Canada; maclower@uwaterloo.ca

* Correspondence: mfowler@uwaterloo.ca; Tel.: +1-519-888-4567 (ext. 33415)

Received: 1 June 2020; Accepted: 26 June 2020; Published: 27 June 2020

Abstract: Emissions and pollution from the transportation sector due to the consumption of fossil fuels by conventional vehicles have been negatively affecting the global climate and public health. Electric vehicles (EVs) are a cleaner solution to reduce the emission and pollution caused by transportation. Lithium-ion (Li-ion) batteries are the main type of energy storage system used in EVs. The Li-ion battery pack must be considerably large to satisfy the requirement for the vehicle's range, which also increases the cost of the vehicle. However, considering that most people use their vehicles for short-distance travel during daily commutes, the large pack is expensive, inefficient and unnecessary. In a previous paper, we proposed a novel EV powertrain design that incorporated the use of a zinc–air (Zn–air) battery pack as a range-extender, so that a smaller Li-ion pack could be used to save costs. The design and performance aspects of the powertrain were analyzed. In this study, the environmental and economic benefits of the proposed dual-battery powertrain are investigated. The results from the new powertrain were compared with values from a standard EV powertrain with one large Li-ion pack and a conventional internal combustion engine vehicle (ICEV) powertrain. In addition, an air pollution model is developed to determine the total amount of pollution released by the transportation sector on Highway 401 in Ontario, Canada. The model was then used to determine the effects of mass passenger EV rollout on pollution reduction.

Keywords: electric vehicles; range extenders; powertrain design; zinc–air battery; lithium-ion battery; electric vehicle transition

---

1. Introduction

A recent report by the international panel on climate change (IPCC) highlighted anthropogenic climate change as a major threat to humanity and life on earth [1]. The IPCC determined that climate change, if not addressed, would cause an increase in mean temperature in many regions. Some regions would also see an increase in the frequency and intensity of droughts, while some would be affected by a rise in sea levels. The main cause of anthropogenic climate change is the ever-increasing amount of greenhouse gas (GHG) emissions. Therefore, GHG emissions must be drastically reduced in order to prevent the negative consequences of climate change [2]. Currently, the transportation sector accounts for a large portion of GHG (mainly CO_2) emissions. In Canada, Clean Energy Canada found that the transportation sector was responsible for 23% of Canada's GHG emissions as of 2016 [3]. The Canadian transportation sector is dominated by internal combustion engine vehicles (ICEVs). Aside from CO_2,

ICEVs also emit other hazardous gases including carbon monoxide (CO), sulfur dioxide (SO$_2$), nitrous oxides (NO$_x$) and volatile organic compounds (VOCs). These air pollutants were found to increase one's risk of dementia, diabetes, asthma and premature death.

A viable alternative to ICEVs is electric vehicles (EVs). EVs use a motor, often powered by a lithium-ion (Li-ion) battery pack, to propel the vehicle [4]. EVs have lower environmental and health impacts compared to ICEVs because they do not emit exhaust, thus, they do not directly release any emissions or pollutants [5]. EVs can help prevent climate change, thus protecting the environment and public health [6]. Although electricity production is usually much cleaner than oil and gas consumption, the degree of which depends on the method that the electricity was produced. Hence, EVs are more favorable in regions that use nuclear and renewable sources to generate electricity [7]. For example, a study done in Texas showed that EVs powered by coal increased air pollution by 350%, while EVs powered by renewable energy or natural gas decreased air pollution by 50% and 70%, respectively [8]. In Ontario, Canada, there is a surplus of electricity produced from different energy sources, and therefore, the market penetration of EVs would be beneficial [9].

However, there are several issues that prevent the widespread adoption of EVs, including the high cost of the vehicles, the lack of charging infrastructure, and the limited range associated with EVs [10,11]. A solution to the limited range and high cost of EVs is to utilize a range extender, which is an auxiliary energy storage system (ESS) that provides the vehicle with additional energy to complement the primary battery in propelling the vehicle [12]. There are several different types of range extenders, including ICE, fuel cell, free-piston linear generator (FPLG) and micro gas turbine (MGT) [13]. An ICE range extender converts gasoline into electricity using a fuel converter. The ICE unit is comprised of an electronically controlled combustion engine connected to a generator, which supplies the required electricity to the motor [14]. A fuel cell extended-range EV has a tank of hydrogen fuel, which gets converted into usable electricity by the fuel cell to support the battery in powering the vehicle [15]. The fuel cell range extender can increase the range and performance of the vehicle while still drastically reducing emissions. An FPLG range extender uses a combustion and linear generator to convert chemical energy into electrical energy [16]. An MGT range extender draws in clean air, compresses it, and passes it through a turbine at extremely high revolutions to generate energy [17]. Despite their advantages, all of the aforementioned range-extending solutions have some drawbacks. ICE is not environmentally friendly, fuel cell and FPLG are not cost-effective, and MGT is not efficient. In recent years, the concept of using a zinc–air (Zn–air) battery pack as a range extender for EVs has been investigated [18].

Metal–air batteries have gained some attention for their potential use in EVs because of their high energy density and low cost [19]. The feasible cell chemistries include lithium–air (Li–air), aluminum–air (Al–air) and Zn–air. Li–air batteries have the highest energy density, but there are many potential deficiencies and safety concerns hindering their commercialization [20]. Al–air batteries have an aqueous electrolyte, making them safer than Li–air batteries. The drawback of Al–air batteries is that they are not easily recharged and need to be replaced once they are drained [19]. Zn–air batteries have a lower energy density than Li–air and Al–air batteries. However, Zn–air batteries have attracted significant commercial and research interest for several reasons. The raw materials required to manufacture Zn–air batteries are abundant, which lowers the cost of the pack significantly [21]. Zn–air batteries also have a greater energy density and specific energy than Li-ion batteries. They have good resistance to aging degradation but are susceptible to cycling degradation [19]. This means that they can last a long time if left unused but cannot be cycled as many times as Li-ion batteries.

Recent breakthroughs in Li-ion battery research have allowed for increased range of EVs. However, the current solution to the range problem involves a larger Li-ion battery pack, which is expensive and redundant for daily use of EVs, as most people only use their vehicles to commute short distances between work and home. In our previous work, we proposed a novel powertrain design, incorporating a Li-ion battery pack as the primary ESS and a Zn–air battery pack as the range extender, with the necessary electronic controls to facilitate their coordination and optimization [18]. The Li-ion battery pack would be used for daily commutes while the Zn–air battery pack would serve as a range extender for longer trips. By utilizing these two ESS units in one powertrain, a smaller Li-ion pack can be used, which would lower the cost of the vehicle. In addition, the Zn–air pack is only used on longer trips, which reduces the cycling of the pack and eliminates its exposure to cycling degradation. Therefore, this design allows each battery to contribute its strengths while minimizing its weaknesses.

In this study, the environmental and economic implications of the proposed novel dual-battery-pack powertrain are examined, since only the design and performance aspects of the powertrain were investigated in our previous study. The analysis should provide a more complete picture of the feasibility and benefits of the proposed powertrain design. The results from the dual-pack powertrain are compared with values from a standard single-pack Li-ion battery EV and a conventional ICEV. In addition, an air pollution model is used to determine the amount of certain pollutants released annually by the transportation sector on Highway 401 in Ontario, Canada. The benefits of the pollution reduction from mass EV rollout are then analyzed for this case study. The rest of this study is organized as follows. Section 2 provides the analysis of the environmental and economic benefits of the proposed powertrain design, while Section 3 presents the benefit analysis of pollution reduction from the potential market penetration of EVs. Finally, Section 4 outlines some concluding remarks.

2. Environmental and Economic Benefits of the Novel Powertrain Design

2.1. Background of the Dual-Battery Powertrain Design

This section outlines the summary of the specifications of the novel dual-battery powertrain design, which was configured to optimize its range, cost and longevity in our previous study [18]. The powertrain design is shown in Figure 1, where the range extender is the Zn–air battery pack and the primary ESS is the Li-ion battery pack. The two ESS were connected in series instead of in parallel since the Zn–air pack may not have the required power density to satisfy the motor power demand during acceleration. Therefore, it can only directly provide electricity to the Li-ion pack which then effectively powers the motor. Figure 2 shows the energy management strategy of the powertrain. The strategy is implemented to ensure the range extender is only used when necessary since the Zn–air battery pack has poor cycling life and should not be used often.

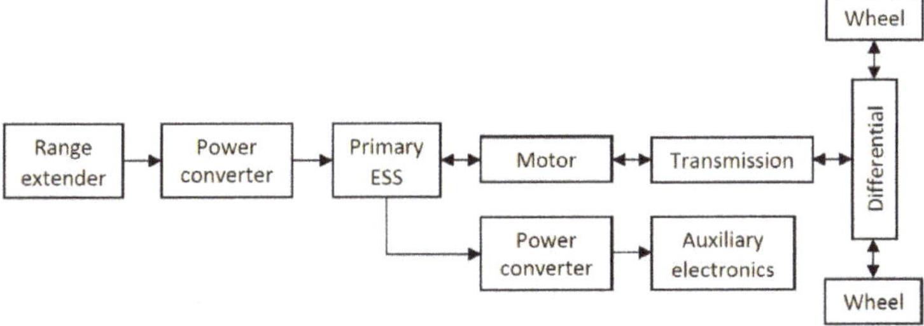

Figure 1. Dual-battery powertrain design.

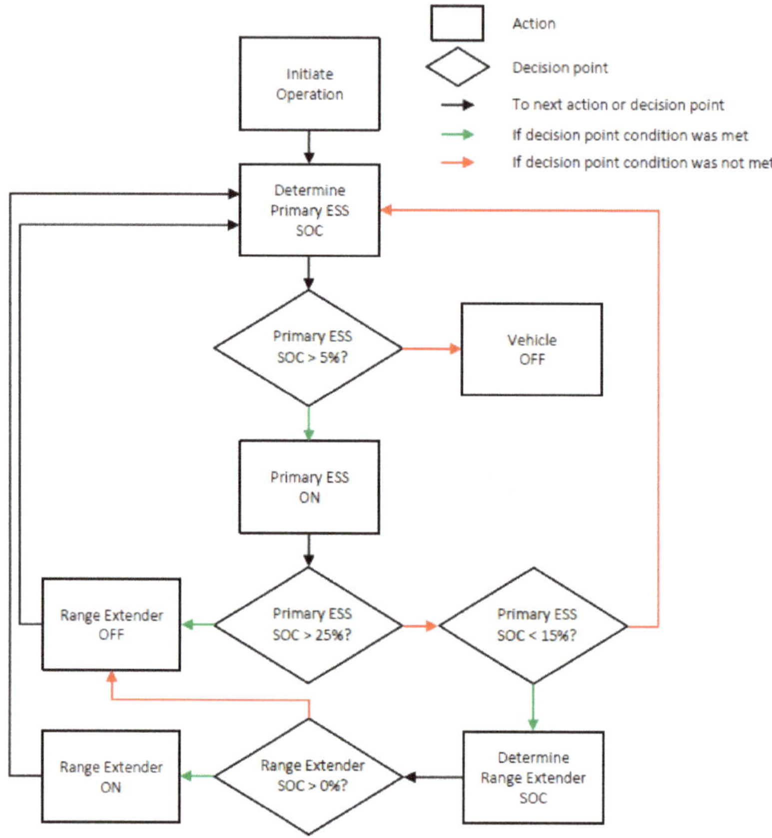

Figure 2. Flow chart of the powertrain energy management strategy.

Several designs of the Zn–air and Li-ion battery packs in the proposed powertrain (2BEV) were examined and the optimal designs are shown in Table 1. A similar EV with one large Li-ion battery pack (1BEV) was also modeled and compared with the proposed design. The modeling and simulation of the powertrain were performed in MATLAB/Simulink (MathWorks, Natick, MA, USA). The software models the performance of individual vehicle components in response to the demands placed on the vehicle by the driver. Because Zn–air batteries are not yet commercially available, the modeling was created based on previously published research with some adjustments to allow for achievable improvements. The Li-ion packs and other vehicle components were modeled based on data from commercial components. The specifications of the Li-ion cells were taken from commercial A123 pouch cells while the Zn–air cells were button cells. The total cell weights in Table 1 were multiplied by a packaging factor, which accounted for the weight of pack packaging components, to determine the pack weight.

Table 2 shows the results of the MATLAB/Simulink simulation runs for the proposed powertrain and the traditional EV powertrain. In the simulation, the 2BEV performance compared favorably to the 1BEV. The traveling distances under various driving conditions were similar, but the cost of the dual packs was 30% lower than that of the large Li-ion pack. The fuel economy of the 1BEV was found to be slightly better, but the overall vehicle performance was not significantly compromised as seen by the similar acceleration time in both powertrains. In addition, the simulation also demonstrated that the Zn–air pack can last up to 13 years if used optimally.

Table 1. Summary of specifications of different battery packs in the standard one large Li-ion battery pack (1BEV) and proposed powertrain (2BEV).

Parameter	Units	Li-ion Pack (Small) in 2BEV	Zn–air Pack in 2BEV	Li-ion Pack (Large) in 1BEV
Cells per module		15	4	15
Modules in series		7	75	7
Arrays in parallel		2	22	10
Weight per cell	g	496	55.6	496
Total cell weight	kg	104	367	521
Packaging factor		1.25	1.25	1.25
Pack weight	kg	130	459	651
Nominal pack voltage	V	347	360	347
Pack capacity	Ah	39	147	196
Pack energy	kWh	13.6	52.9	67.9
Maximum SOC		1	1	1
Minimum SOC		0.05	0	0.05
Pack cost	CAD *	$3610	$9150	$18,070

* currency exchange (2019): 1 CAD = 0.7536 USD.

Table 2. Performance and cost comparison between 1BEV and 2BEV.

Parameter	Units	2BEV	1BEV
Vehicle weight	kg	2005	2014
City range	km	383	375
Highway range	km	432	437
Combined average range	km	405	403
City fuel economy	kWh/100 km	21.7	20.2
Highway fuel economy	kWh/100 km	20.1	17.4
Combined average fuel economy	kWh/100 km	21.4	18.8
Battery pack cost	CAD	$12,760	$18,070
Zn–air pack maximum lifetime	years	13.2	N/A
Time from 0 to 100 km/h	s	15.06	15.23

2.2. Environmental and Economic Comparison of Different Powertrain Designs

We performed further analyses regarding the environmental and economic benefits of the novel powertrain design to complement the findings in our previous study shown in the preceding section. In this section, a commercial ICEV, the 2019 Honda Civic, was used to compare the economic and environmental impact of the 2BEV and the 1BEV in Ontario, Canada. The ICEV data were taken from its real-world cost and performance. The EVs data were estimated based on some assumptions as well as some real-life data. The comparison results between the powertrains are shown in Table 3.

Table 3. Economic and environmental comparison between the 2BEV, 1BEV and internal combustion engine vehicle (ICEV).

Cost	Units	2BEV	1BEV	ICEV
Estimated MSRP	CAD	$26,321	$31,625	$17,890
Harmonized sales tax	CAD	$3422	$4111	$2326
Fuel cost	CAD/yr	$416	$373	$1686 *
Maintenance cost	CAD/yr	$584	$584	$720
Combined cost	CAD	$36,680	$42,380	$36,930
CO_2 emissions	kg/yr	75	56	3354

* based on 7.1 L/100 km.

The manufacturer's suggested retail price (MSRP) of the EVs was determined by removing the cost of certain conventional powertrain components and adding the cost of the corresponding EV powertrain components [22]. The Civic maintenance cost was based on a Canadian Automobile Association (CAA) estimate [23]. The EVs maintenance costs were assumed to be 19% lower [24]. Fuel cost was calculated from a price of $1.20/L for gasoline and $0.10 kW/h for electricity in Ontario. The combined cost for each vehicle was calculated using Equation (1). The combined cost is the total amount the vehicle would cost over a ten-year lifespan with an interest rate of 7.25% when converting the maintenance and fuel costs to the 2019-equivalent dollar amount [23]. The electricity generated in Ontario has an emission intensity of 0.0296 kg CO_2/kWh [25]. This value was used to determine the total annual CO_2 emissions from the EVs.

$$CC = MSRP + HST + (FC + MC)\sum_{n=1}^{10} 1.0725^{-n} \tag{1}$$

where CC is the combined cost, MSRP is the manufacturer's suggested retail price, HST is the harmonized sales tax (13%), FC is the fuel cost and MC is the maintenance cost.

There are several noteworthy findings from this analysis. The higher fuel cost and higher emissions of the 2BEV compared to the 1BEV can be linked to the fact that Zn–air batteries have lower efficiency than Li-ion batteries. However, despite the higher fuel cost, the combined cost of the 2BEV is lower than the 1BEV, which can be attributed to the low capital cost of the Zn–air battery pack. Overall, the 2BEV is observed to be the most cost-effective option of the three. In addition, even though the CO_2 emissions value from the 2BEV is slightly larger than the 1BEV, both emissions values of the EVs are significantly smaller than the ICEV. This is because Ontario only has a small amount of its electricity generated from natural gas while the majority is produced through cleaner renewable energy sources or nuclear power. Overall, the proposed 2BEV design realizes significant environmental and economic benefits, as it outperforms the ICEV in terms of emissions and outperforms the 1BEV in terms of costs. It should be noted that the environmental impact of battery manufacturing is not discussed in this study, and it is possible that the impacts of producing batteries may outweigh the benefits of using EVs.

3. The Benefits of Pollution Reduction from EV Rollout in Canada

In the previous section, the environmental and economic benefits of the 2BEV were discussed in comparison to the 1BEV and ICEV. It was observed that the design of the proposed powertrain can be comparable to other commercial designs. In order to further understand the feasibility and benefits of this technology, the reduction of air pollution from EV market penetration should also be investigated. Emissions of certain pollutants can be reduced by adopting EVs [26]. This section studies the economic benefits of pollution reduction from EV rollout in the case of Toronto, Ontario, Canada—specifically on Highway 401, one of the major highways in the province.

3.1. Air Pollution in Canada

The reduction of air pollution is a major incentive for the rollout of EVs. Anthropogenic pollution in Canada is responsible for approximately 144,000 deaths per year as stated by Health Canada in 2017 [27]. Fine particulate matter ($PM_{2.5}$), nitrogen dioxide (NO_2) and ozone (O_3) are some of the main chemicals that contribute to air pollution. The economic cost of $PM_{2.5}$ and O_3 is 42 billion dollars annually, with incidents caused by these air pollutants accounting for $39 billion. Health care and missed work also cost $2 billion and $861 million, respectively [28]. Through a study that investigated the impact of transportation specifically, it was found that $PM_{2.5}$, sulfur dioxide (SO_2), nitrous oxides (NO_x) and volatile organic compounds (VOCs) emissions cost about $7.9 billion [29]. Pollution from light passenger vehicles excluding road dust cost $2 billion, with 52% of the emissions being attributed to NO_x. Another study found that in Toronto, 1300 premature deaths and 3550 hospitalizations annually were caused by acute exposure and chronic exposure to air pollution, costing the city $878 million [30]. Of these, air pollution produced from traffic accounted for 280 deaths and

1090 hospitalizations. Toronto traffic was also stated to be the cause of a significant number of acute respiratory disease cases.

Major air pollutants from ICEVs include NO_x, SO_x, VOCs, carbon monoxide (CO) and particulate matter ($PM_{2.5}$ and PM_{10}). NO_x and VOCs react in still air and sunlight to form the pollutant O_3 [31]. In traffic, EVs do not emit any pollutants except for the inevitable road dust (particulate matter) from braking, as do all on-road vehicles. Therefore, in locations where traffic is a significant contributor to air pollution, EV adoption would play an important role in improving the air quality within the region.

3.2. Air Pollution Modeling

A simple method to quantitatively determine the benefits EVs have on reducing air pollution is to estimate the amount of pollution being emitted per passenger vehicle. The associated costs per ton of pollutants are then calculated and multiplied by the number of passenger vehicles in the specified region. However, this simple approach does not account for population characteristics, traffic patterns or pollutant distribution. A more intensive method would be to model an entire traffic network. This method would create a more accurate economic assessment based on traffic patterns, airflow patterns and population characteristics. Despite its accuracy, this model would be arduous and require an unfeasibly large amount of data. Therefore, we developed a model with moderate complexity that contains traffic data, but not population nor weather patterns. The model only considers Highway 401 in Ontario and does not incorporate any other Toronto traffic. The model has a reasonable scope, utilizes readily available data, and can deliver a good estimation of the amount of certain air pollutants.

3.2.1. Modeling Software and Model Overview

The model was created through the modeling software transportation air quality system (TRAQS). TRAQS combines motor emissions vehicle simulator (MOVES) and American Meteorological Society/Environment Protection Agency Regulatory Model (AERMOD) into one user-friendly software that simplifies large simulations covering varying conditions over long periods of time. MOVES determines the emission rate based on its modeled traffic patterns, while AERMOD processes the emission rate through air dispersion modeling to determine the pollutant distribution. The schematic of the structure and functions of TRAQS can be seen in Figure 3.

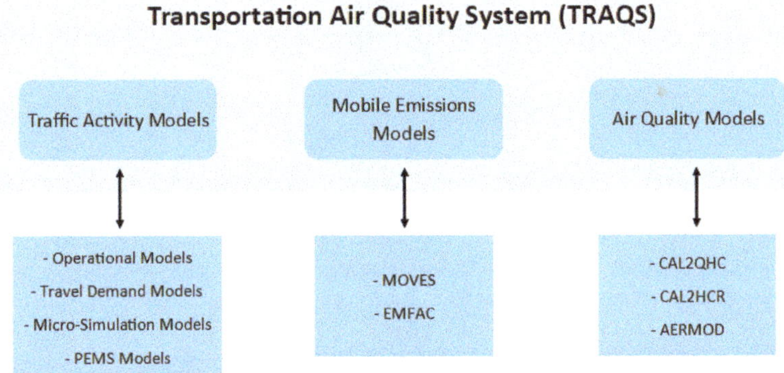

Figure 3. Schematic of transportation air quality system (TRAQS) software structure and functions.

The TRAQS model represents the pollution source, Highway 401, as a sequence of connecting segments in between major on- and off-ramps. This representation is valid because traffic flow along these segments is mostly constant as vehicles are unable to enter or exit the highway. In each segment, the average speed and traffic volume were recorded on a seasonal basis and over four intervals during the day. The day was divided into morning peak (6 A.M. to 9 A.M.), midday (9 A.M. to 4 P.M.), afternoon peak (4 P.M. to 7 P.M.) and overnight (7 P.M. to 6 A.M.). Figure 4 shows the defined route in our TRAQS model. The green line represents the Highway 401 from Renforth Drive (West) to Kingston Road (East). The non-shaded areas in figure represent the 500-m distance from the highway to its sides. The average pollution released was calculated for this specific area near the highway.

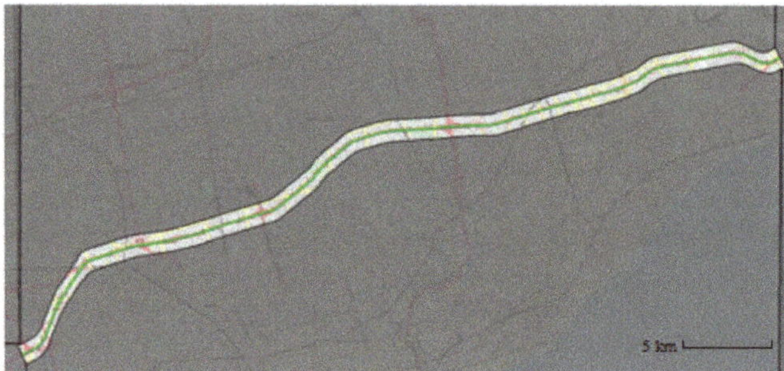

Figure 4. TRAQS study area—Highway 401 from Kingston Road to Renforth Drive.

Once the area and traffic flow patterns were defined in the model, the next set of parameters was established. These parameters included the vehicle–fuel combinations (e.g., diesel-powered truck, gasoline-powered passenger car, etc.) and pollutants to model. Several additional traffic characteristics were also specified for each season and portion of the day. These additional characteristics include the fraction of vehicles coinciding with each vehicle–fuel combination (e.g., 8% diesel-powered trucks, 59% gasoline-powered passenger cars, etc.), fuel formulation and supply, vehicle age distribution and climatic conditions. It should be noted that the climatic conditions were assumed to be the average temperature and humidity by the time of day and year. Once all the components in the model were specified, the model generated an emissions report containing an estimate of emissions in tons per year.

3.2.2. Data Overview

TRAQS provides some default data of vehicle age distribution, fuel supply and formulation and climatic conditions, which were determined by the United States Environmental Protection Agency (EPA). The traffic patterns and distribution were collected from the Ontario Ministry of Transportation (MTO). MTO publishes the annual average daily traffic (AADT) volumes for all of Ontario's major highways, including Highway 401 [32]. These data are categorized by highway segment, but not by season or hour. Equation (2) provided by the MTO must be used to determine the seasonal average traffic volumes.

$$AADT = f_i \cdot ADT_i \tag{2}$$

The average daily traffic for a specified period i of the year (ADT_i) is derived from the AADT and a conversion factor, f_i, provided by the MTO. The conversion factor corresponds to a certain period in the year and a given section on Highway 401. An hourly traffic distribution was created based on sample hourly data from the MTO for the first week of March in 2016. The generated distribution had slight variations over some sections (standard deviation < 6%), but this was considered negligible. Thus, the same hourly distribution was used for all sections of the highway. From the seasonal ADT and the hourly distribution, the hourly traffic volumes were estimated for morning peak, midday, afternoon peak and overnight and for all seasons and sections of the highway. In addition to the hourly traffic volumes, the hourly average speed data for the highway segments was also provided by the MTO.

Data from the MTO and the Bureau of Transportation Services (BTS) was also used to estimate the number of each type of vehicle–fuel combination. According to the BTS, 56% of the vehicle fleet are gasoline-powered cars and 39% are light diesel trucks [33]. The remaining 5% of the fleet consist of medium and heavy diesel trucks (3%) and diesel cars and gasoline trucks (2%). For this work, diesel cars and gasoline trucks are considered negligible. The MTO provided approximations for the percentage of trucks on Highway 401, which was split evenly between short-haul and long-haul single-unit diesel trucks. The remaining percentage was divided into gasoline-powered cars and light diesel trucks corresponding to the ratio specified by the BTS.

3.3. Methodology of Economic Benefits Estimation

The first step to estimate the economic costs due to air pollution surrounding Highway 401 was to generate a concentration profile of the highway using the air dispersion model and the emissions output from TRAQS. This concentration profile and the population characteristics within its horizon were then used as the basis for determining the economic costs to Canadians due to pollution. Using this method, Transport Canada estimated the economic and social costs of air pollution per unit of pollutant considering three major aspects, including impacts on human health, impacts on agriculture crops and visibility impacts [34]. Since the report referenced data and prices from the year 2000, an inflation rate was calculated to convert the costs to 2019-dollar values, as shown in Equation (3). The Canadian consumer price index (CPI) according to Statistics Canada was used, with CPI 2000 being 94.5 and CPI 2019 being 136.0 [35]. The resulting inflation rate of 43.92% was then applied to the 2000-dollar values to give their corresponding values in terms of a 2019 dollar.

$$Inflaction\ Rate = \frac{CPI2019 - CPI2000}{CPI2000} \times 100 \tag{3}$$

The overall cost from pollution on Highway 401 was estimated using the annual emissions calculated using TRAQS and the cost per unit pollutant shown in Table 4. It should be noted that the values from Transport Canada are estimations for the entire province of Ontario. Therefore, the cost calculated using this method would be a large underestimate compared to the true cost, since Highway 401 is near regions with high population density, which increases the economic and health impacts of air pollution in these communities.

Table 4. Economic cost per unit pollutant in the years 2000 and 2019.

Pollutant	Cost (2000-dollars)	Cost (2019-dollars)
Units	CAD/ton	CAD/ton
NO_x	$5940	$8548
$PM_{2.5}$	$29,100	$41,879
SO_2	$6520	$9383
VOCs	$877	$1262

By modeling the traffic and pollution on Highway 401 in TRAQS, the potential air pollution reduction and benefits to Canada's economy of EVs could be determined. The annual emissions of the pollutants were obtained using TRAQS. The economic costs of pollution were estimated using reported data from Transport Canada. The benefits of EV adoption were then calculated for different levels of market penetration. The schematic of the overall methodology is shown in Figure 5.

3.4. Results of Economic Benefits Estimation

TRAQS was used to model air pollution on Highway 401 using average traffic flow and speed data from the MTO, as well as vehicle distribution information from the BTS and the MTO. The annual emissions of relevant pollutants from the highway and their social and economic costs are shown in Table 5.

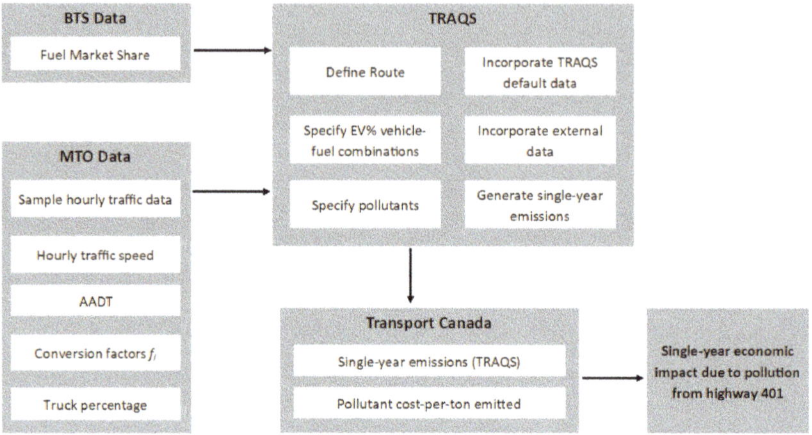

Figure 5. Air pollution cost estimation methodology.

Table 5. Baseline (0% electric vehicles (EV) share) emissions and costs of pollution from Highway 401.

Pollutant	Tons/year	CAD/ton	CAD/year
NO_x	2366	$8548	$20,224,568
$PM_{2.5}$	198	$41,879	$8292,042
SO_2	9.4	$9383	$84,447
VOCs	959	$1262	$1210,258
Total			**$29,811,315**

The TRAQS model was then adjusted for different percentages of the passenger gasoline vehicles being replaced with EVs (20%, 40%, 60%, 80% and 100%). The procedure to obtain emissions and costs for the different EV market penetration scenarios was the same as for the baseline (0%) scenario and is described in Sections 3.2 and 3.3. Figures 6 and 7 present the annual emissions and economic costs at different levels of EV market penetration. The cost values used in these figures are shown in Appendix A.

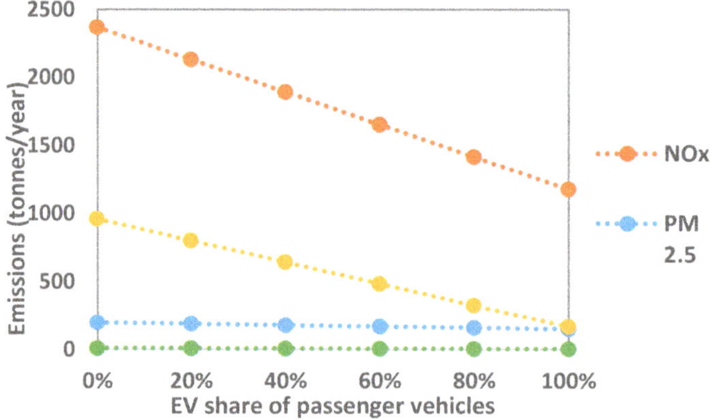

Figure 6. Annual emissions from Highway 401 at different EV penetration levels.

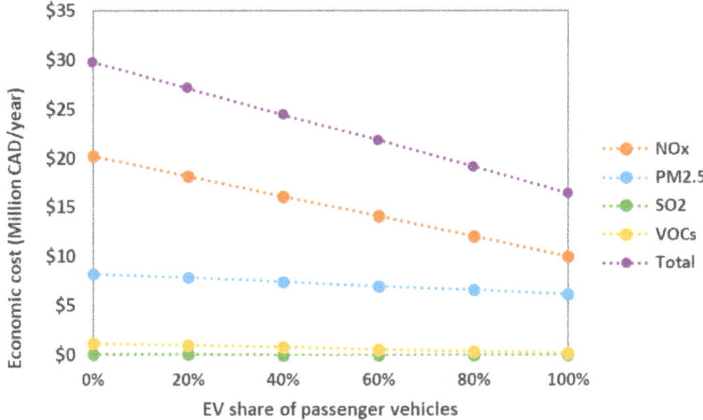

Figure 7. Annual cost of emissions from Highway 401 at different EV penetration levels.

At 100% EV penetration for passenger vehicles, the total cost of emissions decreases by 44.8%, which indicates that trucks account for a significant portion of the costs. In terms of cost and volume, NO_x is the largest contributor, which agrees with the findings from a study done by Sawyer et al. [29]. On the other hand, SO_2 emissions have minimal impact on the total cost, which was also shown by Li et al. [36]. VOCs are significant by mass but have a low overall economic cost. $PM_{2.5}$ emissions have the second-largest cost, but the cost only decreases slightly with increasing EV adoption levels. For every additional 20% EV penetration, $PM_{2.5}$ emissions decrease by only 5%, whereas NO_x emissions decrease by 10%, and SO_2 and VOC emissions decrease by 16%. According to the MTO, trucks account for only 8–10% of traffic volume on Highway 401, but our results show that they contribute a significant amount to air pollution. Therefore, despite the 100% EV market penetration of passenger vehicles, the total economic cost from pollution on Highway 401 was only reduced by 44.8%, while the remaining cost was due to air pollution caused by trucks. Batteries are generally not a feasible option for electric trucks; hence, research works on fuel cells and hydrogen economy have gained attention recently. The combination of battery electric passenger vehicles and fuel cell electric trucks would substantially reduce the economic and social costs of air pollution caused by the transportation sector in Canada.

4. Conclusions

In this study, the environmental and economic benefits of a novel powertrain design for EVs were investigated. The powertrain uses a Li-ion battery pack as its primary ESS and a Zn–air battery pack as a range extender, with an optimal energy management strategy. The novel EV powertrain was compared with a standard EV powertrain with one large Li-ion pack and a conventional ICEV powertrain, in terms of costs and emissions. An air pollution model was also developed to study the benefits of EV adoption in Ontario, Canada. The model used traffic and vehicle fleet data on Highway 401 in Toronto to estimate the total amount of pollution caused by the transportation sector and its corresponding economic and social costs. It was then adjusted to quantitatively determine the effects of varying levels of EV share in the passenger vehicles fleet.

Some of the key discoveries from this work are:

- The total cost of the 1BEV—including the MSRP, maintenance and fuel costs—was 15% higher than the ICEV, while the total cost of the 2BEV was comparable to the ICEV. In addition, both the 1BEV and 2BEV showed significant greenhouse gas emissions reduction compared to the ICEV, as electricity is generated cleanly in Ontario.

- It was found that 29.8 million CAD per year in economic and social costs are attributed to the air pollution associated with the traffic on Highway 401. Increasing the number of EVs in the fleet showed promising results to lower this cost. With potentially 100% of the passenger vehicles being EVs, the costs associated with air pollution decreased by 44.8%—or down to 16.5 million CAD per year.

- NO_x is the pollutant with the highest associated cost, accounting for 20.2 million CAD per year. However, it was shown that this cost could be drastically reduced when the EV market share increases, down to 10 million CAD per year at 100% EV market penetration. At 8.3 million CAD per year, $PM_{2.5}$ is the second most costly pollutant. Unlike NO_x, the cost of $PM_{2.5}$ was not reduced significantly by increasing the number of EVs in the fleet, only down to 6.2 million CAD per year at 100% EV market penetration.

Author Contributions: Conceptualization, S.S. and M.F.; methodology, S.S., M.-K.T., E.S. and M.L.; formal analysis, S.S. and M.-K.T.; writing—original draft preparation, S.S. and M.-K.T.; writing—review and editing, R.V., D.W. and M.L.; visualization, R.V., D.W. and M.L.; supervision, M.F. All authors have read and agreed to the published version of the manuscript.

Funding: Canada Research Chair Tier I - Zero Emission Vehicles and Hydrogen Energy Systems Grant number: 950-232215; and The Natural Sciences and Engineering Research Council of Canada (NSERC), Discovery Grants Program, RGPIN-2020-04149.

Acknowledgments: This work was supported by equipment and manpower from the Department of Chemical Engineering at the University of Waterloo. Special thanks to Danielle Skeba for the contribution in editing the paper.

Conflicts of Interest: The authors declare no conflict of interest.

Abbreviations

1BEV	powertrain with 1 battery pack
2BEV	powertrain with 2 battery packs
AADT	annual average daily traffic
AERMOD	AMS/EPA regulatory model
BTS	Bureau of Transportation Services
CAA	Canadian Automobile Association
CPI	consumer price index
EPA	Environmental Protection Agency
ESS	energy storage system
EV	electric vehicle
FPLG	free-piston linear generator
GHG	greenhouse gas
ICE	internal combustion engine
ICEV	internal combustion engine vehicle
IPCC	international panel on climate change
Li-ion	lithium-ion
MGT	micro gas turbine
MOVES	motor emissions vehicle simulator
MSRP	manufacturer's suggested retail price
MTO	Ontario Ministry of Transportation
PM	particulate matter
TRAQS	transportation air quality system
VOC	volatile organic compound
Zn–air	zinc–air

Appendix A

Table A1. Data for Figure 6 (tons/year).

EV%	NOx	PM2.5	SO2	VOCs
0%	2366	198	9	959
20%	2129	188	8	800
40%	1891	178	6	641
60%	1652	168	5	482
80%	1414	158	3	323
100%	1175	148	2	165

Table A2. Data for Figure 7 (CAD/year).

EV%	NO_x	$PM_{2.5}$	SO_2	VOCs	Total
0%	$20,224,568	$8,292,042	$84,447	$1,210,258	$29,811,315
20%	$18,198,692	$7,873,252	$75,064	$1,009,600	$27,156,608
40%	$16,164,268	$7,454,462	$56,298	$808,942	$24,483,970
60%	$14,121,296	$7,035,672	$46,915	$608,284	$21,812,167
80%	$12,086,872	$6,616,882	$28,149	$407,626	$19,139,529
100%	$10,043,900	$6,198,092	$18,766	$208,230	$16,468,988

References

1. IPCC. Global Warming of 1.5 °C. 2018. Available online: https://www.ipcc.ch/sr15/ (accessed on 25 July 2019).
2. IPCC. Climate Change 2014: Synthesis Report. 2014. Available online: https://www.ipcc.ch/site/assets/uploads/2018/05/SYR_AR5_FINAL_full_wcover.pdf (accessed on 25 July 2019).
3. Clean Energy Canada, David Suzuki Foundation, Environmental Defence, Équiterre and the Pembina Institute. Reducing GHG Emissions in Canada's Transportation Sector. 2016. Available online: https://www.equiterre.org/sites/fichiers/fmm_transportation_recs.pdf (accessed on 30 July 2019).
4. Tran, M.-K.; Fowler, M. Sensor Fault Detection and Isolation for Degrading Lithium-Ion Batteries in Electric Vehicles Using Parameter Estimation with Recursive Least Squares. *Batteries* **2020**, *6*, 1. [CrossRef]
5. Tran, M.-K.; Fowler, M. A Review of Lithium-Ion Battery Fault Diagnostic Algorithms: Current Progress and Future Challenges. *Algorithms* **2020**, *13*, 62. [CrossRef]
6. Buekers, J.; Van Holderbeke, M.; Bierkens, J.; Panis, L.I. Health and environmental benefits related to electric vehicle introduction in EU countries. *Transp. Res. Part D Transp. Environ.* **2014**, *33*, 26–38. [CrossRef]
7. Nealer, R.; Reichmuth, D.; Anair, D. Cleaner Cars from Cradle to Grave: How Electric Cars Beat Gasoline Cars on Lifetime Global Warming Emissions. 2015. Available online: https://www.ucsusa.org/sites/default/files/attach/2015/11/Cleaner-Cars-from-Cradleto-Grave-full-report.pdf (accessed on 26 July 2019).
8. Tessum, C.W.; Hill, J.D.; Marshall, J.D. Life cycle air quality impacts of conventional and alternative light-duty transportation in the United States. *Proc. Natl. Acad. Sci. USA* **2014**, *111*, 18490–18495. [CrossRef] [PubMed]
9. Hajebrahimi, A.; Kamwa, I.; Huneault, M. A novel approach for plug-in electric vehicle planning and electricity load management in presence of a clean disruptive technology. *Energy* **2018**, *158*, 975–985. [CrossRef]
10. Haustein, S.; Jensen, A.F. Factors of electric vehicle adoption: A comparison of conventional and electric car users based on an extended theory of planned behavior. *Int. J. Sustain. Transp.* **2018**, *12*, 484–496. [CrossRef]
11. Berkeley, N.; Jarvis, D.; Jones, A. Analysing the take up of battery electric vehicles: An investigation of barriers amongst drivers in the UK. *Transp. Res. Part D Transp. Environ.* **2018**, *63*, 466–481. [CrossRef]
12. Husain, I. *Electric and Hybrid Vehicles Design Fundamentals*, 2nd ed.; CRC Press Taylor & Francis Group: Boca Raton, FL, USA, 2011.
13. Bassett, M.D.; Hall, J.; Cains, T.; Taylor, G.; Warth, M.; Vogler, C. The development of a range extended electric vehicle demonstrator. *SAE Technical Paper* **2013**. [CrossRef]

14. Martin, N. Application of the internal combustion engine as a range-extender for electric vehicles. *Combust. Engines* **2013**, *154*, 781–786.

15. Jensen, H.C.; Schaltz, E.; Koustrup, P.; Andrasen, S.; Kær, S. Evaluation of Fuel-Cell Range Extender Impact on Hybrid Electrical Vehicle Performance. *IEEE Trans. Veh. Technol.* **2013**, *62*, 50–60. [CrossRef]

16. Virsik, R.; Heron, A. Free piston linear generator in comparison to other range-extender technologies. *World Electr. Veh. J.* **2013**, *6*, 426–432. [CrossRef]

17. Shah, R.M.B.R.A.; McGordon, A.; Amor-Segan, M. Micro Gas Turbine range extender—Validation techniques for automotive applications. In Proceedings of the IET Hybrid and Electric Vehicles Conference 2013, London, UK; 6–7 November 2013.

18. Sherman, S.B.; Cano, Z.P.; Fowler, M.; Chen, Z. Range-extending Zinc-air battery for electric vehicle. *AIMS Energy* **2018**, *6*, 121–145. [CrossRef]

19. Fu, J.; Cano, Z.P.; Park, M.G.; Yu, A.; Fowler, M.; Chen, Z. Electrically Rechargeable Zinc-Air Batteries: Progress, Challenges, and Perspectives. *Adv. Mater.* **2017**, *29*, 1604685. [CrossRef] [PubMed]

20. Wang, C.; Xie, Z.; Zhou, Z. Lithium-air batteries: Challenges coexist with opportunities. *APL Mater.* **2019**, *7*, 40701. [CrossRef]

21. Mainar, A.R.; Iruin, E.; Colmenares, L.C.; Kvasha, A.; de Meatza, I.; Bengoechea, M.; Leanet, O.; Boyano, I.; Zhang, Z.; Blazquez, J.A. An overview of progress in electrolytes for secondary zinc-air batteries and other storage based on zinc. *J. Energy Storage* **2018**, *15*, 304–308. [CrossRef]

22. An Overview of Costs for Vehicle Components, Fuels, Greenhouse Gas Emissions and Total Cost of Ownership Update 2017. Available online: https://www.semanticscholar.org/paper/An-Overview-of-Costs-for-Vehicle-Components-%2C-Fuels-Friesa-Kerlera/3dfd84187b7192d7ab6a1e71afd6dd5cc3621345 (accessed on 26 June 2020).

23. CAA. Driving Costs. 2013. Available online: https://www.caa.ca/wp-content/uploads/2016/09/CAA_Driving_Cost_English_2013_web.pdf (accessed on 30 July 2019).

24. Propfe, B.; Redelbach, M.; Santini, D.; Friedrich, H. Cost analysis of Plug-in Hybrid Electric Vehicles including Maintenance & Repair Costs and Resale Values. *World Electr. Veh. J.* **2012**, *5*, 886–895.

25. National Energy Board. Provincial and Territorial Energy Profiles-Ontario. 2018. Available online: https://www.cer-rec.gc.ca/nrg/ntgrtd/mrkt/nrgsstmprfls/on-eng.html (accessed on 26 July 2019).

26. Requia, W.J.; Mohamed, M.; Higgins, C.D.; Arain, A.; Ferguson, M. How clean are electric vehicles? Evidence-based review of the effects of electric mobility on air pollutants, greenhouse gas emissions and human health. *Atmos. Environ.* **2018**, *185*, 64–77. [CrossRef]

27. Health Canada. Health Impacts of Air Pollution in Canada. 2017. Available online: http://publications.gc.ca/collections/collection_2018/sc-hc/H144-51-2017-eng.pdf (accessed on 30 July 2019).

28. Smith, R.; McDougal, K. Costs of pollution in Canada. 2017. Available online: https://www.iisd.org/sites/default/files/publications/costs-of-pollution-in-canada.pdf (accessed on 26 July 2019).

29. Sawyer, D.; Seton, S.; Welburn, C. Evaluation of Total Cost of Air Pollution Due to Transportation in Canada. 2007. Available online: http://publications.gc.ca/collections/collection_2008/tc/T22-148-2007E.pdf (accessed on 30 July 2019).

30. Toronto Public Health. Path To Healthier Air: Toronto Air Pollution Burden of Illness Update. Toronto Public Health. 2014. Available online: https://www.toronto.ca/legdocs/mmis/2014/hl/bgrd/backgroundfile-68506.pdf (accessed on 30 July 2019).

31. Union of Concerned Scientists. Cars, Trucks, Buses and Air Pollution. 2018. Available online: https://www.ucsusa.org/clean-vehicles/vehicles-air-pollution-and-humanhealth/cars-trucks-air-pollution (accessed on 30 July 2019).

32. Ontario Ministry of Transportation. *Provincial Highways Traffic Volumes*; Ontario Ministry of Transportation Library: Ottawa, ON, Canada, 2016.

33. Chambers, M.; Schmitt, R. Fact Sheet Diesel-powered Passenger Cars and Light Trucks. 2015. Available online: https://www.bts.gov/sites/bts.dot.gov/files/legacy/DieselFactSheet.pdf (accessed on 30 July 2019).

34. Transport Canada. Estimates of the Full Cost of Transportation in Canada. 2008. Available online: http://publications.gc.ca/collections/collection_2009/tc/T22-165-2008E.pdf (accessed on 26 July 2019).

35. Statistics Canada. *Consumer Price INDEX, Monthly, Not Seasonally Adjusted*; Government of Canada: Ottawa, ON, Canada, 2018.
36. Li, N.; Chen, J.P.; Tsai, I.C.; He, Q.; Chi, S.Y.; Lin, Y.C.; Fu, T.M. Potential impacts of electric vehicles on air quality in Taiwan. *Sci. Total Environ.* **2016**, *566*, 919–928. [CrossRef] [PubMed]

 *vehicles*

Article

Powertrain Control for Hybrid-Electric Vehicles Using Supervised Machine Learning

Craig K. D. Harold *, Suraj Prakash * and Theo Hofman *

Control Systems Technology Group, Mechanical Engineering Department, Eindhoven University of Technology, 5612 AZ Eindhoven, The Netherlands

* Correspondence: c.k.d.harold@student.tue.nl (C.K.D.H.); s.prakash@tue.nl (S.P.); t.hofman@tue.nl (T.H.); Tel.: +31-40-247-2827 (T.H.)

Received: 26 February 2020; Accepted: 11 May 2020; Published: 14 May 2020

Abstract: This paper presents a novel framework to enable automatic re-training of the supervisory powertrain control strategy for hybrid electric vehicles using supervised machine learning. The aim of re-training is to customize the control strategy to a user-specific driving behavior without human intervention. The framework is designed to update the control strategy at the end of a driving task. A combination of dynamic programming and supervised machine learning is used to train the control strategy. The trained control strategy denoted as SML is compared to an online-implementable strategy based on the combination of the optimal operation line and Pontryagin's minimum principle denoted as OOL-PMP, on the basis of fuel consumption. SML consistently performed better than OOL-PMP, evaluated over five standard drive cycles. The EUDC performance was almost identical while on FTP75 the OOL-PMP consumed 14.7% more fuel than SML. Moreover, the deviation from the global benchmark obtained from dynamic programming was between 1.8% and 5.4% for SML and between 5.8% and 16.8% for OOL-PMP. Furthermore, a test-case was conducted to emulate a real-world driving scenario wherein a trained controller is exposed to a new drive cycle. It is found that the performance on the new drive cycle deviates significantly from the optimal policy; however, this performance gap is bridged with a single re-training episode for the respective test-case.

Keywords: machine learning; powertrain control; automatic re-training; hybrid electric vehicles; dynamic programming; transmission; energy management

---

1. Introduction

In the late 1970s, the European Union (EU) established the link between air quality and automotive emissions, thereby setting in motion policies to reduce air pollution. In 1992, the Euro norm for passenger cars was introduced that set a ceiling for concentration of pollutants [1]. These norms are made more stringent with time [2] and enforces companies to adopt more efficient automotive powertrains. This can be illustrated with the growth in hybrid electric vehicle (HEV) market share and the estimated increase in sales over the next decade [3]. In line with the efforts to improve overall powertrain efficiency, significant strides have been made in transmission development. As a result, the continuously variable transmission (CVT) with a steel pushbelt is predicted to achieve an efficiency of 97% [4].

Based on these automotive trends and the superiority of HEV topology P2 over P1 [5], a P2 plug-in hybrid electric vehicle (PHEV) with a CVT is considered as the system, wherein an electric motor (EM) is directly connected to the drive shaft while the internal combustion engine (ICE) is connected in parallel via a clutch, depicted in Figure 1. An energy source, in this case a battery (BAT), supplies power to the EM. The combined power from the EM and the ICE is transmitted by the CVT to the wheels, with an intermediate speed reduction through a fixed differential gear. The addition of an EM introduces a torque-split control variable that is strongly inter-coupled with the control of the transmission. Strategic control of the torque-split, transmission gear ratio, clutches, engine on/off, etc., can reduce the combined energy consumption of the EM and ICE. Several strategies exist that seek to minimize this energy consumption, and these strategies are reviewed in Section 1.1.

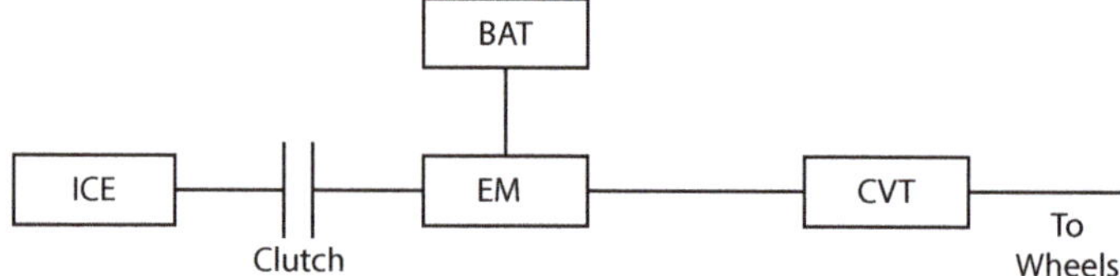

Figure 1. Schematic of a P2 hybrid layout with a CVT.

1.1. Literature Review

The clutch in the P2 configuration disengages the ICE from the powertrain to reduce engine drag; however, it is not considered as a control variable in this study. The clutch is assumed to be activated when there is no demand for ICE power and while activated the ICE is assumed to be idling. Therefore, in this study, only the gear ratio and torque split are considered as control variables and the state of charge of BAT (ζ) is considered as the state of the system.

Accounting for the system dynamics, boundary conditions, constraints on states and control spaces, the two-point boundary value problem can be solved using dynamic programming (DP) that guarantees optimality through an exhaustive search of all control and state grids based on the Bellman's principle of optimality [6,7]. With respect to the considered system, this implies that, for an a priori drive cycle with known boundary conditions, the optimal gear ratio and optimal torque split can be found offline using DP, thereby serving as a benchmark for all other control strategies. This optimal policy is drive cycle dependent, thereby rendering DP unsuitable for online implementation [8]. Online-implementable control strategies exist; however, they are sub-optimal and generally decouple the control variables. Therefore, these online implementable control strategies are reviewed separately as gear ratio control for CVTs in Section 1.1.1 and torque split control in Section 1.1.2.

1.1.1. Gear Ratio Control

CVTs offer a wide range of gear ratios. The classical optimal operating line (OOL) strategy is found to be the most economical method for the conventional drive-train [9]. Subsequently, a modified optimal operating line (M-OOL) strategy that accounts for the CVT system loss is shown to marginally improve over the OOL [10]. A similar method using the equivalent consumption of the electric energy to fuel energy is used to build a hybrid optimal operation Line (H-OOL) [11]. Apart from these standard rule-based methods, another method uses sub-optimal feedback controllers to approximate the optimal control policy and achieves almost optimal performance at a substantially reduced computational effort [12].

1.1.2. Torque Split Control

A comprehensive review of torque split control strategies is given in [13]. Common heuristic based control strategies can be divided into rule-based [14] and fuzzy-logic approaches [15–17]. Fuzzy-logic based approaches are preferred for their robustness and suitability to multi-domain, nonlinear, time-varying systems such as PHEVs [13]. Model predictive control (MPC) methods are also found to be computationally efficient for online implementation [18,19]. However, MPC is heavily dependent on the prediction accuracy and therefore online optimization methods based on Pontryagin's minimum principle (PMP) are preferred [20,21]. An added advantage of PMP is that it is governed by only one costate variable [22]. Similar control strategies, like equivalent consumption minimization strategy (ECMS) first introduced in [23], utilizes the efficiency of the battery and the operating mode to determine the equivalence factor. This ECMS was adapted to PMP [24] and a comparative study with DP was done in [25]. In principle, a good estimation of the costate or equivalence factor can result in near optimal performance [26,27] and therefore online optimization methods are preferred to heuristics [28].

Meanwhile, machine learning (ML) techniques have gained popularity for their ability to control complex tasks by deriving patterns or rules from a data-set or through experience [29,30]. These techniques have also been extended to automotive applications, for example, drive cycle prediction [31], drive cycle recognition [32], training the torque split controller from DP using supervised machine learning (SML) [33], reinforcement learning (RL) for power distribution between the battery and the capacitor [34], etc. In certain tasks, controllers trained using ML have outperformed the controllers based on classical control theory [35].

In the case of supervisory control strategies for an HEV, certain learning based strategies are shown to be comparable to the commonly used control strategies [31]. For continuous-spaces, the actor–critic method was used for the power management in a PHEV [36]. A qualitative study on RL techniques on HEVs and PHEVs shows potential for RL controllers to replace rule based controllers [37]. Similarly, learning based techniques have been used to train neural networks to predict the driving environment and generate an optimal torque split, achieving fuel savings [33]. However, further improvement can be made by customizing the strategy to a specific driving behavior as driving behavior can influence vehicle fuel consumption [32,38]. This driving behavior could be based on the geographic-location, traffic congestion, personal style, etc. In practice, automotive companies offer driving modes such as eco, sport, normal, etc., to address these driver preferences but cannot fully encapsulate the driving behavior. Therefore, the potential of ML can be exploited to bridge the gap to the global optimal without human intervention and thus forms the basis of this research.

Research Question: How can ML be incorporated into supervisory powertrain control in order to adapt to a specific driving behavior?

1.2. Objectives

Apart from the main objective of minimizing overall energy consumption, ML can be extended to improve upon existing practices. In the existing practices for HEVs and conventional powertrains, the control strategy is tuned by experienced calibration engineers through iterative real-time vehicle tests (calibration time) before online implementation, resulting in a strategy that caters to the average driver. Therefore, the objectives (*O1* and *O2*) of the study are to combat the drawbacks of conventional practices, i.e., calibration time and the inability to customize the control strategy to a specific driver. Furthermore, as suggested in literature [37], RL methods can improve fuel economy when compared to the rule based methods. However, these RL techniques come at the cost of learning time and this forms the third objective (*O3*), wherein the learning time must be minimized:

- *O1*: customize the control strategy to a specific driver,
- *O2*: reduce the time consumed for calibration,
- *O3*: improve learning efficiency.

In order to address *O1*, the controller must be able to account for the driver behavior. In this study, the vehicle velocity and its acceleration are considered as a representation of the driver behavior. In practice, the throttle position is considered; however, with a backward facing model, it is replaced with vehicle acceleration. In order to address *O2*, a learning algorithm must be present to adapt to this driver behavior. Several algorithms are available that can learn in real-time or from past data and are discussed in Section 1.1. Real-time learning algorithms like RL require an exploration phase (trial and error) to determine the optimal control respective to the vehicle state, which suggests that it needs to repeatedly encounter identical vehicle states in order to determine the best possible control. However, real-world driving will seldom encounter the identical vehicle states, i.e., identical combination of velocity, acceleration, state of charge, etc. Hence, real-time learning solutions could require thousands of kilometers of driving in order to learn a good control strategy. Objective *O3* is to improve learning efficiency thereby reducing training time. In order to address *O3*, it must be understood that for a given driving trajectory, with boundary conditions, there exists an optimal control policy. Therefore, utilizing this optimal control policy for training would reduce the training time drastically, as there is no requirement of an exploration phase. This would entail that the training occurs after the driving task is completed, in order to find and learn the optimal control policy.

1.3. Contributions

In this paper, a framework is presented that consists of three segments; in segment 1, the route planner analyzes the drive cycle and the end-point condition for the state of charge (ζ) is derived. Under the assumption that the drive cycle is representative of the driving behavior, *O1* is addressed. In segment 2, based on the end-point condition on the state of charge (ζ_f), DP finds an optimal control policy for the a priori drive cycle. Finally, in segment 3, the input parameters from segment 1 and the optimal control policy from segment 2 are used to train a controller using SML algorithms. The absence of human intervention to learn a strategy that addresses *O2* and utilizes the optimal control policy that addresses *O3*.

This trained controller is validated by comparing its performance in terms of fuel consumption, to the global optimal solution derived from DP and an online-implementable control strategy based on literature that uses a combination of OOL and PMP. It should be noted that DP in this study refers to the approximate dynamic programming, wherein the state and control spaces are discretized.

Organization: The paper is organized as follows, Section 2 describes the mathematical modeling of the system. Section 3 formulates the control problem and introduces the proposed framework to solve the problem. Section 4 elaborates on the experimental setup, discusses the results, and presents a test-case. Finally, Section 5 concludes this study and suggests future propositions.

2. Modeling of the System

In this section, the HEV powertrain components are described and the energy flow illustrated in Figure 1 is mathematically modeled. The system is modeled as backward quasi-static, which approximates the system to be static at a given time instance, depicted in Figure 2. Only longitudinal dynamics of the vehicle are considered, and it is assumed that the vehicle only moves forward or is stationary. The equations are taken from [39] and the parameter values are given in experimental design setup is Section 4. Energy losses within each component are taken from manufacturer specifications or modeled from test-bench data.

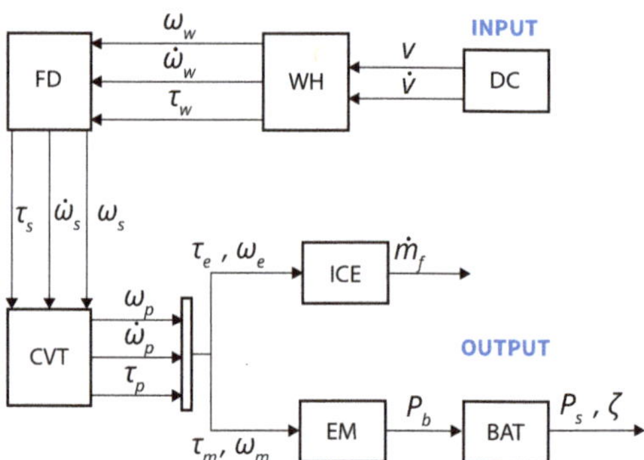

Figure 2. Model overview (DC—Drivecycle, WH—Wheel, FD—Fixed differential, CVT gearbox, ICE—Internal combustion engine, EM—Electric motor, BAT—battery).

Input parameters: The input to the system is the drivecycle, specifically the vehicle speed and the vehicle acceleration, recorded at 1 Hz. To physically achieve this acceleration at a given velocity, the resisting forces must be equal to the driving force applied by the wheel on the road. The resisting forces taken into account are aerodynamic drag (F_a), rolling resistance (F_r), gravity (F_g) and inertia (F_i), given respectively by Equations (1)–(4):

$$F_a = \frac{1}{2} \cdot \rho \cdot c_d \cdot A_f \cdot v^2 \tag{1}$$

$$F_r = m_v \cdot g \cdot \mu_r \cdot cos(\alpha) \quad for \ (v > 0) \tag{2}$$

$$F_g = m_v \cdot g \cdot sin(\alpha) \tag{3}$$

$$F_i = m_v \cdot (1 + m_r) \cdot \dot{v} \tag{4}$$

where ρ is the density of air, c_d is the aerodynamic coefficient, A_f is the frontal surface area, v is the vehicle speed, m_v is the mass of the vehicle, g is the acceleration due to gravity, μ_r is the static rolling coefficient, α is the road inclination, m_r is the mass of rotating parts, and $\dot{v}$ is the acceleration of the vehicle.

Wheel: The driving force (F_w) required at the point of contact of the wheel with the road is the sum of the resisting forces. Subsequently, the torque of the wheel axle is calculated as a factor of the wheel radius. The wheel speed and wheel acceleration can be calculated from the vehicle speed and the vehicle acceleration respectively, as shown in Equations (6) and (7):

$$F_w = F_a + F_r + F_g + F_i$$

$$\tau_w = F_w \cdot r_w \tag{5}$$

$$\omega_w = \frac{v}{r_w} \tag{6}$$

$$\dot{\omega}_w = \frac{\dot{v}}{r_w} \tag{7}$$

where τ_w is the torque at the wheel axle, r_w is the radius of the wheel, ω_w is the rotational speed of the wheel, and $\dot{\omega}_w$ is the rotational acceleration of the wheel.

Differential: The fixed differential factors in the fixed gear ratio, resulting in the required torque and rotational speed at the secondary pulley:

$$\tau_s = \frac{\tau_w}{\gamma_{fd}} \tag{8}$$

$$\omega_s = \omega_w \cdot \gamma_{fd} \tag{9}$$

where γ_{fd} is the ratio of the fixed differential gear, τ_s is the torque at the secondary pulley and ω_s is the rotational speed of the secondary pulley.

Gearbox: The gearbox used in this study is a push-belt CVT type P920 [40], with an under-drive ratio of 0.416 and an over-drive ratio of 2.149. Transmission of speed and torque from the secondary pulley to the primary pulley is dependent on the selected gear ratio; this relation is shown in Equation (11). The loss in transmission of power is attributed to the mechanical loss and pumping loss, modeled from experimental data [40]. An example of the mechanical loss is illustrated in Figure 3 for seven different gear ratios for the vehicle speed of 40 kmph. These losses are measured at the test-bench for the full range of gear ratios at various vehicle speeds and stored in a lookup-table:

$$\tau_p = \tau_s \cdot \gamma_g + \tau_{lo} \tag{10}$$

$$\omega_p = \frac{\omega_s}{\gamma_g} \tag{11}$$

where τ_p is the torque at the primary pulley, γ_g is the selected gear ratio of the CVT, τ_{lo} is the torque loss within the CVT that is the sum of the mechanical and pumping losses, and ω_p is the rotational speed of the primary pulley.

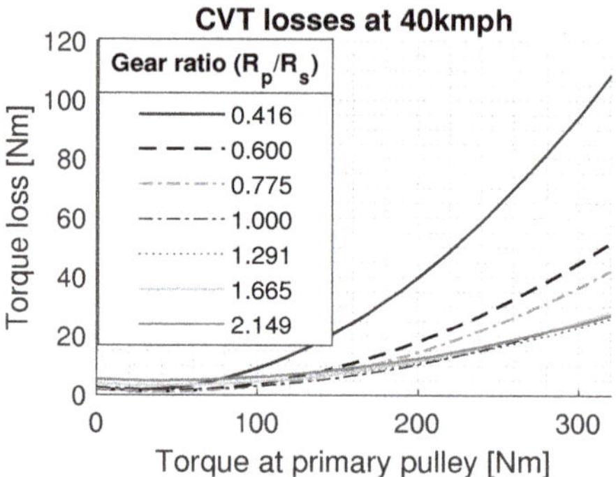

Figure 3. CVT mechanical loss at vehicle speed of 40 kmph [40].

Torque split: The torque at the primary pulley of the CVT is the combined torque delivered by the EM and ICE:

$$\tau_p = \tau_e + \tau_m \tag{12}$$

where τ_e is the ICE torque and τ_m is the EM torque

Engine: The ICE is a 1.6 L, 82-kW unit producing a maximum torque of 143-Nm and is taken from the Peugeot 206 model year 2005. The instantaneous ICE torque and speed are used to determine the fuel consumption from the brake specific fuel consumption (BSFC) map, depicted in Figure 4. The BSFC map expresses the fuel consumed in [g/kWh] that is taken from the manufacturer's specification and is converted to [g/s] using Equation (13):

$$\dot{m}_f = \begin{cases} \dfrac{\tau_e \cdot \omega_e \cdot BSFC(\tau_e, \omega_e)}{3600 \cdot 1000} & \text{if } \tau_e > 0 \\[2ex] m_{f,idle} & \text{if } \tau_e \leq 0 \end{cases} \tag{13}$$

where $\dot{m}_f$ is the instantaneous fuel mass flow in [g/s], BSFC is the fuel consumed in [g/kWh], τ_e is the ICE torque, ω_e is the ICE rotational speed, and $m_{f,idle}$ is the fuel consumption at idling.

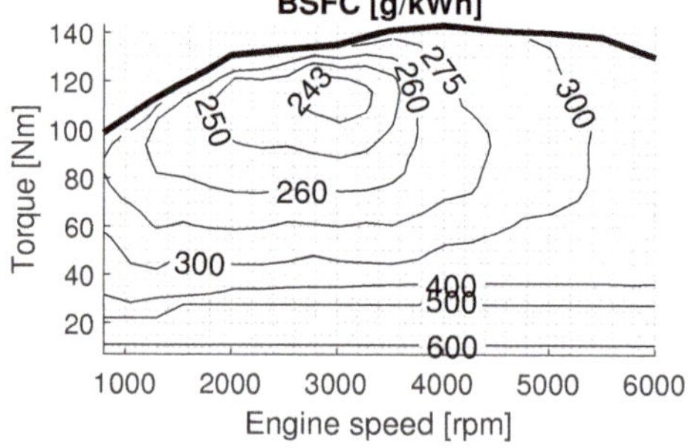

Figure 4. Engine BSFC map.

Electric motor: The 30-kW permanent magnet EM is taken from the 1999 Toyota Prius PHEV (Aichi, Japan). The efficiency of the EM ($\eta_m(\tau_m, \omega_m)$) can be determined from the efficiency map in

Figure 5 and the output power of the battery (P_b) can be calculated as in Equation (14). The data for the efficiency map are taken from test bench measurements [41]:

$$P_b = \begin{cases} \tau_m \cdot \omega_m \cdot \eta_m(\tau_m, \omega_m) & \text{generating} \\[2ex] \dfrac{\tau_m \cdot \omega_m}{\eta_m(\tau_m, \omega_m)} & \text{motoring} \end{cases} \tag{14}$$

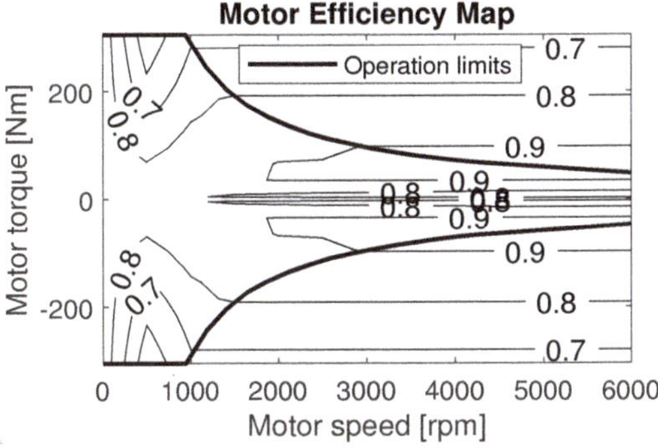

Figure 5. Motor efficiency map [41].

Battery: The 288-V and 6-Ah nickel metal hydride (NiMH) battery pack is taken from the Toyota Prius 2000 model. It is modeled as a voltage input with an internal resistance for simulation and as an equivalent resistance circuit to achieve convexity in the Hamiltonian. The test bench measurements from the Insight battery pack are scaled up to the Prius battery pack [42]. The coulombic efficiency (η_c) is assumed to be 0.905 for charging and discharging. The current within the battery can be calculated from the output power of the battery, and is given in Equation (15):

$$I_b = \frac{\left(U_{oc} - \sqrt{U_{oc}^2 - 4 \cdot \Omega \cdot P_b} \right) \cdot \eta_c}{2 \cdot \Omega} \tag{15}$$

Subsequently,

$$\zeta(t) = \zeta(t-1) - \frac{I_b}{Q_0 \cdot 3600}, \tag{16}$$

where I_b is the battery current, U_{oc} is the open circuit voltage, Ω is the internal resistance of the battery circuit, P_b is the power output of the battery, η_c is the coulombic efficiency of the battery, ζ is the state of charge, t is the time instance, and Q_0 is the nominal battery capacity.

With the system modelled, the next section describes the formulation and solution of the control problem.

3. Methodology

In this section, the control problem is formulated in Section 3.1. Subsequently, the solution for this control problem using the proposed SML framework is described in Section 3.2. In order to evaluate the performance of the proposed SML framework, conventional solutions such as DP is elaborated upon in Section 3.3 and an online implementable control strategy OOL-PMP is introduced and described in Section 3.3.

3.1. Control Problem

The objective of the control problem is to minimize energy consumption over the drive cycle, while satisfying the physical constraints of the system. Considering the practical application of a PHEV, the boundary conditions for state of charge (ζ_i, ζ_f) is fixed for a given drive cycle and the fuel consumption (m_f) is minimized:

$$J = \int_{t_0}^{t_f} \dot{m}_f \, dt \tag{17}$$

where t_0 is the initial time and t_f is the final time.

The control problem can be formulated as:

$$\min_{x,u} J(x,u)$$

subject to

$$h_1 := \dot{x}(t) - f(x(t), u(t), t) = 0$$
$$h_2 := \zeta_i - \zeta(t_0) = 0$$
$$h_3 := \zeta_f - \zeta(t_f) = 0$$
$$h_4 := \gamma_{g,min} - \gamma_g(t_0) = 0$$
$$g_{1,2} := \gamma_{g,min} \leq \gamma(t) \leq \gamma_{g,max}$$
$$g_3 := P_e(t) - P_{e,max} \leq 0$$
$$g_{4,5} := \omega_{e,min} \leq \omega_e(t) \leq \omega_{e,max}$$
$$g_{5,6} := P_{m,min} \leq P_m(t) \leq P_{m,max}$$
$$g_{7,8} := \omega_{m,min} \leq \omega_m(t) \leq \omega_{m,max}$$
$$g_{9,10} := I_{b,min} \leq I_b(t) \leq I_{b,max}$$
$$g_{11,12} := \zeta_{min} \leq \zeta(t) \leq \zeta_{max}$$
$$g_{13,14} := -1 \leq u_{ts}(t) \leq 1$$
$$g_{15} := -p(t_0) \leq 0$$

State space; x = $\{\zeta\}$ where $\zeta = [\zeta_{min}, \zeta_{max}] \subset R$ with $\zeta_{min} = 0.1$ (10% of battery capacity) and $\zeta_{max} = 0.9$ (90% of battery capacity). Control space; u = $\{u_{ts}, u_g\}$ where $u_{ts} = [-1, 1] \subset R$ and $u_g = [\gamma_{g,min}, \gamma_{g,max}] \subset R$ with $\gamma_{g,min} = 0.416$ and $\gamma_{g,max} = 2.149$. $p(t_0)$ is the costate that satisfies the boundary conditions for PMP.

The following subsection introduces the machine learning framework that is proposed as a solution for the control problem.

3.2. Solution Using Supervised Machine Learning

A few comments are in order, and it is assumed that a robust baseline strategy exists while training data are being accumulated. Secondly, in order to satisfy the objective (O1), the controller training is performed from scratch. Thirdly, in the specified framework, training occurs on completion of the driving task. Therefore, it is assumed that the vehicle is equipped with sufficient memory to store vehicle states.

Proposed Framework: The framework is divided into three segments, namely, Route Planner (RP), Dynamic Programming (DP), and Supervised Machine Learning (SML). The flow of events and parameters are illustrated in Figure 6:

- **Route Planner** records the drive cycle and the initial condition (ζ_i) for the respective drive cycle. The velocity trajectory depicted in the route planner segment in Figure 6 is an example of the recorded drive cycle. Based on this drive cycle, an end-point condition ζ_f is determined. In this study, ζ_f is calculated assuming 1.1% battery charge is available for the distance of 1 km. The assumption is made based on the average driving and charging cycles of HEVs in the Netherlands [43]. The requirement of the route planner is to set the boundary condition for the a priori drive cycle. There are more sophisticated planners based on traffic congestion, terrain, charging stations, etc. but do not add value to this study, hence neglected.
- Secondly, **Dynamic Programming** solves the two-point boundary value problem satisfying ζ_f resulting in the optimal control policy (u_{ts}^*, u_g^*) and optimal state trajectory (ζ^*, γ_g^*), for the given drive cycle. The discretized state and control spaces are elaborated in Section 3.3.
- Thirdly, **Supervised Machine Learning** segment develops a control strategy by mapping the input parameters from the drive cycle to the optimal control policy from DP, using SML algorithms. The rules derived from this mapping represent the control strategy and make predictions for a new input as shown in Figure 6. No universal algorithm exists to model the system; therefore, an SML algorithm is selected based on an exhaustive search. The SML algorithm is selected with a five-fold cross validation based on its accuracy of predictions, deviation of false predictions from the optimal value, and computational time for each prediction. The various algorithms are shown in Table 1 along with their prediction accuracy and the number of predictions the algorithm is capable of every second. Both characteristics are desired to be as high as possible and based on this, the selected algorithm is highlighted. Additionally, a memory module is used to store previously recorded data for the purpose of re-training the controller.

Table 1. Performance of Supervised Machine Learning Algorithms for the NEDC case.

Algorithm	Gear-Ratio Control		Torque-Split Control	
	Accuracy (%)	Prediction/Sec	Accuracy (%)	Prediction/Sec
Decision Tree-Fine	96.8	6700	98.0	68,000
Decision tree-medium	94.3	24,000	96.6	100,000
Linear discriminant	70	16,000	78.7	75,000
Quadratic discriminant	78.0	30,000	-	-
SVM-cubic	93.4	6500	94.9	3800
SVM-fine Gaussian	96.5	3500	93.8	3200
KNN-fine	94.8	15,000	93	41,000
KNN-medium	94.6	12,000	90.9	27,000
KNN-cubic	94.5	12,000	91.1	39,000
KNN-weighted	97.0	18,000	93.3	39,000
Ensemble-bagged trees	97.0	3300	98.0	7800
Ensemble boosted trees	96.7	3300	97.3	9100

The learning algorithms used for individual control strategies are discussed in Section 3.2.1 and the parameters with which the algorithm achieved the accuracy and prediction speed highlighted in Table 1 are introduced.

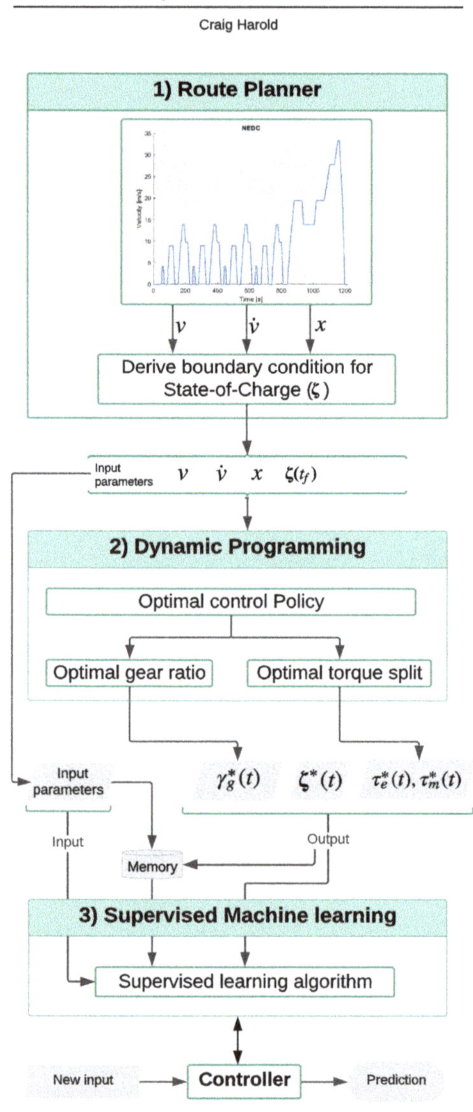

Figure 6. Proposed framework.

3.2.1. Supervised Machine Learning Algorithms

Decision Tree: As the name suggests, decision tree (DT) builds a model for the data to go from observation to prediction through branches. It is representative of a root system beneath a tree and is also representative of human decision-making. Each node represents binary logic and filters down to a prediction based on this set of binary logic gates. The nodes of the tree are split based on impurity gain (δI), given by Equation (18):

$$\delta I = P(T)i_t - P(T_L)i_l - P(T_R)i_R \tag{18}$$

where $P(T)i_t$ is the probability of the splitting candidate or node t in the set of all observations T, $P(T_L)i_{t_L}$ is the probability that left child node (t_L) is present in the left observation set T_L and $P(T_R)i_{(t_R)}$ is the probability that the right child node (t_R) is present in the right observation set (T_R). In essence, a node is selected and all the observations are partitioned at the node. The impurity gain checks the number of instances of a class that are common on both sides of the partition, thereby the impurity of the class.

For this study, the primary pulley speed (ω_p), the torque demand at the primary pulley (τ_p), and the state of charge of the battery (ζ) were used as input features, and the optimal torque split was the desired output from the DT. The properties of the decision tree used for training are as follows: maximum number of splits was set to 100 and k-fold cross validation is set to 5. An example with relevance to the system model is depicted in Figure 7, wherein the decision tree is trained for torque split control, but limited to 10 nodes. It is intuitive that, for the negative torque demand (power flow from the wheels to the energy source), the resulting torque split is closer to 1 indicating complete regeneration.

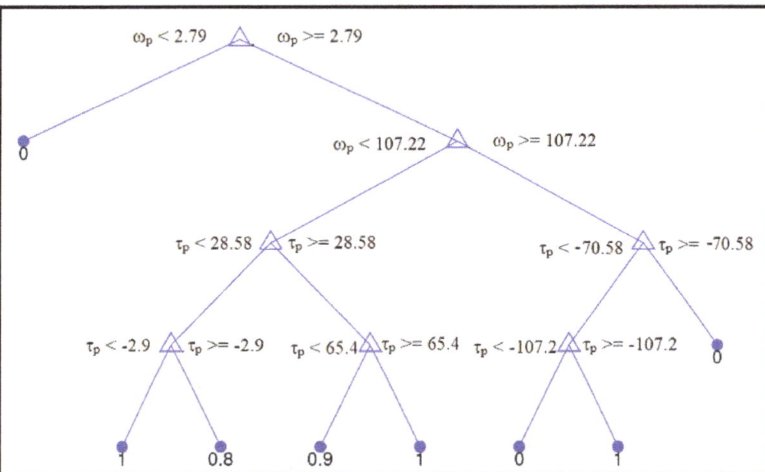

Figure 7. Decision tree trained for torque split control (u_{ts}) with 10 nodes.

K-Nearest Neighbors: The K-Nearest Neighbor (KNN) algorithm is non-parametric, i.e., no model is fitted to the data and all the work is done when a prediction is required. In principle, the KNN algorithm takes a vote of the closest neighbors to predict an output. The 'K' in KNN represents the number of neighbors to consider. Therefore, the 'K' should be an odd number to ensure a majority in the vote. KNN is used as the gear ratio control algorithm, wherein the input features are the vehicle speed (v) and vehicle acceleration ($\dot{v}$) and the desired output is the optimal gear ratio (γ_g^*). It must be noted that $\dot{v}$ is used as a feature since the drivecycles considered do not include elevation profiles. In case of non-horizontal drivecycles, the torque required at the secondary pulley (τ_s) will substitute $\dot{v}$ and in case of a forward facing model or practical applications, the throttle input will substitute $\dot{v}$. The properties of the KNN algorithm used are as follows; the closest neighbors are determined by the Euclidean distance, the number of votes accounted for is 7, equal weighting given to all neighbors and the k-fold cross validation is set to 5. It must be noted that to ensure effective learning with the limited available data points from each drive cycle, the CVT was discretized into seven equally spaced classes for the SML case—ergo limiting the performance of the SML control strategy. However, with the abundance of data from real-world driving, this limitation can be overcome and in turn the full potential of the CVT can be exploited.

In order to evaluate the performance of the proposed framework, the SML controller is compared to conventional solutions that are elaborated in Section 3.3.

3.3. Solutions Using Classical Control

To validate the SML controller, its performance is compared to the global benchmark set by DP and the online implementable benchmark set by OOL-PMP.

Global Benchmark, DP: DP results in the global optimal control policy for the a priori drive cycle with boundary conditions ζ_i and ζ_f. A cost matrix is built with all possible state-action combinations at each timestep through the drive cycle. All infeasible states or actions are penalized with a high cost. Finding a path with minimal cost through the cost matrix results in the optimal control policy.

In order to build the cost matrix, the state and control space is discretized. The CVT is discretized into 100 gear ratios to exploit the full potential of the CVT. The torque split control ($u_{ts} \in [-1, 1]$) is discretized into intervals of 0.05, and the implication of the torque split variable is shown in Equations (19) and (20). For cases where $\omega_p > 0$,

$$\tau_e = (1 - u_{ts}) \cdot \tau_p \tag{19}$$

$$\tau_m = (u_{ts}) \cdot \tau_p \tag{20}$$

where ω_p is the rotational speed of the primary pulley and τ_p is the torque required at the primary pulley. The state space $x = \{\zeta\}$ is discretized as follows: $\zeta \in [0.1, 0.9]$ in intervals of 0.005 and $\gamma_g \in [0.416, 2.149]$ discretized to 100 ratios. The time interval of 1 s is chosen since the difference in ζ is very small for any smaller intervals of time.

Online Benchmark, OOL-PMP: The combination of OOL-PMP is based on a decoupled approach from the literature review in Section 1.1. The OOL is constructed by using the most efficient points of ICE operation over the entire ICE speed range and is used to control the gear ratio (u_g), while the PMP method is used to control the torque split (u_{ts}). Since the system is modeled as backward facing, the power required at the primary pulley (P_p) is estimated and subsequently a gear ratio is selected by OOL. It is counter-intuitive to use a torque split controller in combination with OOL, since OOL inherently determines the operating point of the ICE and ergo the operating point of the EM based on the power request at the primary pulley. However, to ensure that the torque split is optimal and online-implementable, a PMP approach is used to control the torque split. The formulation of the Hamiltonian and application of PMP is taken from [22] and given in Equation (21). The losses in the EM are modelled as a second degree polynomial, while the ICE losses and BAT open-circuit voltage are approximated by a linear fit. The control variable chosen is the power of the battery (P_b):

$$H = F - p \cdot \dot{x} = P_f + p \cdot P_s \tag{21}$$

where P_f is the fuel power, x is the state of charge (ζ), $\dot{x} = -P_s$ is the time derivative of ζ, and (p) is the costate:

$$H = \gamma_{p_1}\left(P_p - \frac{-\gamma_{m_1} + \sqrt{\gamma_{m_1}^2 + 4 \cdot \gamma_{m_2} \cdot (P_b - \gamma_{m_0})}}{2 \cdot \gamma_{m_2}}\right) + \gamma_{p_0} \cdots$$

$$\cdots + p \cdot \frac{U_{oc} \cdot (U_{oc} - \sqrt{U_{oc}^2 - 4 \cdot \Omega \cdot P_b})}{2 \cdot \Omega} \tag{22}$$

where P_p is the power required at the primary pulley of the CVT, ($\gamma_{p_1}, \gamma_{p_0}$) are the coefficients of the linear fit that models the ICE losses, ($\gamma_{m_2}, \gamma_{m_1}, \gamma_{m_0}$) are the coefficients of the second degree polynomial used to model the EM losses, U_{oc} is the open-circuit voltage of the BAT, and Ω is the resistance of the BAT.

The necessary conditions of PMP are as follows:

$$\frac{\partial H}{\partial P_b} = 0$$

$$\frac{\partial H}{\partial \zeta} = \dot{p}$$

Solving the first condition results in the optimal P_b^*, shown in Equation (23):

$$P_b^* = \frac{U_{oc}^2 (\gamma_{p1}^2 - p^2 \cdot \gamma_{m1}^2 + 4 \cdot p^2 \cdot \gamma_{m2} \cdot \gamma_{m0})}{4 \cdot (\Omega \cdot \gamma_{p1}^2 + p^2 \cdot U_{oc}^2 \cdot \gamma_{m2})} \tag{23}$$

Solving this as an initial value problem using a bisection algorithm results in the initial costate $p(t_0)$ that satisfies the boundary conditions ζ_i and ζ_f. In order to the solve this boundary value problem, the exact velocity and acceleration profile from the respective drivecycle are considered.

With the proposed SML solution described along with the conventional solutions in order to compare the performance of the SML control strategy, the following Section 4 discusses the results of the study.

4. Results

This section outlines the results obtained by solving the control problem defined in Section 3.1. The experiment setup is outlined in Section 4.1, results of which are discussed in Section 4.2 wherein the performance of the SML controller is compared to the OOL-PMP and DP strategies. Furthermore, a test-case is elaborated in Section 4.3 wherein the effects of re-training the SML controller are observed.

4.1. Experimental Design

As mentioned in Section 3.2, the real-world drive cycle of the user is used in the framework. However, for experimental purposes, the real-world driving data are replaced by standard drive cycles and each drivecycle is assumed to represent the respective individual driving behavior. The drive cycles NEDC, EUDC, FTP75-highway, WLTP, and JP10-15 mode are used in simulation. To make a valid comparison between the control strategies, the boundary conditions are fixed respectively for each drive cycle as per the assumption made in the RP in Section 3.2. The parameters for the system modeled in Section 2 are given in Table 2.

4.2. Numerical Results

The results summarized in Table 3 show the fuel consumption in liters per 100 kilometers and the percentage loss in fuel when compared to the optimal solution. In Table 3, the proposed machine learning solution is denoted as SML, the globally optimal solution is denoted as DP, and the online implementable solution is denoted as OOL-PMP. It is evident from Table 3 that the SML control strategy consistently performs better than the OOL-PMP control, thereby bridging the gap to DP. The OOL-PMP performs well with steady state drive cycles while deviates significantly from the optimal solution with instantaneous drive cycles. The NEDC case is elaborated to give a better understanding of the performance figures achieved in Table 3 by illustrating the differences in control strategy operation over the drivecycle.

NEDC Results: The NEDC case is elaborated with graphical illustrations for the comparison of controllers. Due to legibility concerns with graphical representations, the more prominent drive cycle, WLTP, is not illustrated. However, numerical results for all the drive cycles are tabulated in Table 3. Figure 8 illustrates the working of the gear ratio controllers over the complete drive cycle while Figure 9 showcases the differences in the control strategies over the last quarter of the drive cycle. Subsequently, ICE and EM operating points are illustrated in Figures 10 and 11, and the resulting fuel consumption is illustrated in Figure 12. It is evident from this comparison that the SML controller almost perfectly tracks the DP control strategy, hence resulting in a 2.6% loss in optimality for the NEDC. While the OOL-PMP controller significantly deviates from the optimal control and results in 11.5% increase in fuel consumption.

Table 2. Vehicle parameters.

Parameter	Symbol	Value	Unit
Base vehicle weight	m_v	1300	kg
Wheel radius	r_w	0.316	m
Final drive ratio	γ_{fd}	4.695	-
Air drag coefficient	c_d	0.36	-
Rolling resistance coefficient	μ_r	0.01	-
Air density	ρ	1.18	kg/m^3
Frontal area	A_f	2.5813	m^2
Mass of rotating parts	m_r	0.05	-
Gravitational constant	g	9.81	m/s^2
CVT under-drive ratio	$\gamma_{g,min}$	0.416	-
CVT over-drive ratio	$\gamma_{g,max}$	2.149	-
Max. Engine power	$P_{e,max}$	82	kW
Max. Electric motor power	$P_{m,max}$	30	kW

Table 3. Fuel consumption in liters per 100 kms.

	DP	OOL-PMP	SML
NEDC (11.7 km)	4.25	4.74 (+11.5%)	4.36 (+2.6%)
EUDC (7.0 km)	4.68	4.95 (+5.77%)	4.93 (+5.34%)
JP10-15 (4.2 km)	4.05	4.33 (+6.9%)	4.21 (+3.8%)
FTP (17.7 km)	3.88	4.53 (+16.75%)	3.95 (+1.8%)
WLTP (23.3km)	4.75	5.31 (+11.8%)	4.84 (+1.9%)

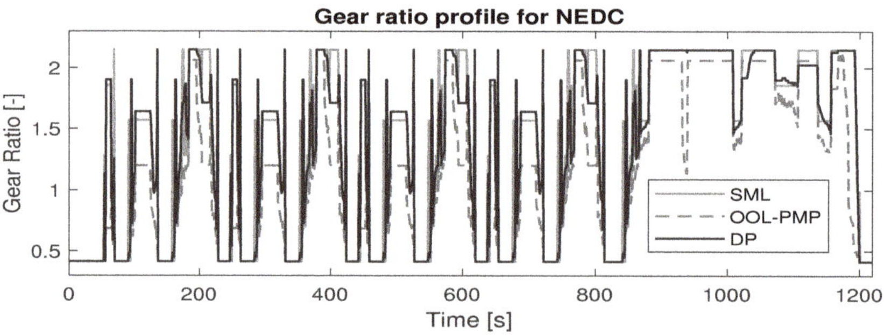

Figure 8. Gear Ratio Controller comparison on the NEDC.

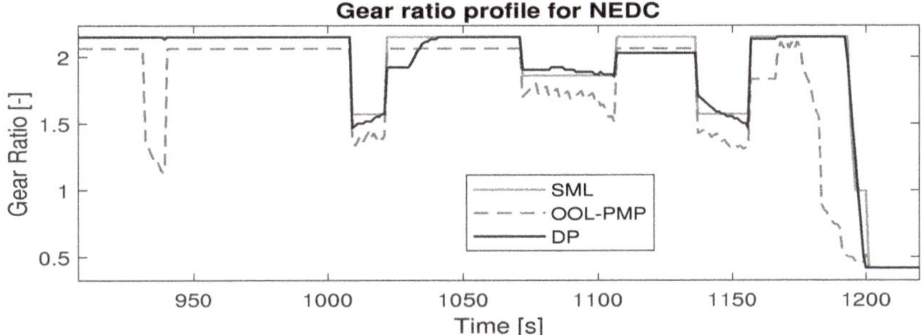

Figure 9. Gear ratio controller comparison over the last quarter of the NEDC.

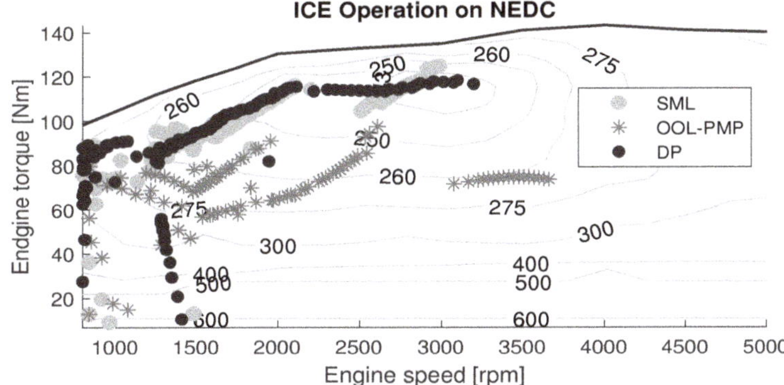

Figure 10. Engine operating points on the NEDC.

4.3. Test Case

Evident from Section 4.2, the near perfect tracking by the SML controller is dependent on the quality of the dataset. Additionally, the good performance is also attributed to the fact that the trained controller was tested with the same input drive cycle. However, it seldom occurs wherein the identical training conditions are experienced in practice. Therefore, it is critical for the proposed framework to efficiently adapt the control strategy to the newly experienced drivecycle. A test case is conducted wherein the control strategy trained with the EUDC is utilized on the FTP-75 (representing the newly experienced drivecycle), and the effect of re-training is observed. It is evident from Figure 13a that the SML controller trained with EUDC largely deviates from the optimal control strategy. In line with the principle of the proposed framework to adapt to driving behavior, the control strategy was then re-trained with the FTP-75 drivecycle—thereby significantly bridging the gap of SML control strategy to the optimal strategy with a single episode of re-training. Further re-training of the control strategy improved the performance and is illustrated in Figure 14a,b. However, as mentioned earlier, the exact combination of vehicle states and driving conditions is not likely to reoccur in practice and therefore the effect of re-training with a single episode is more relevant to real-world application. It is important to note that, in this test case, the new drivecycle forms a significant portion of the training dataset and hence resulted in a drastic change; this is analogous to early stages of re-training in the real-world. Intuitively, a less drastic change in the control strategy occurs when the new drivecycle is a small fraction of the training dataset, thereby capturing only the essence of the new drivecycle while maintaining the desired control from the previous learning episodes.

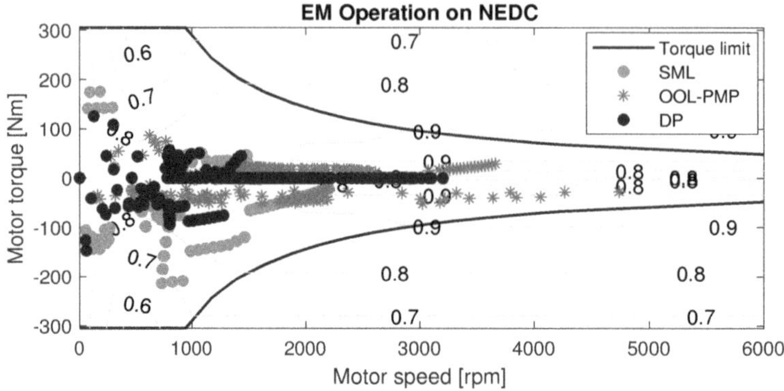

Figure 11. Electric Motor operating points on the NEDC.

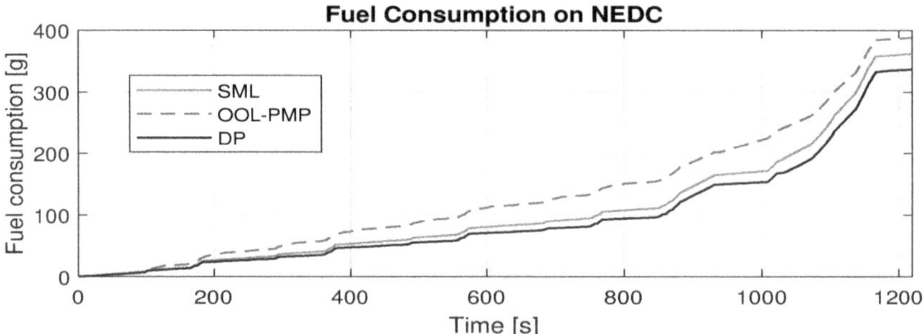

Figure 12. Fuel consumption comparison for identical ζ_f on the NEDC.

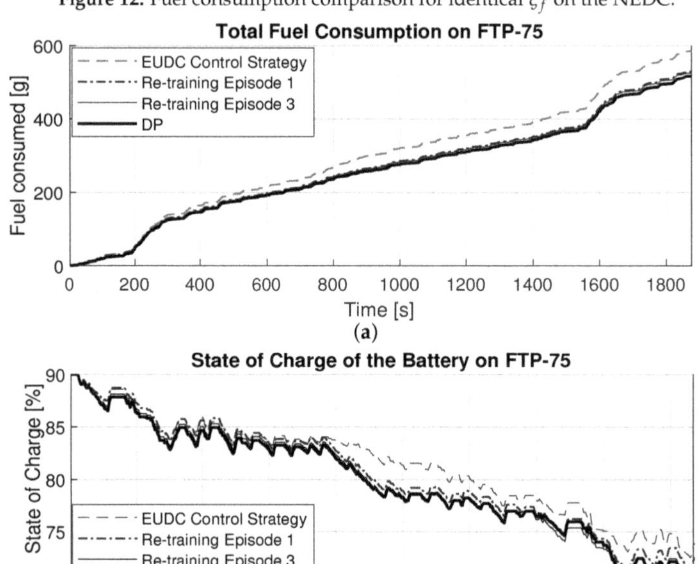

Figure 13. Comparison between EUDC trained controller and re-trained controllers on the FTP75 drive cycle. (**a**) fuel consumption; (**b**) state-of-charge.

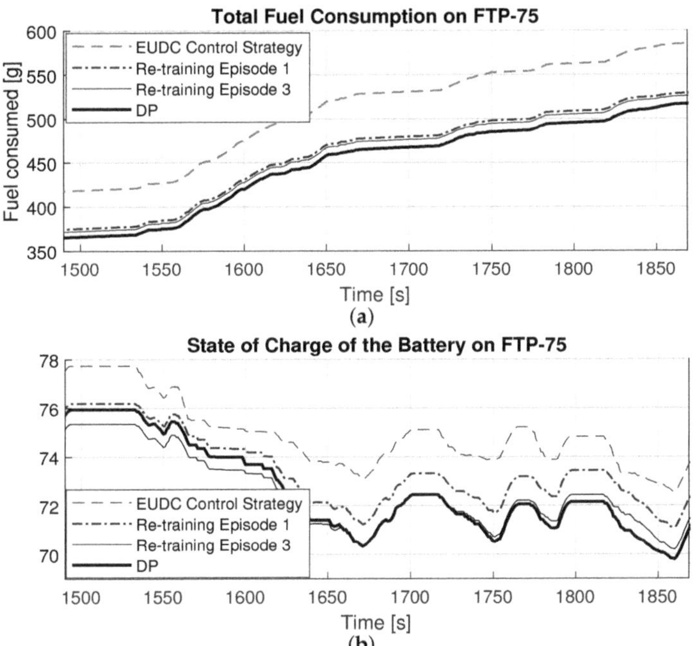

Figure 14. Controller comparison for the last 20% of FTP75 drive cycle. (**a**) fuel consumption; (**b**) state-of-charge.

5. Conclusions

The results show that the learned control strategy (SML) outperforms the conventional online implementable strategy (OOL-PMP), tested over several standard drivecycles in terms of fuel economy. This indicates that the proposed framework could effectively learn from past driving experiences to reproduce close to optimal results for the identical training conditions, i.e., the same drivecycle. However, since real-world driving will seldom experience identical training conditions, it is critical to study the efficacy of the proposed framework in adapting the control strategy to newly experienced drivecycles. Therefore, a test-case is conducted wherein the control strategy trained for EUDC is utilized on the FTP-75 and the effect of re-training the control strategy with the new drivecycle is observed. The EUDC strategy applied to the FTP-75 showed a large deviation from the optimal trajectory and a single episode of re-training significantly bridged the gap to optimality. Therefore, the proposed framework could be a viable alternative to the existing control strategies that adapts efficiently to a specific driver behavior.

Downsides of this proposed framework are briefly highlighted, the requirement of on-board computational power to perform DP is a real-world limitation but off-loading the computational burden is a prospective solution, since modern cars already record vehicle data and upload it to a cloud in the presence of internet connectivity. Furthermore, the algorithm properties, size of the dataset, and quality of dataset have shown to affect the performance and require more in-depth research.

As for future applications and improvements, the current framework using past data can be replaced with a predicted drive cycle based on geographic location, traffic congestion, terrain, weather, etc. that is easily available with current map technology. Additional states can be added, such as a slope sensor to consider the elevation of the road, in turn providing a more extensive control strategy capable of accounting for dynamic environments.

Vehicles **2020**, *2*, 267–286

Author Contributions: Conceptualization, C.K.D.H., S.P. and T.H.; methodology, C.K.D.H. and T.H.; software, C.K.D.H.; writing—original draft preparation, C.K.D.H.; writing—review and editing, C.K.D.H. and T.H.; supervision, S.P. and T.H. All authors have read and agreed to the published version of the manuscript.

Funding: This research received no external funding.

Conflicts of Interest: The authors declare no conflict of interest.

Abbreviations

Acronym	Name
BAT	Battery
CVT	Continuously Variable Transmission
DC	Drive Cycle
DT	Decision Tree
DP	Dynamic Programming
ECMS	Equivalent Consumption Minimization Strategy
EM	Electric Motor
EU	European Union
EUDC	Extra Urban Driving Cycle
FD	Fixed Differential
FTP75	Federal Test Procedure 75
HEV	Hybrid Electric Vehicle
H-OOL	Hybrid Optimal Operating Line
ICE	Internal Combustion Engine
JP10-15	Japanese 10–15 driving cycle
KNN	K Nearest Neighbors
ML	Machine Learning
M-OOL	Modified Optimal Operating Line
NEDC	New European Driving Cycle
OOL	Optimal Operating Line
PHEV	Plugin Hybrid Electric Vehicle
PMP	Pontryagin's Minimum Principle
RL	Reinforcement Learning
RP	Route Planner
SML	Supervised Machine Learning
SVM	Support Vector Machines
WH	Wheel
WLTP	Worldwide Harmonized Light Vehicle Test Procedure

References

1. Bennett, G. *Air Pollution Control in the European Community*; U.S. Department of Energy Office of Scientific and Technical Information: Oak Ridge, TN, USA, 1992.
2. Bielaczyc, P.; Woodburn, J.; Szczotka, A. An assessment of regulated emissions and CO_2 emissions from a European light-duty CNG-fueled vehicle in the context of Euro 6 emissions regulations. *Appl. Energy* **2014**, *117*, 134–141. [CrossRef]
3. IEA. *Global EV Outlook*; IEA: Paris, France, 2019.
4. Sluis, F.; Noll, E.; Leeuw, H. Key technologies of the pushbelt CVT. *Target* **2013**, *25*, 50.
5. Yang, Y.; Hu, X.; Pei, H.; Peng, Z. Comparison of power-split and parallel hybrid powertrain architectures with a single electric machine: Dynamic programming approach. *Appl. Energy* **2016**, *168*, 683–690. [CrossRef]
6. Bellman, R. Dynamic Programming. *Science* **1957**, *153*, 34–37. [CrossRef] [PubMed]
7. Bertsekas, D.; Tsitsiklis, J. *Parallel and Distributed Computation: Numerical Methods*; Prentice Hall: Englewood Cliffs, NJ, USA, 1989; Volume 23.
8. Kermani, S.; Delprat, S.; Guerra, T.; Trigui, R.; Jeanneret, B. Predictive energy management for hybrid vehicle. *Control Eng. Pract.* **2012**, *20*, 408–420. [CrossRef]

9. Bonsen, B.; Steinbuch, M.; Veenhuizen, P. CVT ratio control strategy optimization. In Proceedings of the 2005 IEEE Conference Vehicle Power and Propulsion, Chicago, IL, USA, 7 September 2005.

10. Ryu, W.; Kim, H. CVT ratio control with consideration of CVT system loss. *Int. J. Automot. Technol.* **2008**, *9*, 459–465. [CrossRef]

11. Kim, C.; NamGoong, E.; Lee, S.; Kim, T.; Kim, H. Fuel economy optimization for parallel hybrid vehicles with CVT. *SAE Trans.* **1999**, *108*, 2161–2167.

12. Pfiffner, R.; Guzzella, L. Optimal operation of CVT-based powertrains. *Int. J. Robust Nonlinear Control* **2001**, *11*, 1003–1021. [CrossRef]

13. Huang, Y.; Wang, H.; Khajepour, A.; Hongwen, H.; Ji, J. Model predictive control power management strategies for HEVs: A review. *J. Power Sources* **2017**, *341*, 91–106. [CrossRef]

14. Hofman, T.; Steinbuch, M.; Druten, R.; Serrarens, A. Rule-based energy management strategies for hybrid vehicle drivetrains: A fundamental approach in reducing computation time. *IFAC Proc. Vol.* **2006**, *39*, 740–745. [CrossRef]

15. Pusca, R.; Ait-Amirat, Y.; Berthon, A.; Kauffmann, J. Fuzzy-logic-based control applied to a hybrid electric vehicle with four separate wheel drives. *IEE Proc. Control Theory Appl.* **2004**, *151*, 73–81. [CrossRef]

16. Shi, G.; Jing, Y.; Xu, A.; Ma, J. Study and simulation of based-fuzzy-logic parallel hybrid electric vehicles control strategy. In Proceedings of the Sixth International Conference on Intelligent Systems Design and Applications, Jinan, China, 11 December 2006; Volume 1, pp. 280–284.

17. Koo, E.; Lee, H.; Sul, S.; Kim, J. Torque control strategy for a parallel hybrid vehicle using fuzzy logic. In Proceedings of the Conference Record of 1998 IEEE Industry Applications Conference, Thirty-Third IAS Annual Meeting (Cat. No. 98CH36242), St. Louis, MO, USA, 12–15 October 1998; Volume 3, pp. 1715–1720.

18. Opila, D.; Wang, X.; McGee, R.; Grizzle, J. Real-time implementation and hardware testing of a hybrid vehicle energy management controller based on stochastic dynamic programming. *J. Dyn. Syst. Meas. Control* **2013**, *135*, 021002. [CrossRef]

19. Joševski, M.; Abel, D. Distributed predictive control approach for fuel efficient gear shifting in hybrid electric vehicles. In Proceedings of the 2016 European Control Conference (ECC), Aalborg, Denmark, 29 June–1 July 2016; pp. 2366–2373.

20. Zheng, C.; Xu, G.; Cha, S.; Liang, Q. Numerical comparison of ECMS and PMP-based optimal control strategy in hybrid vehicles. *Int. J. Automot. Technol.* **2014**, *15*, 1189–1196. [CrossRef]

21. Kim, N.; Cha, S.; Peng, H. Optimal Control of Hybrid Electric Vehicles Based on Pontryagin's Minimum Principle. *IEEE Trans. Control Syst. Technol.* **2010**, *19*, 1279–1287.

22. Jager, B.; Keulen, T.; Kessels, J. *Optimal Control of Hybrid Vehicles*; Springer: Berlin, Germany, 2013.

23. Paganelli, G.; Delprat, S.; Guerra, T.; Rimaux, J.; Santin, J. Equivalent consumption minimization strategy for parallel hybrid powertrains. In Proceedings of the IEEE 55th Vehicular Technology Conference, VTC Spring 2002 (Cat. No. 02CH37367), Birmingham, AL, USA, 6–9 May 2002; Volume 4, pp. 2076–2081.

24. Serrao, L.; Onori, S.; Rizzoni, G. ECMS as a realization of Pontryagin's minimum principle for HEV control. In Proceedings of the 2009 American Control Conference, St. Louis, MO, USA, 10–12 June 2009; pp. 3964–3969.

25. Serrao, L.; Onori, S.; Rizzoni, G. A comparative analysis of energy management strategies for hybrid electric vehicles. *J. Dyn. Syst. Meas. Control* **2011**, *133*, 031012. [CrossRef]

26. Musardo, C.; Rizzoni, G.; Guezennec, Y.; Staccia, B. A-ECMS: An adaptive algorithm for hybrid electric vehicle energy management. *Eur. J. Control* **2005**, *11*, 509–524. [CrossRef]

27. Onori, S.; Serrao, L. On Adaptive-ECMS strategies for hybrid electric vehicles. In Proceedings of the International Scientific Conference on Hybrid and Electric Vehicles, Malmaison, France, 6–7 December 2011; Volume 67.

28. Pisu, P.; Rizzoni, G. A Comparative Study Of Supervisory Control Strategies for Hybrid Electric Vehicles. *IEEE Trans. Control Syst. Technol.* **2007**, *15*, 506–518. [CrossRef]

29. Silver, D.; Schrittwieser, J.; Simonyan, K.; Antonoglou, I.; Huang, A.; Guez, A.; Hubert, T.; Baker, L.; Lai, M.; Bolton, A.; et al. Mastering the game of go without human knowledge. *Nature* **2017**, *550*, 354. [CrossRef]

30. Mnih, V.; Kavukcuoglu, K.; Silver, D.; Graves, A.; Antonoglou, I.; Wierstra, D.; Riedmiller, M. Playing atari with deep reinforcement learning. *arXiv* **2013**, arXiv:1312.5602.

31. Geulen, S.; Josevski, M.; Nellen, J.; Fuchs, J.; Netz, L.; Wolters, B.; Abel, D.; Ábrahám, E.; Unger, W. Learning-based control strategies for hybrid electric vehicles. In Proceedings of the 2015 IEEE Conference on Control Applications (CCA), Sydney, NSW, Australia, 21–23 September 2015; pp. 1722–1728.

32. Jeon, S.; Jo, S.; Park, Y.; Lee, J. Multi-Mode Driving Control of a Parallel Hybrid Electric Vehicle Using Driving Pattern Recognition. *J. Dyn. Syst. Meas. Control* **2000**, *124*, 141–149. [CrossRef]

33. Murphey, Y.; Park, J.; Kiliaris, L.; Kuang, M.; Masrur, M.; Phillips, A.; Wang, Q. Intelligent Hybrid Vehicle Power Control—Part II: Online Intelligent Energy Management. *IEEE Trans. Veh. Technol.* **2013**, *62*, 69–79. [CrossRef]

34. Xiong, R.; Cao, J.; Yu, Q. Reinforcement learning-based real-time power management for hybrid energy storage system in the plug-in hybrid electric vehicle. *Appl. Energy* **2017**, *211*, 538–548. [CrossRef]

35. Juang, J.; Lin, R.; Liu, W. Comparison of classical control and intelligent control for a MIMO system. *Appl. Math. Comput.* **2008**, *205*, 778–791. [CrossRef]

36. Li, Y.; Hongwen, H.; Peng, J.; Zhang, H. Power Management for a Plug-in Hybrid Electric Vehicle Based on Reinforcement Learning with Continuous State and Action Spaces. *Energy Procedia* **2017**, *142*, 2270–2275. [CrossRef]

37. Hu, X.; Liu, T.; Qi, X.; Barth, M. Reinforcement Learning for Hybrid and Plug-In Hybrid Electric Vehicle Energy Management: Recent Advances and Prospects. *IEEE Ind. Electron. Mag.* **2019**, *13*, 16–25. [CrossRef]

38. Ericsson, E. Independent driving pattern factors and their influence on fuel-use and exhaust emission factors. *Transp. Res. Part D Transp. Environ.* **2001**, *6*, 325–345. [CrossRef]

39. Guzzella, L.; Sciarretta, A. *Vehicle Propulsion Systems: Introduction to Modeling and Optimization*; Springer: Berlin, Germany, 2013; Volume 1.

40. Vroemen, B. *Component Control for the Zero Inertia Powertrain*; Control Systems Technology: Osborne Park, WA, USA, 2001; pp. 85–89.

41. National Renewable Energy Laboratory. *NREL's Testing of Prius Japanese Motor at Unique Mobility 4/1999*; Technical Report; National Renewable Energy Laboratory: Golden, CO, USA, 2001.

42. National Renewable Energy Laboratory. *NREL Test Data From Testing Entire Insight Battery Pack Model Year 2000*; Technical Report; National Renewable Energy Laboratory: Golden, CO, USA, 2001.

43. Franke, T.; Krems, J. Understanding charging behavior of electric vehicle users. *Trans. Res. Part F Traffic Psychol. Behav.* **2013**, *21*, 75–89. [CrossRef]

 *vehicles*

Article

Investigation on the Impact of Degree of Hybridisation for a Fuel Cell Supercapacitor Hybrid Bus with a Fuel Cell Variation Strategy

Julius S. Partridge *, Wei Wu and Richard W. G. Bucknall

Dept. of Mechanical Engineering, University College London, Torrington Place, London WC1E 7JE, UK; w.wu.11@alumni.ucl.ac.uk (W.W.); r.bucknall@ucl.ac.uk (R.W.G.B.)
* Correspondence: julius.partridge.09@ucl.ac.uk; Tel.: +44-207-679-7063

Received: 19 November 2019; Accepted: 17 December 2019; Published: 19 December 2019

Abstract: This paper presents the development of a control strategy for a fuel cell and supercapacitor hybrid power system for application in a city driving bus. This aims to utilise a stable fuel cell power output during normal operation whilst allowing variations to the power output based on the supercapacitor state-of-charge. This provides flexibility to the operation of the system, protection against over-charge and under-charge of the supercapacitor and gives flexibility to the sizing of the system components. The proposed control strategy has been evaluated using validated Simulink models against real-world operating data collected from a double-decker bus operating in London. It was demonstrated that the control strategy was capable of meeting the operating power demands of the bus and that a wide range of degrees of hybridisation are viable for achieving this. Comparison between the degree of hybridisation proposed in this study and those in operational fuel cell (FC) hybrid buses was carried out. It was found that the FC size requirement and FC variation can be significantly reduced through the use of the degree of hybridisation identified in this study.

Keywords: degree of hybridization; energy management; hybrid propulsion; proton exchange membrane fuel cell; simulink, supercapacitor

---

1. Introduction

The London bus network is the largest road transportation network in the UK and is an essential part of the public transportation network [1]. This, however, results in significant contributions to both Greenhouse Gas (GHG) and local pollutant emissions [2–4], with strategies such as the ultra-low emission zone implemented as a means of reducing these emissions through deployment of hybrid and zero emissions technologies [5]. One of the more promising potential zero emissions solutions for bus applications is the proton exchange membrane (PEM) fuel cell technology. The PEM fuel cell (which will be referred to as FC in this paper) uses hydrogen as its fuel and produces electricity and water as a waste product through an electrochemical process [6]. Hybridisation of FCs with some form of energy storage is a promising solution to solving the problems of over sizing the FC stack and the FC's poor transient response [6]. Much work has been carried out in the field of PEM FC hybrids for vehicular applications, where hybridisation with battery and/or supercapacitor (SC) technologies has been considered. This covers FC/battery [7–10], FC/SC, and FC/battery/SC hybrids [11–14], with some examples given. The literature review that follows focuses on FC/SC hybrids as these are most relevant to the work presented here.

In the work of [15] a comparison between fuel cell hybrid configurations and Energy Storage System (ESS) technologies is presented for use in a vehicle. Of the available configurations it was found that connecting the FC and SC via DC/DC converters provides the best solution in terms of reducing the stress on the fuel cell and achieving a high hydrogen economy because of the optimal fuel cell operation.

A number of examples of this configuration have been presented in the literature, such as [16–24]. A control strategy based on reducing the transient changes on the FC load has been developed and experimentally tested in [16]. It was shown that the developed system avoids fuel starvation of the FC whilst using the SC to meet transient power changes. In the work of [17] a control strategy based on reducing the transient response of the FC is considered. This was tested against the ECE15 EU drive cycle and performed acceptably. An energy management strategy utilising short-term future energy demand prediction was developed and tested through both simulation and experimentation in [18]. It was found that this strategy offers improved performance, owing partly to the better management of the SC for regenerative braking. Components sizing and development of a control strategy based on Pontryagin's minimum principle has been developed with cost functions of hydrogen consumption, SC supercapacitor state-of-charge (SoC) and fuel cell durability in [19]. The control strategy maintained a fairly stable FC output but did exhibit a large range of FC outputs. An equivalent consumption minimisation strategy (ECMS) is employed to assess the sizing of system components against different driving cycles in [20]. The results suggest a significant variation to the FC output is beneficial in terms of the hydrogen consumption. In the work of [21] the energy management is achieved by using only the SC for transient responses and only the FC for stable load conditions. This however necessitates large transient changes to the FC output. In [22], a control system aimed at providing voltage regulation on the busbar, tracking of the SC reference current and asymptotic stability of the closed-loop system was developed. In the work of [23], the control strategy focused on a differential flatness control that offers a simple and effective means of reducing the transient power demand changes on the FC. In the work of [24], an interleaving technique was successfully used to improve the voltage and current control in the FC/SC hybrid system. This focussed primarily on the short-term system performance. Along with work of [25,26] that each proposed control strategies to mitigate the stress applied on the FC from a step response to the output power demand. In real world application, step response is rarely required for a vehicle application while frequent variation is often required. In the work of [27–29] representative duty cycles such as New European Drive Cycle (NEDC) was used to evaluate their proposed control strategies to control the FC output power. The work in the literature highlights that there are numerous methods of controlling the balance of power in a FC/SC hybrid system. Most of the proposed designs have however focussed on the short-term operation of the system and/or have also resulted in significant variations in the FC power output. The aim of the work presented here is to limit the transient response of the FC power output and to assess the possible sizing solutions for a FC/SC hybrid power plant against real-world load profiles. This approach additionally allows for the assessment of the potential for downsizing the FC stack.

The work detailed in this paper is a continuation of the research presented in [30–32] and further considers the evaluation of the degree of hybridisation through improvement of the control strategy. In the previous work, a FC/SC hybrid propulsion system had been developed, constructed and simulated. The FC was used as a fixed output power source to eliminate the dynamic stress applied to the FC. The SC was used to supplement the FC output power and meet the dynamic power demands. A stabilised FC control strategy was designed and demonstrated to be capable of maintaining the FC output constant while enabling the propulsion system to meet the dynamic load demands of a bus. A strategy to identify the FC output power and required SC size was proposed and shown to perform as expected, although a number of limitations of the control strategy were highlighted. These are mainly the required prior knowledge of the required FC output, the lack of flexibility and protection against over and under charge.

The limitations lead to another question. Would it be best practice to maintain the FC and boost converter power output at a predefined and constant setting throughout the entire journey? Hence, this research aims to:

1. Investigate a strategy to facilitate variation of the FC output control operation to eliminate or mitigate the identified limitations.
2. Investigate the impact of the degree of hybridisation for a FC/SC hybrid bus with the proposed control strategy.

Within this paper the outline and development of the updated control strategy is detailed. The performance of the control strategy is compared against the stabilised control strategy previously employed against real-world performance data collected from a city driving bus. Finally, an assessment of the degree of hybridisation is carried out for variations to the control strategy parameters. The novel contributions of this paper are as follows. The development of the control strategy to include protection offers novelty in its application to real-world data and the impact this has on the sizing of the system components. This highlights the viability of using SCs as the energy storage medium even for long drive cycles and for significant downsizing of the FC used. Further to this the wide range of possible sizing solutions shows the flexibility available to the designer.

2. Data Collection

Operational performance data collected from an ADL Enviro 400H diesel hybrid bus (Alexander Dennis, Larbert, UK) operating in London was used as the basis to test and compare the control strategies. This comprised of data for a whole day of operation of the bus whilst in operation on the 388-bus route, comprising roughly 18 hour of operation, as shown in Figure 1. For the purposes of this study the data collected was used to provide power profiles of the traction motor power demand upon which the control strategies could be tested. For this the data collected regarding the motor input power were used directly as the power profile and were based on the assumption that the traction motor used on the Enviro 400H would be kept the same with the proposed FC/SC hybrid system. The power profiles used to compare and assess the control strategies implemented are detailed in Table 1. The purpose of these driving cycles was to test the system under a variety of operating conditions which provide high power, low power and long duration performance requirements.

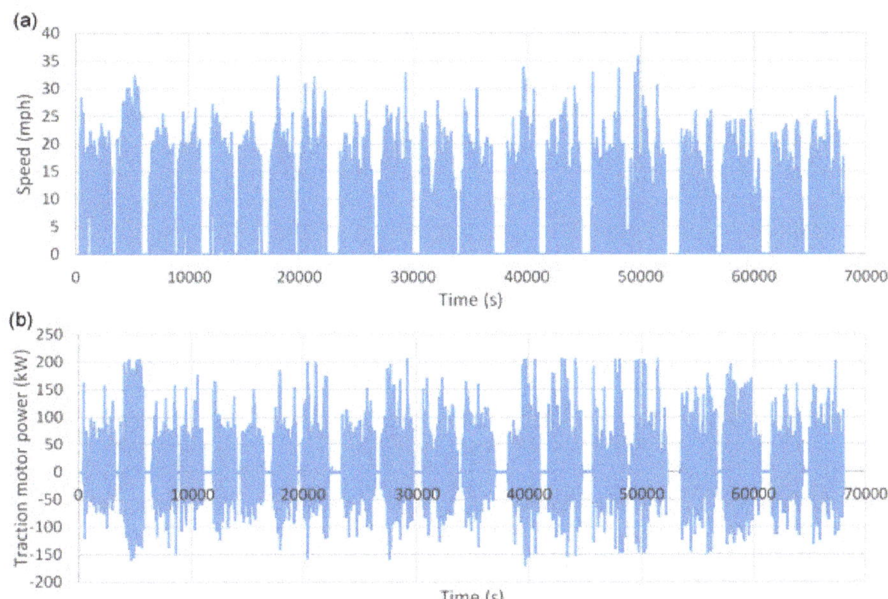

Figure 1. Profile of the real-world bus performance data of an Enviro 400H bus on the 388-bus route. (**a**) Bus speed, (**b**) traction motor power.

Table 1. The route data used to assess the fuel cell (FC) variation control strategy.

Profile No.	Average Power	Duration	Notes
1	16.03 kW	1860 s	Highest power route
2	5.65 kW	3000 s	Lowest power route
3	10.70 kW	6400 s	First three routes
4	9.45 kW	69,100 s	Whole day

3. FC/SC Operation Strategy

The FC/SC hybrid configuration is shown in Figure 2. The originally proposed FC/SC operation strategy is to keep the FC at a constant pre-defined output power while using the SC to cover any transient power demand, as detailed in [32]. In this system the balance of power between the FC, SC and load is controlled on the common busbar linking these components. This method has been validated and tested in [31,32] and was shown to perform well under transient conditions whilst maintaining a stable busbar voltage (630 V in this case). Since the voltage is maintained at a constant value, the power balance is directly controlled by controlling the magnitude of the current and can simply be written as:

$$I_{load} = I_{fc\_out} + I_{SC\_out} \tag{1}$$

where each of the current values are defined on the 630 V busbar and I_{load} is the current to/from the load, I_{fc_out} is the current from the FC and I_{sc_out} is the current to/from the SC. The balance of power provided by Equation (1) remains the default control for the proposed control strategy detailed in this paper. The output power of the FC and boost converter (P_{fc_out}) is defined as 110% of the average power requirement of the bus duty cycle (P_{load}) with the additional 10% included to account for the losses in the SC buck/boost converter and is maintained at a constant value. The SC was sized by considering the cumulative energy change over the course of the drive cycle, with a 20% margin over the magnitude of cumulative energy change chosen as the SC size, further details of this can be found in [32]. This strategy has been proven capable of providing a reasonable estimation of the required degree of hybridisation for a certain duty cycle. A more detailed description of the system can be found in [32]. However, the strategy has been proven to lack the flexibility required to work effectively across a range of different load profiles and offers no protection against under charge and overcharge of the SC module. To address this, a simple overcharge and undercharge protection strategy is introduced which aims to both provide protection to the energy system and provide greater operational flexibility. Whilst the protection of the system is an important consideration it is also worth considering how the presence of such a protection system will impact upon the sizing of system components.

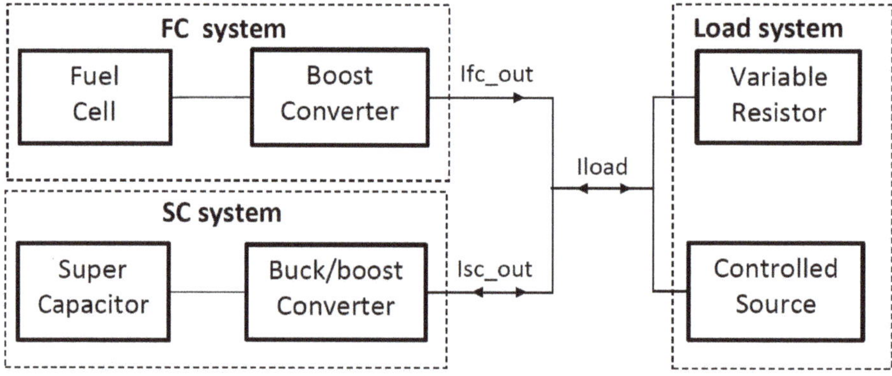

Figure 2. Schematic of the FC/supercapacitor (SC) hybrid system.

3.1. Overcharge Protection Design

To prevent the SC overcharging, a higher threshold value (HTV) was assigned. The HTV is the threshold of the SC SoC, which when exceeded, the value of $I_{\text{fc\_out}}$ begins to ramp down as a means of preventing the SC from overcharging. The intent is to calculate a new $I_{\text{fc\_out}}$ reference (and thus a new $P_{\text{fc\_out}}$) based on the SoC of the SC. The calculation for overcharge protection was carried out using the equation:

$$I_{fc\_out\_new} = \left(\frac{100 - SoC}{100 - HTV}\right) \times I_{fc\_out} \tag{2}$$

The change in $I_{\text{fc\_out}}$ decreases linearly with the SC SoC, such that when SC SoC reaches 100%, the value of $I_{\text{fc\_out}}$ is 0 A. The value of HTV of the overcharge protection was selected to be 90% SoC. Hence if the SC SoC exceeds 90% the value of $I_{\text{fc\_out}}$ will decrease, reducing the charging rate of the SC during charge operation and also increasing the discharge rate during discharge operation. Limiting power transients on the FC has also been proved to be very important in [33–35]. Hence a rate limiter was added to control the rate of change requirement applied on the FC. It takes at least 30 s at a constant rate to increase from no load power (0 kW) to full load power (85 kW) and with the same rate of change limit when power output needs to be decreased.

3.2. Undercharge Protection Design

To prevent the SC from becoming fully discharged, a lower threshold value (LTV) was assigned. In this case, the value of $I_{\text{fc\_out}}$ will ramp up if the SC SoC falls below the LTV and acts as a means of protecting against the SC SoC becoming depleted. The calculation for undercharge protection was carried out using the equation:

$$I_{fc\_out\_new} = I_{fc\_out} + \frac{(Ifc_{max} - Ifc_{out}) \times (LTV - SoC)}{LTV - LL} \tag{3}$$

where $I_{\text{fc\_max}}$ is the maximum output current of the FC and boost converter can provide and is set as 120 A, amounting to a maximum power output of 76 kW (85 kW at the FC). For the initial tests, the value of the LTV is set at 60%. Additionally, a lower limit (LL) is introduced and acts as the value of the SC SoC at which $I_{\text{fc\_max}}$ is reached and assigned as 30%. An increased $I_{\text{fc\_out}}$ will charge the SC at a higher rate during charge operation and also reduce the power demand placed on the SC during discharge operation. The new $I_{\text{fc\_out}}$ will be increased by an amount determined by the SC SoC until $I_{\text{fc\_out}}$ reaches the maximum value of 120 A. A rate limiter has also been added to ensure the change in FC output is gradual.

3.3. Control Strategy Overview

The overall control strategy implemented for the hybrid system is based on a defined value of $I_{\text{fc\_out}}$ and the SoC of the SC. This can be summarised as follows,

$$\begin{cases} I_{fc\_out\_new} = I_{fc\_out} + \frac{\left(I_{fc\_max} - I_{fc\_out}\right) \times (LTV - SoC)}{LTV - LL} & LL < SoC < LTV \\ I_{fc\_out\_new} = I_{fc\_out} & LTV < SOC < HTV \\ I_{fc\_out\_new} = \left(\frac{1 - SOC}{1 - HTV}\right) \times I_{fc\_out} & HTV < SOC \end{cases}$$

The basis of the strategy is to control the value of $I_{\text{fc\_out}}$ based on the SOC of the SC. Under normal operation the value of $I_{\text{fc\_out}}$ is taken as the user defined value. If the SOC is less than the LTV then $I_{\text{fc\_out}}$ is increased to prevent the SC from becoming depleted. The increase in $I_{\text{fc\_out}}$ is limited in accordance with the maximum output of the fuel cell. If the SoC is greater than the HTV, then $I_{\text{fc\_out}}$ decreases to prevent the SC overcharging. This will be referred to as the FC variation strategy, with the Simulink model of the hybrid system and control strategy shown in Figure 3.

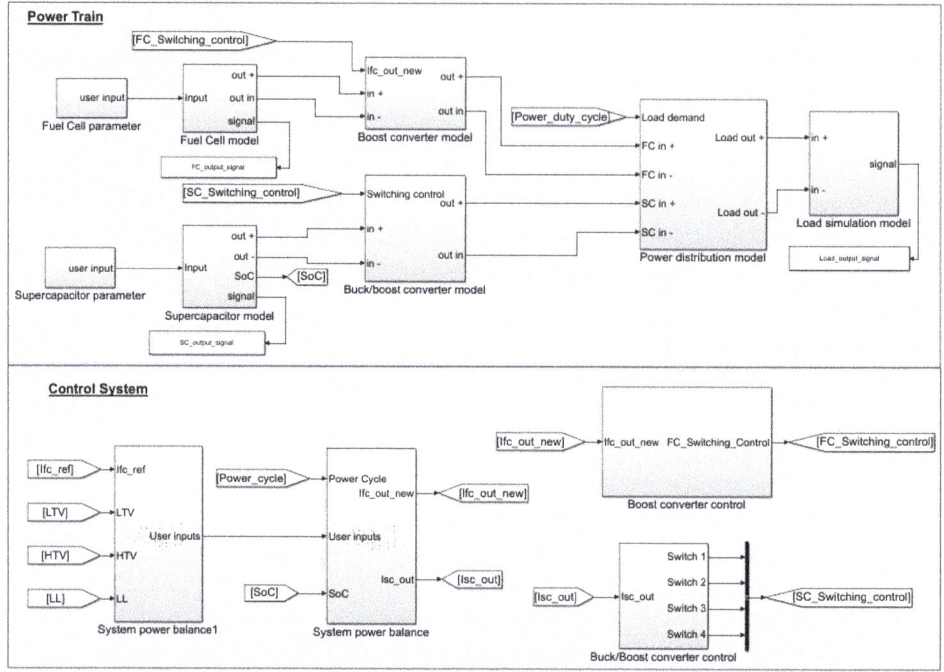

Figure 3. Simulink model of the proposed configuration and control structure.

4. Performance with FC Variation Strategy

The degree of hybridisation is identified for each of the profiles detailed in Table 2, where the SC size is defined by calculating the required cumulative energy from the SC based on the duty cycle. The FC variation strategy is applied to the power profiles detailed in Table 2 and compared with the performance of the constant FC output strategy. It should be noted that the FC rated power for each of the simulations is 85 kW, however the results of the simulations determines the minimum size of FC that would be required based on the maximum output observed for each simulation.

Table 2. Table detailing the parameters used for each of the load profiles.

Profile No.	I_{fc_out}	P_{fc_out}	SC	HTV	LTV
1	28.0 A	17.63 kW	65 F (2.08 kWh)	90%	60%
2	9.9 A	6.22 kW	44 F (1.41 kWh)	90%	60%
3	18.7 A	11.77 kW	124 F (3.97 (kWh)	90%	60%
4	16.5 A	10.39 kW	506 F (16.2 kWh)	90%	60%

For power profile 1 (Figure 4a), the initial FC and boost converter output power was determined to be 17.63 kW and the SC was sized at 2.08 kWh. These settings matched those used for the tests carried out previously without the FC variation strategy. The final SoC at the end of the driving cycle was reasonably close to the initial SoC when using the FC variation strategy at the same degree of hybridisation as that of the base line comparison tests without the FC variation strategy. It should also be noted that the FC and boost converter output reference was reduced a number of times between 100 s and 500 s when the SoC attained 90%. Additionally, the FC and boost converter output reference was increased multiple times to prevent undercharge triggered at the 60% threshold, particularly between 1200 s and 1380 s. The peak FC and boost converter output power is 31.52 kW for this 32-min journey. This requires a FC power output of 35 kW when a 90% average boost converter efficiency is considered. Hence the required degree of hybridisation for profile 1 (high power journey with the highest average power) would be 35 kW FC/2.08 kWh SC.

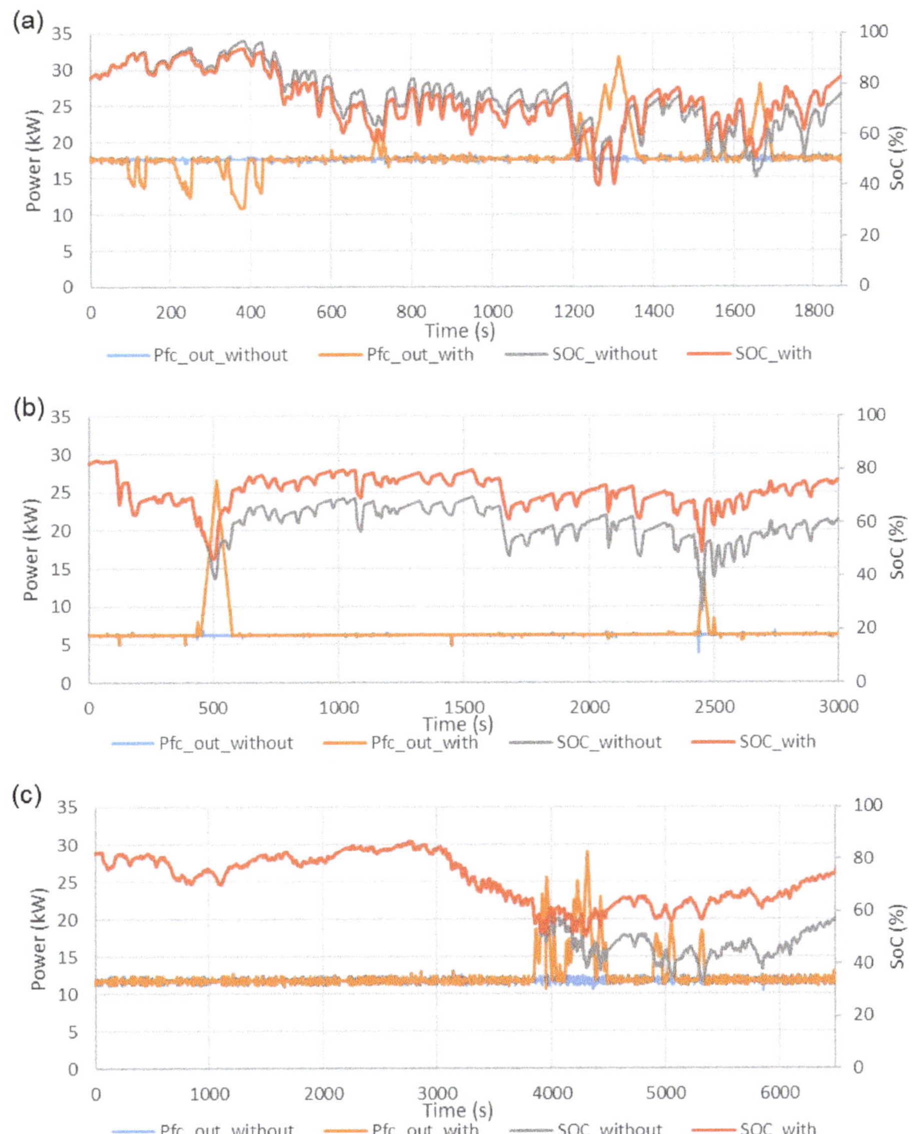

Figure 4. Boost converter output power and supercapacitor state-of-charge (SoC) variation of different bus journeys with and without FC variation strategy for (**a**) journey with highest average power, (**b**) journey with lowest average power and (**c**) first three journeys of the day.

For profile 2, the bus journey with the lowest average power (Figure 4b), the initial FC and boost converter output power was set as 6.22 kW and the SC was sized at 1.41 kWh. In this scenario, the SoC of the SC at the end of the test was higher, with the FC variation strategy in operation. The FC and boost converter output was regulated depending on the SoC of the SC, with two occasions when the undercharge protection was engaged. The peak FC and boost converter output power is 26.2 kW which equates to a required FC maximum output power of 29.1 kW assuming a 90% boost converter efficiency. Hence the required degree of hybridisation for this low power journey would be 29.1 kW FC/1.41 kWh SC.

The driving cycle comprising three completed bus journeys, profile 3 (Figure 4c), used the initial FC and boost converter output power setting of 11.77 kW and a 3.97 kWh SC. As expected, the variations in SoC are identical for both the models with and without the FC variation strategy until the SoC drops to the lower threshold value. Once beyond the FC variation trigger point, the SoC was sustained at an overall higher level as would now be expected. The FC and boost converter output power setting clearly increased during the second part of this bus journey where higher power operations occurred. This results in a significant increase in the SoC of the SC at the end of the journey for the FC variation strategy. The peak power output of the FC and boost converter output power in this driving cycle is 28.7 kW which requires a FC capable of delivering up to 31.9 kW rated output power. Hence the required degree of hybridisation for this longer driving cycle is 31.9 kW FC/3.97 kWh SC. It can be seen that the calculated degrees of hybridisation for all three driving cycles functioned as expected with the inclusion of the FC variation strategy.

The strategy to identify the degree of hybridisation was validated against a number of driving cycles with the inclusion of the FC variation strategy. The model will be used to identify the required degree of hybridisation for the entire day of route 388. The average power of the entire day (without driver breaks) has been measured at 9.45 kW based on the operation power measurements. That gives the required initial FC and boost converter output power base reference as 10.39 kW. Based on the load average and FC initial power output setting, the minimum capacity for the SC was determined to be 505 F, which equates to a maximum 16.2 kWh of stored energy. The model has been tested with the entire day's power profile (profile 4). The FC and boost converter output power and the SoC variation have been plotted in Figure 5.

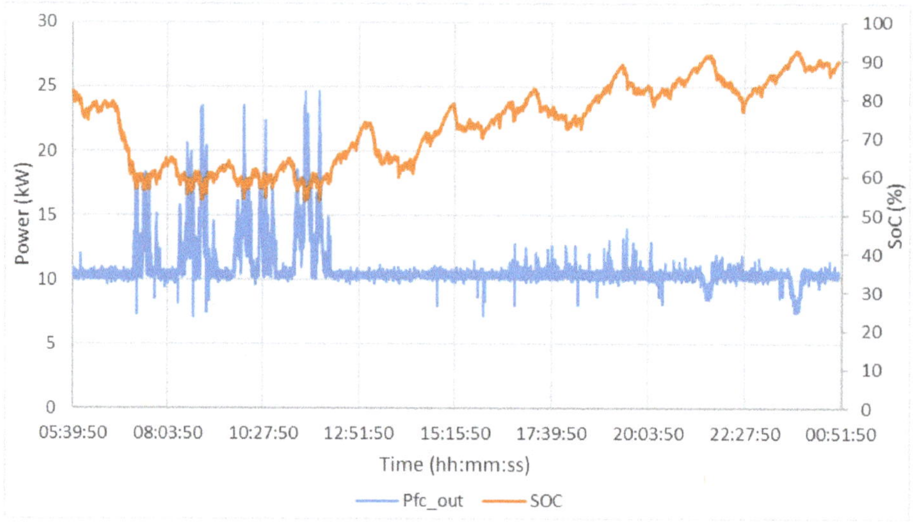

Figure 5. Boost converter output power and SoC variation of the entire operation day of route 388.

It is evident that the SC was generally at low SoC having delivered large amounts of energy initially to propel the bus during the morning portion of the driving cycle and largely absorbed excess energy during the afternoon and evening portions of the driving cycle. As a result, the FC and boost converter output was increased significantly by the FC variation strategy in the morning operations and then decreased on two occasions in the afternoon and evening operations. This is because morning (rush hour) driving requires a lot of starts which are high load events and rarely will the bus attain appreciable speeds which would also compromise regenerative energy capture. It was found the average charge efficiency of the SC throughout the entire day was 82.7%, while the discharge efficiency was 90.3%. The SoC was maintained within the prescribed operational range. The proposed degree of hybridisation proved capable of delivering effective bus operation for the entire day. Since the highest power of the FC and boost converter output is 24.2 kW, this equates to a required FC power of 26.9 kW with a 90% average boost converter efficiency. Therefore, the degree of hybridisation on route 388 bus for the operating day can be identified to be 26.9 kW FC/16.2 kWh SC.

The variable FC output control strategy showed to limit the variation of the SC SoC and thus allow the system to provide for the long-term transient power demands of the bus without either depleting of over-charging the SC. It has been seen that the FC variation strategy results in significantly less variation in the SC SoC during operation, this leads to the situation where the calculated SC size can potentially be reduced by utilising the FC variation strategy. This also allows for a greater flexibility in the degree of hybridisation of the system and will now be explored.

The response of the system during over- and under-charge is highlighted in Figure 6. The over-charge and under-charge protection response is taken from profiles 1 and 3 respectively. It can be seen in Figure 6(a) that the SC SoC rises above the HTV (90%) at 346 s as a result of a regenerative braking event. This causes the value of Ifc_out to decrease. This is followed by a period with no load power requirements. During this period the FC continues to charge the SC but at a decreasing rate. At 376 s an acceleration event occurs, resulting in the SC discharging before SC SoC falls below the HTV at 396 s. During the period of over-charge protection the SC is still able to meet all of the transient demands whilst the FC output is able to slowly ramp down. Similarly, for under-charge protection (Figure 6(b)), a period of relatively high-power demand occurs at around 4885 s. This causes the SC SoC to fall below the LTV. At this point the FC output begins to ramp up, and limits the rate of discharge SC. A regenerative braking event starting at 4962 s acts to recharge the SC with the FC output ramping down as a result. The SC SoC rises above the LTV at 4978 s and coincides with the FC output returning to the reference value. Again the SC is able to meet the transient load demands whilst the FC is able to ramp slowly.

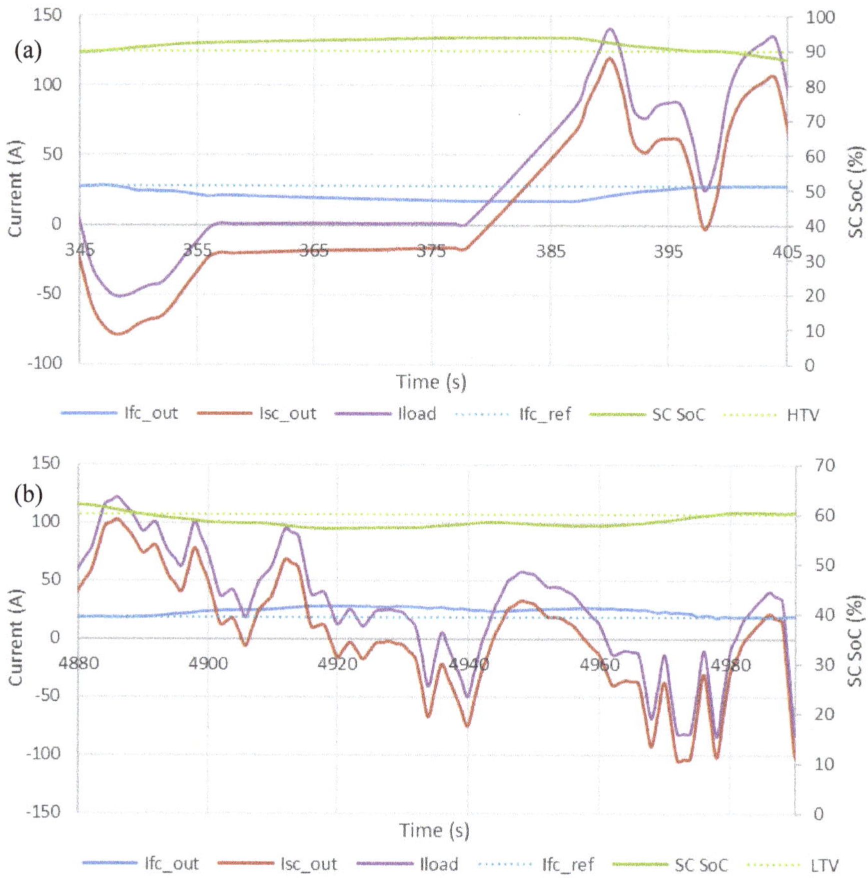

Figure 6. System response of the current to over- and under-charge protection. (**a**) Shows the over-charge protection during profile 1 and (**b**) shows the under-charge protection during profile 3.

5. Degree of Hybridisation Investigation

This section aims to investigate the impact on the degree of hybridisation of the FC variation strategy and the impact that changes to the SC sizing and control strategy parameters has on this. The tests are all carried out on profile 4, the whole day operating profile on route 388, with an initial FC and boost converter output base reference (10.39 kW) utilised for the tests. The SC size utilised for the previous test for the full day driving cycle was a 505 F SC (16.2 kWh) with a 60% lower threshold undercharge protection. The same tests have been carried out with different SC sizes to investigate the impact of degree of hybridisation applied on the same driving cycle. The SC size has been decreased while running the same duty cycle simulation. The required FC power has also been determined by using the highest required power from the FC. The tests have also been run for different values of the LTV, with values of 50%, 60%, and 70% utilised to determine the impact of this on the performance of the system and resulting degree of hybridisation. Hence a degree of hybridisation ratio between the required FC size and SC size can be obtained. The obtained results have been plotted in Figure 7.

It can be seen that reducing the SC size results in an increase in the FC power required, since a smaller SC will experience quicker variations to the SoC. It was also found that further reducing the SC size beyond 3.2 kWh will cause the system to fail for this particular profile. The failure was caused by the SC SoC dropping to quickly for the FC to be able to respond sufficiently and is a result of the SC being too small to effectively act as a damper for the transient power demands of the power profile. It is clear from Figure 6 that the SC size can be reduced significantly but that this comes at the cost of a larger required FC power. Variations to the value of the LTV had a significant impact on the viable values of the degree of hybridisation of the system. It can be seen that reducing the LTV to 50% would increase the required size of both the FC and SC, whereas increasing the LTV to 70% would reduce the required size of both the FC and the SC. It has been found there is a trade-off relationship between the SC size reduction and FC size increase.

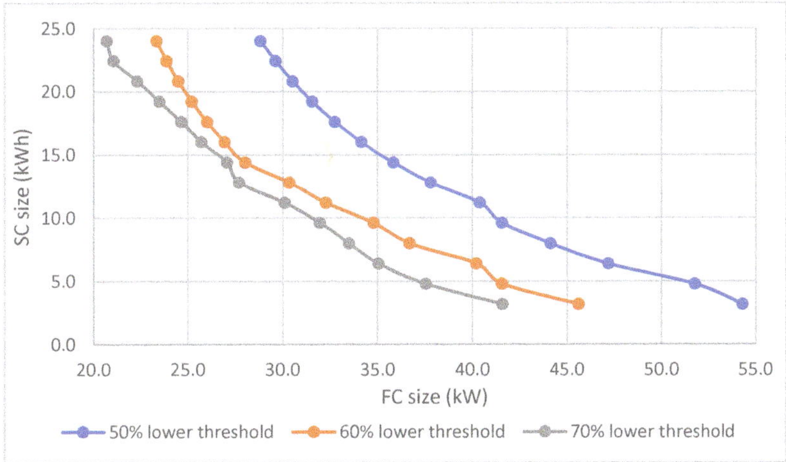

Figure 7. The required FC power and SC size for the full day driving cycle with different lower thresholds.

Although it has been shown that the system having the lowest required FC and SC size occurs for a lowest threshold setting at 70%, this results in more frequent adjustments to the FC output. Utilising a high value of LTV results in a change in the dynamic of the system, where the SC is acting to meet the transient demands but the change in fuel cell output varies more readily to adapt to the power profile. In the cases with the smallest SC size, this variation occurs more frequently and rapidly because of the increase in the rate of change of the SC SoC. This increase in variation comes at the cost of using the FC over a wider dynamic power band. To investigate the FC variation frequency, the percentage of time the FC output varied from the initially defined value of I_{fc_out} has been calculated. The results are plotted with different lower thresholds as shown in Figure 8.

It can be seen from Figure 8 that the 70% lower threshold was subject to the most FC variation for a given SC size, where the FC varied its output for nearly 47% of the day for the worst-case scenario. The variation includes the FC and boost converter output being increased to prevent SoC depletion or being decreased to prevent overcharge. It was found that the average power of the FC and boost converter output for each case is nearly the same with less than 1% variation in results. Since the net power profile of the load are the same for each of the degrees of hybridisation, varying the SC size and LTV will not affect the energy delivery to or from the SC. The minor difference is caused by the charge/discharge efficiency and differing values of the final SC SoC. The same average FC output power also means the total energy delivered by the FC is always the same.

It can be seen that the degree of hybridisation can be optimised with respect to a number of parameters. However, there is always a trade-off relationship for the parameter that is being controlled. There will be a number of factors involved and is about finding the "best balance" amongst those factors.

All the degrees of hybridisation in Figure 7 have been shown to be capable of suitably delivering the service for a complete operating day of route 388. The three hybrid option results in minimum variation, minimum FC size and minimum SC size have been highlighted. One of these parameters can be maximised for each case, but this would also consequently change the other parameters. It can be seen selecting the degree of hybridisation is not simply finding a "best" number. The factors would depend on the requirements of the bus designer.

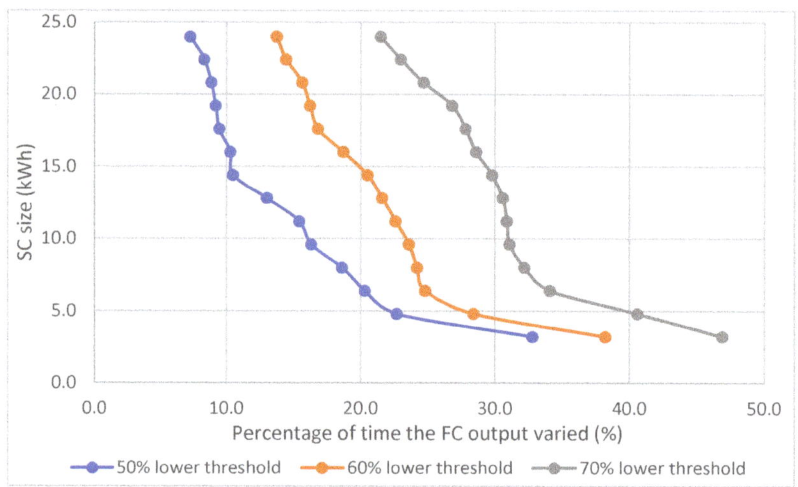

Figure 8. Percentage of time the FC output varied against SC size.

Finally, the degrees of hybridisation proposed for route 388 in this research were compared with other operating FC buses. All the operating buses used for comparison have been in passenger service in commercial use for a relatively long period of time and represent the majority of commercially available FC buses. Information for the operating buses in terms of FC power and energy storage system size was obtained from a number of literatures sources [36–57]. The comparison is plotted in Figure 9. The three options controlled in terms of minimum FC size, SC size and FC variation have been plotted in the FC/SC ratio plot.

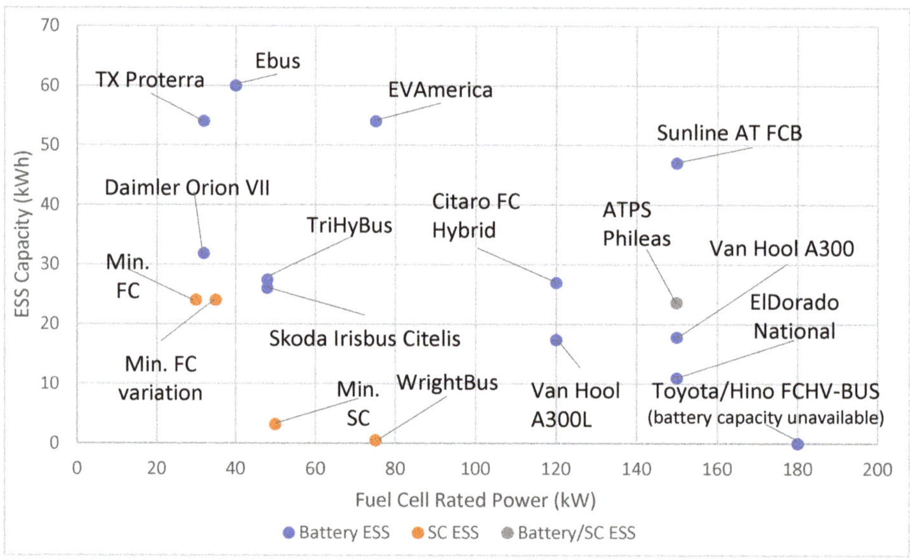

Figure 9. Ratio comparison between route 388 degrees of hybridisation and existing FC buses.

From the FC point of view, it can be seen that the FC size proposed in this research is significantly smaller when compared with those for most existing FC buses. However, there is an important point that needs to be addressed for the FC size comparison. The required FC size used in the proposed degree of hybridisation are the operating power which defines the minimum rated power required from the FC. This is not necessarily the same as the rated power of the FC. Additionally, the degrees of hybridisation proposed in this research were mainly based on the driving cycle of one operating day. The driving cycle is subject to change based on a variety of factors such as season, weather and other events. Although the proposed FC variation strategy will provide some flexibility for the model to be operated under different driving cycles, the required FC size could be increased to be prepared for possible worst-case scenarios. As a result, the degrees identified in this research are more likely to be appropriate for route 388 on that day instead of for route 388 generally. Further information regarding the operating load profile on different days are needed to make an assessment of whether the operating profile collected is representative of normal operation.

From the energy storage point of view, the energy storage size proposed in this research varies over a wider range compared with those installed in existing buses. Most existing FC hybrid bus models utilise Li-ion batteries as the energy storage technology, with the exception of the WrightBus FC bus (Wrightbus, Ballymena, UK) used on the RV1 Bus route in London. The capacities of the battery used in the existing FC buses are generally larger than proposed in this research. The reason for this is the lower power density of the Li-ion batteries [58]. More batteries need to be integrated to provide for the high transient power outputs required. The SC used on the WrightBus (0.5 kWh) is significantly smaller than the proposed SC capacity in the degree of hybridisation for route 388. There are three reasons for this. First, route RV1 is a relatively flat route which was specifically selected for the FC bus demonstration. As a result, the power variations in the route RV1 in terms of magnitude and frequency are expected to be significantly smaller than for the same bus on route 388. Second, RV1 is a single decker bus while the route 388 bus is a double decker bus. This would further reduce the power demand for RV1 compared to the route 388 bus selected for this research. Third, the FC on RV1 is significantly more powerful than the FC proposed for route 388. As shown earlier this has the potential to reduce the size of the required energy storage system.

6. Conclusions

This research evaluated and investigated the control strategy for a FC/SC hybrid power system for city bus applications. This is built on a previously proposed stabilised FC output control strategy and degree of hybridisation identification strategy. Based on the limitations identified with the stabilised operating strategy, a FC variation strategy was applied that offers the facility to adjust the FC and boost converter output reference through monitoring the SC SoC. It was found that the model with the inclusion of the FC variation strategy can not only eliminate the limitations for the initial proposed operation strategy, but also bring potential benefits of further optimising the identified degree of hybridisation. A power profile of a complete day of bus operation was used to test the control strategy and explore the viable range of FC and SC sizing. It was found that the system operated as expected in terms of managing the balance of power and SC SoC throughout the bus journey. It can be concluded that the degree of hybridisation identification strategy can be used to assign an appropriate degree for any FC/SC hybrid bus system and the inclusion of the FC variation strategy is an important feature to add flexibility to power system of the bus.

It was found that there are a wide range of degrees of hybridisation that can fulfil the operating performance requirements. It was found that reducing the size of the SC resulted in the need for a larger required FC power to compensate for the increased rate of change of the SC SoC. Additionally, increasing the value of the LTV resulted in a reduction in both the FC and SC size requirements. However, a greater value of the LTV significantly increased the frequency and magnitude of the variation to the FC power output. It has been found that all of the parameters in the degree of hybridisation are interlinked. As a result, the selection of the degree of hybridisation would be dependent on the requirements of the bus designer. Three controlled degrees of hybridisation namely minimum FC size, minimum SC size and minimum FC variation have been proposed for the route 388. The proposed degrees of hybridisation have been compared with degrees of hybridisation of existing FC buses. It has been found that the FC can be significantly downsized from those used in commercial FC buses.

This research further improved the degree of hybridisation identification strategy by implementing a'FC variation strategy. Although the degrees of hybridisation proposed in this research are more designed for the specific profile, the most important contribution of this research is the strategy to identify and explore the feasible degree of hybridisation options. The degree of hybridisation identification method can be applied on any other route.

Author Contributions: For research articles with several authors, a short paragraph specifying their individual contributions must be provided. The following statements should be used "conceptualization, W.W. and J.S.P.; methodology, W.W. and J.S.P.; software, W.W. and J.S.P.; validation, W.W. and J.S.P.; formal analysis, W.W. and J.S.P.; investigation, W.W. and J.S.P.; resources, W.W. and J.S.P.; data curation, W.W. and J.S.P.; writing—original draft preparation, W.W. and J.S.P.; writing—review and editing, W.W. and J.S.P.; visualization, W.W. and J.S.P.; supervision, R.B.; project administration, R.B.; funding acquisition, R.W.G.B.", please turn to the CRediT taxonomy for the term explanation. Authorship must be limited to those who have contributed substantially to the work reported. All authors have read and agreed to the published version of the manuscript.

Funding: This research was funded by the EPSRC, grant number EP/K021192/1 as part of the HyFCaP project.

Acknowledgments: The authors would like to thank Konrad Yearwood for reviewing this work.

Conflicts of Interest: The authors declare no conflict of interest.

Abbreviations

HTV	Higher threshold value of the supercapacitor SoC (%)
I_{fc_in}	Current output from the Fuel Cell (A)
I_{fc_out}	Current output from the boost converter on the common busbar (A)
I_{fc_max}	Maximum current limit of the fuel cell and boost converter (A)
I_{fc_ref}	Reference value for the boost converter current output on the common busbar (A)
I_{load}	Current to/from the traction motor (A)
I_{SC_in}	Current to/from the Supercapacitor (A)
I_{SC_out}	Current to/from the Buck/Boost converter on the common busbar (A)
LL	Lower limit of the supercapacitor State-of-charge (%)
LTV	Lower threshold value of the supercapacitor SoC (%)
HTV	Higher threshold value of the supercapacitor SoC (%)
P_{fc_out}	Power output from the boost converter on the common busbar (W)
P_{load}	Power to/from the traction motor load (W)
P_{sc_out}	Power to/from the buck/boost converter on the common busbar (W)
SoC	Supercapacitor state-of-charge (%)
V_{fc_in}	Voltage across the Fuel Cell (V)
V_{fc_out}	Voltage across the Boost converter on the busbar (V)
V_{load}	Voltage across the traction motor controller on the busbar (V)
V_{SC_in}	Voltage across the supercapacitor (V)
V_{SC_out}	Voltage across the Buck/Boost converter on the busbar (V)

References

1. London Assembly. London Bus Network Statistics TfL Surface Transport Buses Directorate 2016. Available online: https://www.london.gov.uk/sites/default/files/bus_network_report_final.pdf (accessed on 20 July 2019).
2. TfL. *Health, Safety and Environment Report 2014/15*; Transport for London: London, UK, 2015.
3. Anderson, D. *Transport Statistics Great Britain 2015*; Department for Transport: London, UK, 2015.
4. Datastore, L. *Average Air Quality in London*; King's College London: London, UK, 2019.
5. Carrol, S. *Green Fleet Technology Study for Public Transport*; CENEX: Leicestershire, UK, 2015.

6. Hoffmann, P. *Tomorrow's Energy: Hydrogen, Fuel Cells and the Prospects for a Cleaner Planet*; MIT Press: Cambridge, MA, USA, 2012.

7. Napoli, G.; Micari, S.; Dispenza, G.; Di Novo, S.; Antonucci, V.; Andaloro, L. Development of a fuel cell hybrid electric powertrain: A real case study on a Minibus application. *Int. J. Hydrog. Energy* **2017**, *42*, 28034–28047. [CrossRef]

8. Li, J.; Hu, Z.; Xu, L.; Ouyang, M.; Fang, C.; Hu, J.; Cheng, S.; Po, H.; Zhang, W.; Jiang, H. Fuel cell system degradation analysis of a Chinese plug-in hybrid fuel cell city bus. *Int. J. Hydrog. Energy* **2016**, *41*, 15295–15310. [CrossRef]

9. Fares, D.; Chedid, R.; Panik, F.; Karaki, S.; Jabr, R. Dynamic programming technique for optimizing fuel cell hybrid vehicles. *Int. J. Hydrog. Energy* **2015**, *40*, 7777–7790. [CrossRef]

10. Melo, P.; Ribau, J.; Silva, C. Urban Bus Fleet Conversion to Hybrid Fuel Cell Optimal Powertrains. *Procedia-Soc. Behav. Sci.* **2014**, *111*, 692–701. [CrossRef]

11. Marzougui, H.; Amari, M.; Kadri, A.; Bacha, F.; Ghouili, J. Energy management of fuel cell/battery/ultracapacitor in electrical hybrid vehicle. *Int. J. Hydrog. Energy* **2017**, *42*, 8857–8869. [CrossRef]

12. Hames, Y.; Kaya, K.; Baltacioglu, E.; Turksoy, A. Analysis of the control strategies for fuel saving in the hydrogen fuel cell vehicles. *Int. J. Hydrog. Energy* **2018**, *43*, 10810–10821. [CrossRef]

13. Torreglosa, J.; Garcia, P.; Fernandez, L.; Jurado, F. Predictive Control for the Energy Management of a Fuel Cell-Battery-Supercapacitor Tramway. *IEEE Trans. Ind. Inform.* **2013**, *10*. [CrossRef]

14. Garcia, P.; Torreglosa, J.P.; Fernandez, L.M.; Jurado, F. Control strategies for high-power electric vehicles powered by hydrogen fuel cell, battery and supercapacitor. *Expert Syst. Appl.* **2013**, *40*, 4791–4804. [CrossRef]

15. Zhao, H.; Burke, A.F. Fuel cell powered vehicles using supercapacitors-device characteristics, control strategies, and simulation results. *Fuel Cells* **2010**, *10*, 879–896. [CrossRef]

16. Thounthong, P.; Raël, S.; Davat, B. Control strategy of fuel cell/supercapacitors hybrid power sources for electric vehicle. *J. Power Sources* **2006**, *158*, 806–814. [CrossRef]

17. Behdani, A.; Naseh, M.R. Power management and nonlinear control of a fuel cell e supercapacitor hybrid automotive vehicle with working condition algorithm. *Int. J. Hydrog. Energy* **2017**, *42*, 24347–24357. [CrossRef]

18. Roda, V.; Nigro, N.M.; Carignano, M.G.; Costa-Castell, R.; Junco, S.; Feroldi, D. Energy management strategy for fuel cell-supercapacitor hybrid vehicles based on prediction of energy demand. *J. Power Sources* **2017**, *360*. [CrossRef]

19. Li, T.; Liu, H.; Zhao, D.; Wang, L. Design and analysis of a fuel cell supercapacitor hybrid construction vehicle. *Int. J. Hydrog. Energy* **2016**, *41*, 12307–12319. [CrossRef]

20. Feroldi, D.; Carignano, M. Sizing for fuel cell/supercapacitor hybrid vehicles based on stochastic driving cycles. *Appl. Energy* **2016**, *183*, 645–658. [CrossRef]

21. Allaoua, B.; Asnoune, K.; Mebarki, B. Energy management of PEM fuel cell/supercapacitor hybrid power sources for an electric vehicle. *Int. J. Hydrog. Energy* **2017**, *42*, 21158–21166. [CrossRef]

22. El Fadil, H.; Giri, F.; Guerrero, J.M.; Member, S. Modeling and Nonlinear Control of a Fuel Cell/Supercapacitor Hybrid Energy Storage System for Electric Vehicles. *IEEE Trans. Veh. Technol.* **2014**, *63*, 3011–3018. [CrossRef]

23. Thounthong, P.; Pierfederici, S.; Martin, J.; Hinaje, M.; Davat, B. Modeling and Control of Fuel Cell/Supercapacitor Hybrid Source Based on Differential Flatness Control. *IEEE Trans. Veh. Technol.* **2010**, *59*, 2700–2710. [CrossRef]

24. Benyahia, N.; Denoun, H.; Zaouia, M.; Rekioua, T.; Benamrouche, N. Power system simulation of fuel cell and supercapacitor based electric vehicle using an interleaving technique. *Int. J. Hydrog. Energy* **2015**, *40*, 15806–15814. [CrossRef]

25. Ziaeinejad, S.; Sangsefidi, Y.; Mehrizi-Sani, A. Fuel Cell-Based Auxiliary Power Unit: EMS, Sizing, and Current Estimator-Based Controller. *IEEE Trans. Veh. Technol.* **2016**, *65*, 4826–4835. [CrossRef]

26. Geng, Z.; Hong, T.; Qi, K.; Ambrosio, J.; Gu, D. Modular regenerative emulation system for DC-DC converters in hybrid fuel cell vehicle applications. *IEEE Trans. Veh. Technol.* **2018**, *67*, 9233–9240. [CrossRef]

27. Snoussi, J.; Elghali, S.B.; Benbouzid, M.; Mimouni, M.F. Optimal sizing of energy storage systems using frequency-separation-based energy management for fuel cell hybrid electric vehicles. *IEEE Trans. Veh. Technol.* **2018**, *67*, 9337–9346. [CrossRef]

28. Odeim, F.; Roes, J.; Heinzel, A. Power Management Optimization of a Fuel Cell/Battery/Supercapacitor Hybrid System for Transit Bus Applications. *IEEE Trans. Veh. Technol.* **2016**, *65*, 5783–5788. [CrossRef]

29. Depature, C.; Lhomme, W.; Sicard, P.; Bouscayrol, A.; Boulon, L. Real-Time Backstepping Control for Fuel Cell Vehicle Using Supercapacitors. *IEEE Trans. Veh. Technol.* **2018**, *67*, 306–314. [CrossRef]
30. Wu, W.; Partridge, J.S.; Bucknall, R.W.G. Stabilised control strategy for PEM fuel cell and supercapacitor propulsion system for a city bus. *Int. J. Hydrog. Energy* **2018**. [CrossRef]
31. Wu, W.; Partridge, J.S.; Bucknall, R.W.G. Simulation of a stabilised control strategy for PEM fuel cell and supercapacitor hybrid propulsion system for a city bus. *Int. J. Hydrog. Energy* **2018**, *3*. [CrossRef]
32. Wu, W.; Partridge, J.; Bucknall, R. Development and Evaluation of a Degree of Hybridisation Identification Strategy for a Fuel Cell Supercapacitor Hybrid Bus. *Energies* **2019**, *12*, 142. [CrossRef]
33. O'Hayre, R.; Cha, S.-W.; Colella, W.; Prinz, F.B. *Fuel Cell Fundamentals*; John Wiley & Sons: Hoboken, NJ, USA, 2016.
34. Xie, C.; Xu, X.; Bujilo, P.; Shen, D.; Zhao, H.; Quan, S. Fuel cell and lithium iron phosphate battery hybrid powertrain with an ultracapacitor bank using direct parallel structure. *J. Power Sources* **2015**, *279*, 487. [CrossRef]
35. Barbir, F. *PEM Fuel Cells Theory and Practice*; Elsevier Academic Press: Cambridge, MA, USA, 2005.
36. HyFLEET:Cute. *The HyFLEET: CUTE Project Partners: Uniting for Progress*; Clean Urban Transport for Europe: Ulm, Germany, 2009; pp. 1–52.
37. Tyler, T.; Core, R.D. Present at the Fuel Cell Bus Workshop 2011, San Francisco, CA, USA, 24 February 2011.
38. Hua, T.; Ahluwalia, R.; Eudy, L.; Singer, G.; Jermer, B.; Asselin-Miller, N.; Wessel, S.; Patterson, T.; Marcinkoski, J. Status of hydrogen fuel cell electric buses worldwide. *J. Power Sources* **2014**, *269*, 975–993. [CrossRef]
39. Binder, M.; Faltenbacher, M.; Kentzler, M.; Schuckert, M. *CUTE—Deliverable No. 8: Final Report*; Clean Urban Transport for Europe: Ulm, Germany, 2006; pp. 1–85.
40. London Fuel Cell Bus Trial 2007. Available online: https://www.yumpu.com/en/document/view/27649118/london-hyfleetcute-experience-international-fuel-cell-bus (accessed on 20 July 2019).
41. Trihybus, T. TriHyBus and Triple Hybrid. 2010, pp. 1–24. Available online: www.h2fc-fair.com (accessed on 23 July 2019).
42. Doucek, A.; Janík, L. TriHyBus: The First Fuel Cells Hydrogen Bus in New EU Countries. *ESC Trans.* **2011**, *32*, 49–53. [CrossRef]
43. Eudy, L.; Post, M. *Fuel Cell Buses in U.S. Transit Fleets: Current Status 2014*; National Renewable Energy Laboratory: Golden, CO, USA, 2014.
44. Eudy, L.; Renewable, N.; Chandler, K. *National Fuel Cell Bus Program: Proterra Fuel Cell Hybrid Bus Report, Columbia Demonstration*; United States. Federal Transit Administration: Washington, DC, USA, 2011.
45. Mohrdieck, C. Next Generation Fuel Cell Technology for Passenger Cars and Buses. In Proceedings of the 24th Electric Vehicle Symposium, Stravanger, Norway, 13–16 May 2009; pp. 209–213.
46. Eudy, L.; Chandler, K. *SunLine Transit Agency Advanced Technology Fuel Cell Bus Evaluation: Fourth Results Report*; National Renewable Energy Laboratory: Golden, CO, USA, 2013.
47. Pistoria, G. *Lithium-Ion Batteries*, 1st ed.; Elsevier: Amsterdam, The Netherlands, 2014; ISBN 9780444595133.
48. Slavík, J. Electric Buses in Urban Transport—The Situation and Development Trends. *J. Traffic Transp. Eng.* **2014**, *2*, 45–58.
49. Lipman, T.E.; Gray-Stewart, A.L.; Lidicker, J. Driver Response to Hydrogen Fuel Cell Buses in a Real-World Setting. *Transp. Res. Rec.* **2015**, *2502*, 48–52. [CrossRef]
50. Eudy, L.; Post, M. *Zero Emission Bay Area (ZEBA) Fuel Cell Bus Demonstration Results: Fourth Report*; NREL: Golden, CO, USA, 2015.
51. Zaetta, R.; Madden, B. Hydrogen Fuel Cell Bus Technology State of the Art Review. 2011. Available online: http://s3.amazonaws.com/zanran_storage/nexthylights.eu/ContentPages/2481166193.pdf (accessed on 20 July 2019).
52. Presentation of Emerging Conclusions Detailed Project's Presentation 2014. Available online: https://www.eltis.org/sites/default/files/case-studies/documents/chic_emerging_conclusions_update_december_2014_fv.pdf (accessed on 20 July 2019).
53. Ally, J.; Pryor, T.; Pigneri, A. The role of hydrogen in Australia's transport energy mix. *Int. J. Hydrog. Energy* **2015**, *40*, 4426–4441. [CrossRef]
54. Chandler, K.; Eudy, L. *Cover Photo Credit*; NREL: Golden, CO, USA, 2010.
55. Costa, C. *AC Transit Demos Three Prototype Fuel Cell Buses*; NREL: Golden, CO, USA, 2006.

56. Eudy, L.; Post, M. *American Fuel Cell Bus Project Evaluation: Second Report*; NREL: Golden, CO, USA, 2015.
57. Collaborative, I.F. All Active Demonstrations. 2014. Available online: http://www.gofuelcellbus.com/index.php/project/cologne-apts (accessed on 8 June 2018).
58. Lu, K. *Materials and Energy Conversion, Harvesting and Storage*; John Wiley & Sons: Hoboken, NJ, USA, 2014.

Article

An Approach for Estimating the Reliability of IGBT Power Modules in Electrified Vehicle Traction Inverters

Animesh Kundu [1], Aiswarya Balamurali [1], Philip Korta [1], K. Lakshmi Varaha Iyer [2] and Narayan C. Kar [1,*]

[1] Centre for Hybrid Automotive Research and Green Energy (CHARGE) Labs, University of Windsor, Windsor, ON N9B 3P4, Canada; anik@uwindsor.ca (A.K.); abala@uwindsor.ca (A.B.); kortap@uwindsor.ca (P.K.)

[2] Corporate Engineering and R&D, Magna International Inc., Troy, MI 48098, USA; lakshmiVaraha.Iyer@magna.com

* Correspondence: nkar@uwindsor.ca

Received: 1 June 2020; Accepted: 24 June 2020; Published: 28 June 2020

Abstract: The reliability analysis of traction inverters is of great interest due to the use of new semi-conductor devices and inverter topologies in electric vehicles (EVs). Switching devices in the inverter are the most vulnerable component due to the electrical, thermal and mechanical stresses based on various driving conditions. Accurate stress analysis of power module is imperative for development of compact high-performance inverter designs with enhanced reliability. Therefore, this paper presents an inverter reliability estimation approach using an enhanced power loss model developed considering dynamic and transient influence of power semi-conductors. The temperature variation tracking has been improved by incorporating power module component parameters in an LPTN model of the inverter. A 100 kW EV grade traction inverter is used to validate the developed mathematical models towards estimating the inverter performance and subsequently, predicting the remaining useful lifetime of the inverter against two commonly used drive cycles.

Keywords: Arrhenius model; losses; mission profile; inverter; powertrain; Rainflow algorithm; reliability; thermal network; electric vehicle

1. Introduction

1.1. Background Literature of Inverter Reliability Estimation

Research shows that the power converter is the most vulnerable component in the electric vehicle (EV) powertrain since failure or maintenance of the converter is challenging due to inaccessibility [1]. Thus, research in the areas of inverter reliability estimation and condition monitoring towards design improvement, material optimization and control methods are vital for developing a high performance traction inverter. Towards inverter reliability estimation, developing an accurate model for analyzing temperature variation over a mission profile that affect the power module parameters is of prime importance [2–5]. Temperature variation due to random load profile causes thermal coefficient mismatch that leads to power module material degradation and subsequently towards critical failures like bond-wire lift off and solder fatigue in the power module. However, it is to be noted that the temperature variation within various components of the inverter depends on the load variation based on the mission profile.

In the literature, extensive research has been conducted to analyze and improve inverter reliability through improved control methodology, design parameter identification, and online and offline temperature tracking methods by considering specific mission profiles [5–7]. In order to perform reliability analysis on the inverter, accurate inverter loss model is important for device characterization, performance analysis and temperature estimation. The accuracy of the inverter loss model depends on the parasitic components in the circuit [8], which are difficult to determine experimentally. Hence, it is necessary to develop an accurate analytical model considering the effects of the parasitic components. The most fundamental inverter losses are the conduction and switching losses due to the current flow and switching speed during operation. In [9–11], inverter fundamental loss models have been developed with the device voltage and current characteristics. The energy dissipation map depending on the switching delay is used to determine switching loss of the inverter power switches. For further improvement, considering transient loss model improves the accuracy of inverter reliability estimation. In literature, the transient model has been developed to include turn -on voltage and resistance variation, parasitics components such as terminal capacitances and stray inductances [11]. However, all these influences have not been considered in the loss model towards reliability estimation. Hence, this paper aims at studying the influence of mission profile on reliability estimation considering the transient loss model. Mostly, transient model has been analyzed based on traditional double pulse test (DPT). The transient analysis is developed based on the behavior of parasitic capacitance on the terminals considering physical insights of the device, which increases the complexity of the analytical model. In [11], the parasitic influence considering terminal capacitances and stray induces have been developed based on the datasheet information and DPT test. However, it was challenging to consider temperature changes in the model with the conventional method. Therefore, this paper develops a transient model considering temperature in an iterative loop.

The inverter lifetime considering an EV powertrain has been estimated using power cycling test with a constant load or data driven load profile-based methods for maximum thermal stress analysis [12]. A constant heat generation model is developed to identify hotspot in the power module and bonding strength of the inverter [12–16]. Consequently, variable thermal cycling model is being to identify material degradation based on the average and maximum temperature swing. Finally, the reliability of the inverter estimated is directly related to the change in temperature due to the load variation in the mission profile. In a grid or EV powertrain, it is necessary to consider the data profile for actual lifespan analysis. Previous research on inverter lifetime analysis [17–20], considers the mission profile from the collected load curve based on consumers use. The conventional reliability analysis has been done with standard drive cycle which hardly match with practical scenario.

1.2. Contribution of This Paper

Considering the drawbacks of reliability estimation model in existing literature, this paper develops: (a) an advanced power loss model considering the dynamic and transient effect on the inverter 3-phase output voltage and current, (b) an extensive thermal network model based on inverter material properties to track temperature accurately during load variation and (c) detailed thermal cycle counting method for calculation of lifetime degradation incorporating bonding structure and solder layer dimension. Figure 1 shows the flowchart of the proposed inverter reliability estimation approach. In this paper, an improved inverter lifetime estimation model is developed incorporating an advanced power loss model with dynamic and transient effect such as device saturation, device parasitics on the 3-phase voltage and current for accurate thermal analysis. Cauer network based inverter thermal model is implemented to track temperature changes due to variation of power losses for any mission profile. Thermal network model considering device material information shows the temperature difference at each layer of the power module and accurate mean temperature due to consideration of transient effect in the power loss model. Also, the power loss model and thermal model are dependent on each other, which is updated in an iterative process. This improved power loss model is used to analyze the impact of mission profile for inverter reliability estimation. Finally, a Rainflow algorithm-based cycle counting algorithm is used to estimate remaining lifetime of the inverter. The main objective of this research is to develop an accurate reliability analysis and remaining useful lifetime estimation model of a traction inverter.

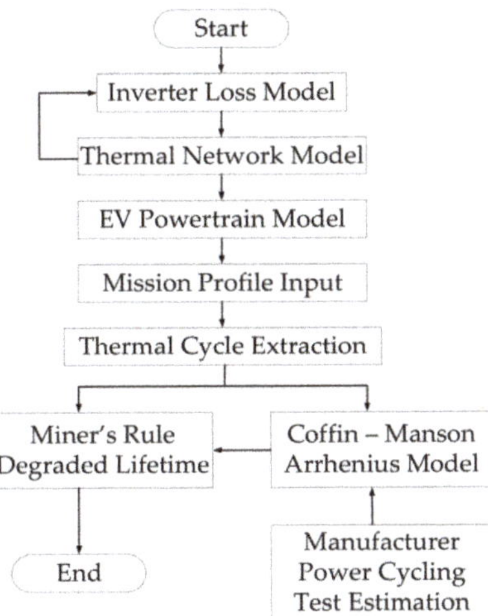

Figure 1. Flowchart of the proposed lifetime estimation procedure.

2. Electro—Thermal Model of 3-Phase VSI for Traction Application

The flowchart of the proposed lifetime estimation procedure is shown in Figure 1. First, the inverter power loss for every load condition is calculated. Following, the calculated power loss is used to observe corresponding temperature changes using the developed thermal network model. The extracted temperature is used as a feedback to the analytical loss model to improve accuracy. This paper considers a detailed inverter loss model considering saturation and non-linearity with parasitic components and an improved thermal network model for dynamic load conditions.

2.1. Improved Power Loss Model Considering Dynamic and Transient Effects

The inverter loss model is generalized for any power switching device. Also, it facilitates fast computation and considers variation of inverter parameters with the temperature changes. The developed model under dynamic operating conditions includes voltage saturation with changing load current, which elaborates conduction and switching losses [7]. Offline data from the device manufacturer has been used for initial condition. In this paper, the conduction loss of the Si-IGBT and Si-diode are modeled separately.

The overall conduction loss combines the losses from six switches. The conduction loss model depends on the current flowing through the device in the operation. Also, the collector—emitter terminal voltage shown in Figure 2a is considered. Both the diode and IGBT voltage and current characteristics have been fused together to simplify the calculation and limit the computational time. The graph also indicates the temperature dependency of the terminal voltage.

With the increase in temperature, the graph shifts in the upward direction. The voltage graph has been stored for room temperature and device maximum temperature. Additionally, linear interpolation method is applied to estimate voltage characteristics at any other temperature. Similarly, the conductive resistance is being calculated using the *V–I* characteristics with the varying temperature.

The energy dissipation during turn—on and off have been acquired with double-pulse test (DPT) method in Figure 3a. Related parameters of the diode and IGBT that have been used in loss and junction temperature calculations have been considered from the device datasheet in Table 1. A half bridge power module has been selected to conduct the DPT. The tests have been conducted at 400 V DC and settled temperatures of 25 °C and 125 °C. The related parameters such as threshold voltage and gate resistance have been stored at the two selected operating temperatures. The parameters in between the selected temperature range are determined by linear interpolation method. The external resistance has been considered based on the high switching operation, which is used to identify minimum switching loss. The initial parameters at the selected temperature have been obtained from the device datasheet and matched with the developed analytical model. Also, the parasitic elements are provided in Table 2. The change in parasitic elements are determined with respect temperature. Depending on the extracted parameters from the DPT, an extended switching loss model has been developed to determine energy dissipation during turn-on and turn-off of the power switches in dynamic operating condition. The switching loss model in (2), considers the on time and off time delay due to the internal and external gate resistance on the input terminal of the power switch. The internal gate resistance is modeled with respect to load current, applied gate voltage and temperature. In the off stage, diode loss is also added to the IGBT switch for simplification. Energy dissipation map during turn–on and off due to high switching operation is represented in Figure 2b. The power losses due to conduction and switching losses from (1) and (2) are considered for both the IGBT and diode with the non–ideality based on temperature variation and voltage saturation [9]. Where V_{CE0}, I_c and R_{CE} are the voltage, current and conductive resistance of a single IGBT device. Energy dissipation due to device switching is represented in (2), where t_{on} and t_{off} are the IGBT turn–on and off time and t_{rr} is the diode turn–off time. V_f represents the voltage due to terminal parasitics. The conduction loss is defined from the device V–I characteristics. In Figure 2c, the switching loss model considers the energy dissipation during turn–off and on through a look-up-table from the test inverter in Figure 3b. The fundamental parameters of the test inverter are given in Table 1, which is obtained from the device datasheet.

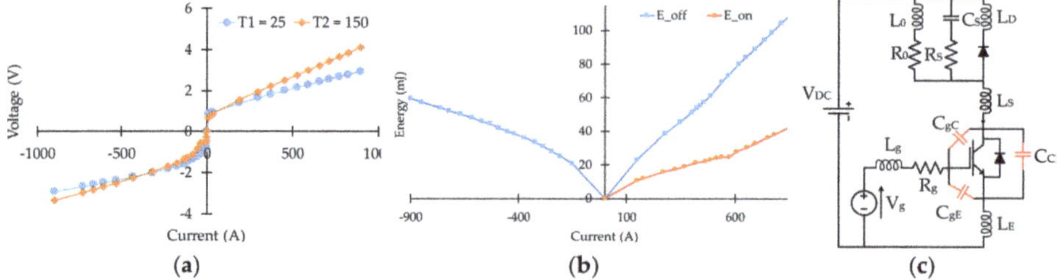

Figure 2. Circuit diagram and device characteristics for parameter extraction towards inverter power loss model, (**a**) *V*–*I* characteristics of IGBT and diode, (**b**) energy dissipation curves for diode and IGBT, and (**c**) Circuit diagram considering terminal parasitics.

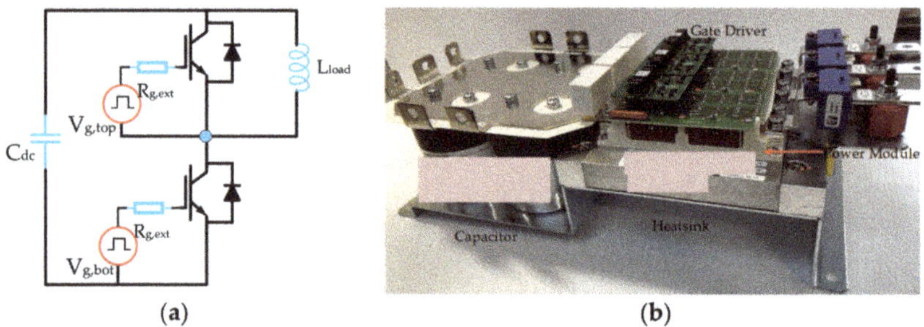

Figure 3. Inverter hardware and IGBT power module layout considered in this paper. (**a**) Inverter DPT circuit diagram, (**b**) 100kW traction inverter.

Table 1. Inverter parameters.

Model	Device Parameters
V_{ce}	1200 V
I_c	556 A
Max. T_j	175 °C
I_F	438 A
R_{ce}	2.3 mΩ
$V_{ge(th)}$	5.8 V
t_{on}	276 ns
t_{off}	538 ns

Table 2. Parasitic elements used to model inverter transient response.

Model	Device Parameters
c_{ies}	26.4 nF
c_{oes}	1.74 nF
c_{res}	1.41 nF
$L_{s/p}$	15 nH

For transient analysis, effective parasitic components among power switch terminal collector, emitter and gate as in Figure 2c have been considered in (3) for accurate voltage measurement towards improved inverter loss model. To improve the accuracy of the inverter power loss model, a comprehensive transient response model has been developed [8,9]. The most significant effect in the transient stage is caused due to collector–gate and gate–emitter capacitance, which is also known as miller capacitance. The transient response of a switch depends on the capacitor charging and discharging factor. Therefore, it is important to monitor capacitor behavior in the switching period. The overall transient loss model is developed in (3)—(6). Threshold voltage is considered during sudden change in junction temperature. The parasitic capacitances are determined in rising and falling interval of the switching period. Previously, derived collector–emitter voltage is included in the model. Collector–emitter resistance, loop and leakage inductances in Table 2, are taken from the datasheet for room temperature, which is being updated in an iterative loop based on temperature.

$$\left.\begin{aligned}
P_{cond,Q}(T_j,t) &= \frac{1}{2}\left[\frac{V_{CE0}(T_j)}{\pi}\hat{i}(t) + \frac{R_{CE}(T_j)}{4}\hat{i}(t)^2\right] + \\
&\quad M\cos\theta\left[\frac{V_{CE0}(T_j)}{8}\hat{i}(t) + \frac{R_{CE0}(T_j)}{3\pi}\hat{i}(t)^2\right] \\
P_{cond,D}(T_j,t) &= \frac{1}{2}\left[\frac{V_{f0}(T_j)}{\pi}\hat{i}(t) + \frac{R_f(T_j)}{4}\hat{i}^2(t)\right] + \\
&\quad M\cos\theta\left[\frac{V_{f0}(T_j)}{8}\hat{i}(t) + \frac{R_f(T_j)}{3\pi}\hat{i}^2(t)\right]
\end{aligned}\right\} \tag{1}$$

$$\left.\begin{aligned}
E_{on} &= \int_0^{t_{on}} V_{CE}(T_j,t)i_c(t)dt \\
E_{off} &= \int_0^{t_{off}+t_{rr}} V_{CE}(T_j,t)i_c(t)dt \\
P_{sw} &= \frac{1}{\pi}f_s[E_{on}(\hat{i}) + E_{off}(\hat{i})]
\end{aligned}\right\} \tag{2}$$

$$V_f = \frac{g_{fs}V_{th} + I_C}{(1+g_{fs}R_g)C_{gC} + C_{CE}}t \tag{3}$$

$$
\left.\begin{array}{r}
C_{ge} = C_{ies} - C_{res} \\
C_{ce} = C_{oes} - C_{res} \\
C_{gc} = C_{res}
\end{array}\right\} \tag{4}
$$

$$
\left.\begin{array}{l}
i_g = C_{ge}\dfrac{dv_{ge}}{dt} + C_{gc}\dfrac{d(v_{ge}-v_{ce})}{dt} \\[2mm]
i_{ce} = \dfrac{v_{ce}}{R_{ce,on}} + C_{ce}\dfrac{dv_{ce}}{dt} - C_{ge}\dfrac{d(v_{ge}-v_{ce})}{dt}
\end{array}\right\} \tag{5}
$$

$$
\left.\begin{array}{l}
v_{ge} = v_c - R_g i_g - (L_g + L_s)\dfrac{di_g}{dt} \\[2mm]
v_{ce} = v_{dc} - v_f - L_p\dfrac{di_{ce}}{dt}
\end{array}\right\} \tag{6}
$$

R_g and g_{fs} represent gate resistance and trans-conductance of the semi-conductor. Gate resistance is stored as a look–up table. In (5) and (6), i_g, i_{ce}, v_{ge} and v_{ce} represent the gate and collector to emitter current and voltage, respectively. L_s and L_g represent the leakage and gate inductances. C_{gC} and C_{ce} represent the parasitic capacitances between the terminal gate–collector and collector–emitter respectively.

2.2. Derivation of Electro—Thermal Model Parameters Considering Inverter Material Information

The reliability of the inverter is dependent on the thermal performance and heat distribution in the device module. The main reason for inverter failure is because of material degradation due to thermal co-efficient mismatch while the temperature changes. The thermal network model has been developed to analyze the thermal stress during load flow in the powertrain (Figure 4). The main objective of the thermal network model is to determine the average temperature and the temperature swing during operation. Further, the possibility of device failure such as solder fatigue and bond-wire lift off are dependent on average temperature and instant temperature variation. Thus, based on the packaging as in Figure 5, the thermal network model is developed for both IGBT and diode.

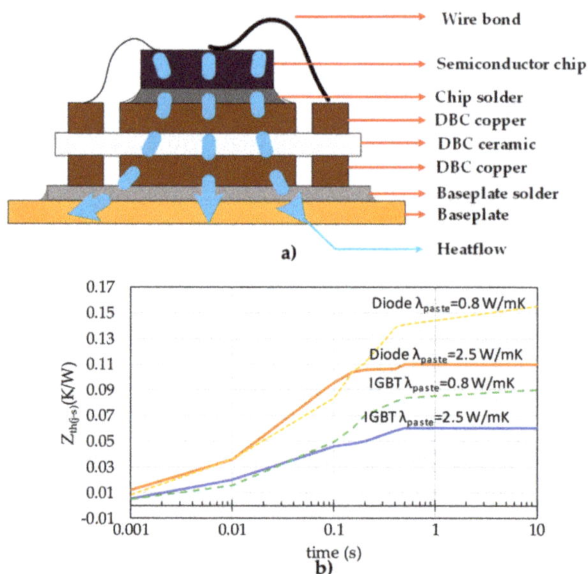

Figure 4. Temperature properties of IGBT model. (**a**) heat flow direction from the chip to heatsink, and (**b**) thermal impedance of inverter module components.

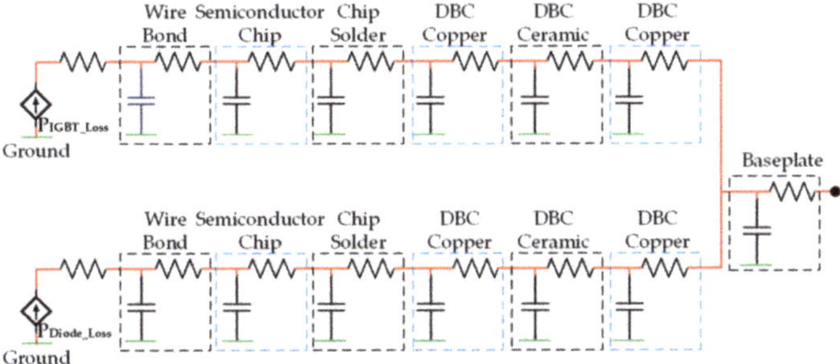

Figure 5. Inverter *RC* thermal network model.

The most common thermal network model is based on Cauer type or Foster network type model. The Foster type model is based on the equivalent circuit model measured from a fitting function. The fitting function is developed from the transient thermal impedance analysis and updated with the linear estimation method. On the other hand, the Cauer type model is based on the material properties and the structure of the power module. The model parameter changes along with the change in material properties, which indicates the fault in the power module effectively. Therefore, this paper focuses on developing the Cauer type inverter thermal network to monitor the temperature changes at each layer of the power module. In this section, the nonuniform heat flow in the power module has been discussed. In the packaging power module heat flows from the top surface IGBT chip to the bottom surface heatsink. The center of the power module carries most of the temperature as the heat flows vertically. In the power module, chip is the most sensitive to heat condition. Also, the aging temperature $T_{j,side}$ is mentioned in (7). The ΔT_j is to monitor the aging process and $T_{j,chip}$ is the chip temperature at the center of the chip. d is the minimum distance:

$$\Delta T_j = \frac{T_{j,chip} - T_{j,side}}{d} \tag{7}$$

In Figure 5, series combination of each thermal resistance and capacitance represent each layer of the device package in Figure 3b. The upper branch represents the heat flow path on the IGBT, and the lower branch shows the heat flow on the diode side. Also, the thermal co-efficient of the device material has been selected from the Table 1. The developed thermal network model represents heat flow at each layer of the device packaging module. The *RC* parameters are identified from the provided material information. The capacitor represents the dynamic temperature changes in dynamic condition. The capacitor represents the temperature rise with the time constant. The thermal model describes the temperature changes at each layer of the module package depending on the loss variation for random dynamic loading. The thermal network model has been developed based on the cross–sectional diagram in Figure 3b. The Z_{jc} defines the sum of the all junction impedances at each layer of the power module. The total impedance combines the seven layers of the IGBT power module. The impedances at the first and the second layer from the top shows the self-heating path due to the die attached [12]. Also, the die in IGBT and diode cause cross heating due to the power loss of its own:

$$Z_{jc} = \sum_{k=1}^{n} \frac{R_k}{1 + s\tau_k} \tag{8}$$

The Cauer-based *RC* thermal network model is expressed in (8). Z_{jc} is the total impedance of the thermal model. k represents number of bond layers in the packaging device. If the power stops flowing into the power switches before it reaches the thermal equilibrium, the device will be steady at the ambient temperature due to the time constant, which depends on the thermal resistance and capacitance value. The time constant is calculated as in [13]. The temperature cooling curve can be represented as an exponential curve in (9):

$$T_{ca}(t) = \sum_{i=1}^{n} \alpha_i e^{\left(-\frac{t}{\tau_i}\right)} \tag{9}$$

The time constant is dependent on the RC parameters in (8). In (9), the time constant is converted into Laplace transfer function. Also, the model is developed for seven layers ($n = 7$) of the power module as in Figure 5:

$$T_{ca}(s) = \sum_{i=1}^{n} \frac{\alpha_i}{s + 1/\tau_i} \tag{10}$$

The inverter thermal analysis is mostly described in 1-D model [12–15]. However, to improve accuracy in this paper third—order analytical model is used to track heat conduction in the power module from the chip to the heatsink. The final Cauer network model in Figure 3b, can be expressed with the Kirchhoff's voltage law the third—order thermal model can be solved as in (11):

$$T_{ca}(t) = \alpha_1 e^{-t/\tau_1} + \alpha_2 e^{-t/\tau_2} + \alpha_3 e^{-t/\tau_3} \tag{11}$$

3. Reliability Performance of VSI under Different Mission Profile

After developing the electro-thermal model of the inverter, a comprehensive stress evaluation can be analyzed with specific load profile based on the inverter design specification for a time duration. This approach is called mission profile. To understand the influence of different mission profiles on reliability estimation, first, a temperature variation profile is estimated using the developed electro-thermal model. Further, using the Rainflow algorithm with extended cycle counting model, the number of temperature cycles is extracted.

3.1. Rainflow Algorithm-Based Inverter Reliability Estimation

For a selected mission profile, the temperature variation of the device module estimated using the developed electro-thermal model is used as an input to the cycle counting algorithm. Subsequently, the offline algorithm is used to find the number of half cycles and full cycles along with their maxima and minima in the temperature profile. Figure 6 shows a sample cycle counting method for a short loading profile. A single 20 s profile has been repeated three times and it increases the temperature by 0.4 °C, which leads towards faulty condition. Figure 6a shows the dynamic performance of the thermal network model. Further, Figure 6b shows the number of half cycle and full cycle count for a single mission profile using the Rainflow algorithm, where a complete temperature swing is considered as a full cycle and each temperature rise or fall is considered as a half cycle [17]. A rising cycle is defined as cycle up and similarly falling half cycle is denoted as cycle down in Figure 6b. A comprehensive Arrhenius model in (12) is used to consider wire bond diameter and solder layer thickness to improve accuracy in reliability analysis. The material geometrical information such as the bonding wire thickness, layer area, height and solder joint material are taken from the manufacturer application note. In (12), N_f defines the number of cycles counted based on the mean temperature and maximum temperature swing in the selected mission profile. N_{Damage} defines the accumulated damage due to the load variation in the mission profile and N_i and N_{fi} are the number of cycles for a specific temperature and the number of temperature cycles in the selected mission profile respectively. Finally, the inverter lifetime is calculated using the damage accumulation factor as in (12) where, K, β_1–β_6 are the curve fitting co-efficient for cycle counting model. The fitting co-efficient are identified with linear curve fitting method. V, I and D represent the conducting voltage, current and the diameter of the wire bond.

$$N_f = K(\Delta T_j)^{\beta_1} e^{\frac{\beta_2}{T_{\max}}} t_{on}^{\beta_3} I^{\beta_4} V^{\beta_5} D^{\beta_6} \tag{12}$$

$$N_{Damage} = \sum \frac{N_i}{N_{fi}} \tag{13}$$

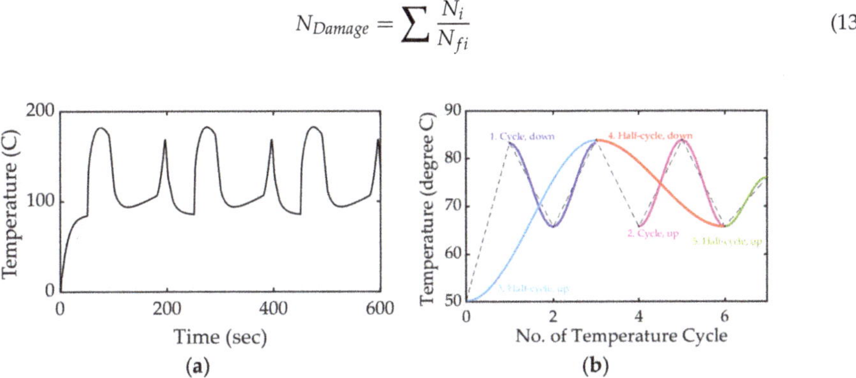

Figure 6. Cycle counting with Rainflow algorithm based on sample data, (a) sample temperature profile, (b) number of cycles counting.

3.2. Impact of Mission Profile on Inverter Remaining Useful Lifetime Estimation

The developed inverter lifetime estimation model has been implemented on an EV grade 100 kW inverter. The damage accumulation in the inverter due to load variation is estimated by the ratio of the initial lifecycle of the inverter and lifecycle after operating under a load profile in (13). The initial inverter state has been determined using device data information provided by the manufacturer. With the provided information ideal inverter condition has been determined and deviation due to load profile has been monitored to identify material degradation. Two most commonly used mission profiles namely Urban Dynamometer Driving Schedule (UDDS) and Worldwide Harmonized Light Vehicles Test Cycle (WLTC3) drive cycles as in Figure 7a,b have been selected for this analysis.

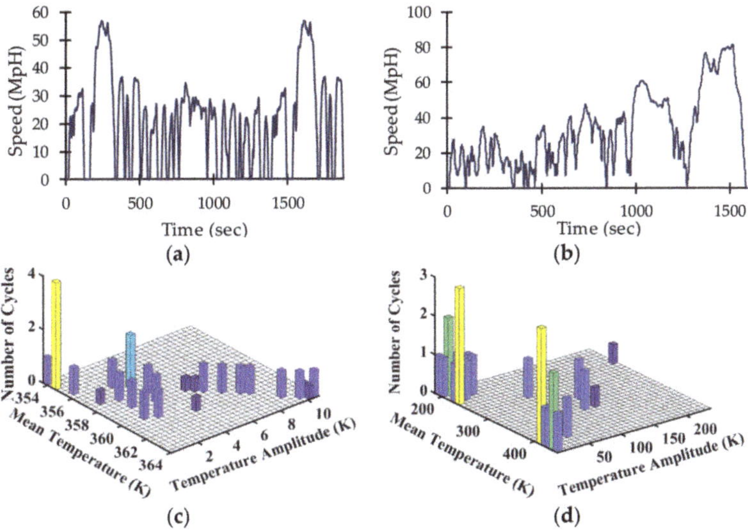

Figure 7. Rainflow based cycle counting for selected mission profiles, (a) urban drive cycle, (b) WLTC3 drive cycle, (c) number of cycles for UDDS, and (d) number of cycles for WLTC3.

To reduce the computational time, cycles with major load changes have been considered for maximum temperature sweep. Also, mean temperature and maximum temperature swing based cycles have been considered to reflect the worst case loading conditions on the inverter. The cycle counting matrix obtained for UDDS and WLTC3 are shown in Figure 7c,d, respectively. First, the UDDS cycle is slightly longer in duration than the WLTC3 cycle. Also, the urban cycle has multiple small cycles, which shows the mean temperature variations in a small range of 354 K to 364 K. On the other hand, WLTC3 drive cycle consists of several low and high load conditions that lead to significant temperature variations. The mean temperature in WLTC3 in between 200 K to 450 K, shows a wide range of load variation, which implies additional stress on the inverter. Similarly, the UDDS has minimum temperature swing observed due to repetitive small cycles and the WLTC3 has maximum temperature swing. Finally, based on the mean temperature and temperature swing, the number of full cycles has been calculated. The number of cycles in UDDS is higher than WLTC3. However, the stress on the inverter is higher in the WLTC3 due to increased mean temperature and maximum temperature swing, which reduces the inverter lifetime by 1.03 years.

4. Conclusions and Future Work

In this paper, a concise inverter reliability prediction model has been developed with improved power loss and thermal models. The power loss model showed a significant effect of transient power loss, which leads to accurate temperature tracking. The thermal network model estimates the temperature difference at each layer of the power module to identify significant material degradation. Finally, the reliability estimation model showed the inverter damage over the selected standard driving range or mission profile. The inverter transient response model incorporating the parasitic elements improves the loss determination model. Also, it improves temperature tracking for random load changes in mission profile. Finally, from the analytical model, it can be concluded that the selected mission profiles have different load variation, which implies different stress level and causes difference in the estimation model by 1.03 years due to a greater number of load cycles in the WLTC3 mission profile, subsequently leading to reduced mileage over time. Hence, online reliability estimation model could be more accurate, which will avoid dependency on the data driven analysis in the conventional analysis. Also, inverter fault prognosis model will be incorporated in the future to observe increased temperature during fault period.

Author Contributions: Concept—A.K., K.L.V.I., N.C.K.; Methodology—A.K., K.L.V.I., N.C.K.; Analysis—A.K., A.B., P.K.; Resources—N.C.K., K.L.V.I.; Preparation Original Draft—A.K.; Writing, Reviewing and Editing—A.K., A.B., P.K., K.L.V.I. and N.C.K.; Supervision—N.C.K. and K.L.V.I. All authors have read and agreed to the published version of the manuscript.

Funding: This research is funded by Magna International, Ontario Center of Excellence (OCE), and Natural Sciences and Engineering Research Council of Canada (NSERC).

Conflicts of Interest: The authors declare no conflict of interest.

References

1. Ni, Z.; Lyu, X.; Yadav, O.P.; Singh, B.N.; Zheng, S.; Cao, D. Overview of Real-Time Lifetime Prediction and Extension for SiC Power Converters. *IEEE Trans. Power Electron.* **2020**, *35*, 7765–7794. [CrossRef]
2. Sangwongwanich, A.; Yang, Y.; Sera, D.; Blaabjerg, F. Mission Profile-Oriented Control for Reliability and Lifetime of Photovoltaic Inverters. *IEEE Trans. Ind. Appl.* **2018**, *56*, 2512–2518.
3. Musallam, M.; Yin, C.; Bailey, C.; Johnson, M. Mission profile-based reliability design and real-time life consumption estimation in power electronics. *IEEE Trans. Power Electron.* **2015**, *30*, 2601–2613. [CrossRef]
4. Sangwongwanich, A.; Yang, Y.; Sera, D.; Blaabjerg, F.; Zhou, D. On the Impacts of PV Array Sizing on the Inverter Reliability and Lifetime. *IEEE Trans. Ind. Appl.* **2018**, *54*, 3656–3667. [CrossRef]
5. Ma, K.; Liserre, M.; Blaabjerg, F.; Kerekes, T. Thermal loading and lifetime estimation for power device considering mission profiles in wind power converter. *IEEE Trans. Power Electron.* **2015**, *30*, 590–602. [CrossRef]
6. Wei, L.; McGuire, J.; Lukaszewski, R.A. Analysis of PWM frequency control to improve the lifetime of PWM inverter. *IEEE Trans. Ind. Appl.* **2011**, *47*, 922–929.

7.   Yang, K.; Guo, J.; Ge, H.; Bilgin, B.; Loukanov, V.; Emadi, A. Transient electro-thermal analysis for a MOSFET based Traction Inverter. In Proceedings of the 2014 IEEE Transportation Electrification Conference and Expo, Detroit, MI, USA, 15–18 June 2014; pp. 1–6.
8.   Li, X.; Jiang, J.; Huang, A.Q.; Guo, S.; Deng, X.; Zhang, B.; She, X. A SiC Power MOSFET Loss Model Suitable for High-Frequency Applications. *IEEE Trans. Ind. Electron.* **2017**, *64*, 8268–8276. [CrossRef]
9.   Xu, Y.; Ho, C.N.M.; Ghosh, A.; Muthumuni, D. A Datasheet-Based Behavioral Model of SiC MOSFET for Power Loss Prediction in Electromagnetic Transient Simulation. In Proceedings of the 2019 IEEE Applied Power Electronics Conference and Exposition (APEC), Anaheim, CA, USA, 17–21 March 2019; pp. 521–526.
10.   Roy, S.K.; Basu, K. Analytical Estimation of Turn on Switching Loss of SiC mosfet and Schottky Diode Pair From Datasheet Parameters. *IEEE Trans. Power Electron.* **2018**, *34*, 9118–9130. [CrossRef]
11.   Peng, H.; Chen, J.; Cheng, Z.; Kang, Y.; Wu, J.; Chu, X. Accuracy-Enhanced Miller Capacitor Modeling and Switching Performance Prediction for Efficient SiC Design in High-Frequency X-Ray High-Voltage Generators. *IEEE J. Emerg. Sel. Top. Power Electron.* **2020**, *8*, 179–194. [CrossRef]
12.   Ye, J.; Yang, K.; Ye, H.; Emadi, A. A Fast Electro-Thermal Model of Traction Inverters for Electrified Vehicles. *IEEE Trans. Power Electron.* **2017**, *32*, 3920–3934. [CrossRef]
13.   An, N.; Du, M.; Hu, Z.; Wei, K. A High-Precision Adaptive Thermal Network Model for Monitoring of Temperature Variations in Insulated Gate Bipolar Transistor (IGBT) Modules. *Energies* **2018**, *11*, 595. [CrossRef]
14.   Tsibizov, A.; Kovačević-Badstübner, I.; Kakarla, B.; Grossner, U. Accurate Temperature Estimation of SiC Power mosfets Under Extreme Operating Conditions. *IEEE Trans. Power Electron.* **2020**, *35*, 1855–1865. [CrossRef]
15.   Nakamura, Y.; Evans, T.M.; Kuroda, N.; Sakairi, H.; Nakakohara, Y.; Otake, H.; Nakahara, K. Electrothermal Cosimulation for Predicting the Power Loss and Temperature of SiC MOSFET Dies Assembled in a Power Module. *IEEE Trans. Power Electron.* **2020**, *35*, 2950–2958. [CrossRef]
16.   Zheng, S.; Du, X.; Zhang, J.; Yu, Y.; Sun, P. Measurement of Thermal Parameters of SiC MOSFET Module by Case Temperature. *IEEE J. Emerg. Sel. Top. Power Electron.* **2020**, *8*, 311–322. [CrossRef]
17.   Ceccarelli, L.; Kotecha, R.M.; Bahman, A.S.; Iannuzzo, F.; Mantooth, H.A. Mission-Profile-Based Lifetime Prediction for a SiC mosfet Power Module Using a Multi-Step Condition-Mapping Simulation Strategy. *IEEE Trans. Power Electron.* **2019**, *34*, 9698–9708. [CrossRef]
18.   Wang, Z.; Wang, H.; Zhang, Y.; Blaabjerg, F. A Viable Mission Profile Emulator for Power Modules in Modular Multilevel Converters. *IEEE Trans. Power Electron.* **2019**, *34*, 11580–11593. [CrossRef]
19.   Choi, U.; Jørgensen, S.; Blaabjerg, F. Advanced Accelerated Power Cycling Test for Reliability Investigation of Power Device Modules. *IEEE Trans. Power Electron.* **2016**, *31*, 8371–8386. [CrossRef]
20.   Xiang, D.; Ran, L.; Tavner, P.; Bryant, A.; Yang, S.; Mawby, P. Monitoring Solder Fatigue in a Power Module Using Case-Above-Ambient Temperature Rise. *IEEE Trans. Ind. Appl.* **2011**, *47*, 2578–2591. [CrossRef]

Article

A Novel Method for Clutch Pressure Sensor Fault Diagnosis

Zhichao Lv and Guangqiang Wu *

School of Automotive Studies, Tongji University, Shanghai 201804, China; lzcren@163.com
* Correspondence: wuguangqiang@tongji.edu.cn

Received: 12 February 2020; Accepted: 28 February 2020; Published: 5 March 2020

Abstract: As a crucial output component, a clutch pressure sensor is of great importance on monitoring and controlling a whole transmission system and a whole vehicle status, both of which play important roles in the safety and reliability of a vehicle. With the help of fault diagnosis, the fault state prediction of a pressure sensor is realized, and this lays the foundation for further fault-tolerant control. In this paper, a fault diagnosis method of Dual Clutch Transmission (DCT) is designed. Firstly, a Variable Force Solenoid (VFS) valve model is established. A feed-forward input system is added to correct the first-order inertial link of the sensor on the second step. Finally, the parameters of the established system model are identified by using the measured data of the actual transmission and the Genetic Algorithm (GA). An identified model is then used for designing a fault observer. The constant output faults of 0, 3, and 5 V, pulse fault, and bias fault that enterprises are concerned with are selected to simulate and verify the fault observer under four different operating conditions. The results show that the designed fault observer has great fault diagnosis performance.

Keywords: fault diagnosis; VFS; GA; input feedforward; fault observation; pressure sensor

1. Introduction

The clutch pressure sensor, as an important signal output component of the drivetrain, plays a very important role in monitoring and controlling the entire drivetrain and its vehicle. Through the acquisition and processing of sensor signals, the optimal control of the clutch under various operating conditions can be completed, and the requirements for vehicle dynamic, economy, safety can also be met. However, in the case of sensor fault, a system cannot accurately monitor the state of the clutch, nor can it effectively control the clutch. This undoubtedly worsens the safety and economy of the vehicle, as well causing irreparable losses to the lifetime of the clutch. Nowadays, the research on the development of automobiles is paying more and more attention to reliability and safety. Research on the fault diagnosis of clutch pressure sensors aims to not only complete the early warning of the sensor fault but also be the basis of the design of fault-tolerant control. Therefore, the diagnosis of a clutch sensor is of great significance.

At present, there are two main methods for fault diagnosis: the model-based fault diagnosis method and the data-based fault diagnosis method. These two ways each have their own pros and cons. The model-based diagnosis method has the advantage of high accuracy, and it can analyze the system and fault from the perspective of mechanism and structure. However, a real system is generally complex, so it is difficult to get an accurate mathematical model of such, which limits model-based diagnosis. The data-based diagnosis method's advantage is that it can easily obtain data, and it can also get expected results by combining intelligent algorithms such as deep learning and artificial intelligence. However, though the data-based method is used to process the output of a system, it cannot get information about the mechanism and structure of the system; as such, the correctness and accuracy of the results largely depend on the output data. It is very difficult to get all the features of a system and its faults through limited amounts of data. Thus, the accuracy of this mode is also limited. It is a practical and wise choice to choose different methods according to different situations.

A method of Fault Detection and Isolation(FDI) control based on statistical process monitoring that is aimed at raising the possibility of implement in industrial systems was proposed in [1]. By recognizing the patterns of measurement data, the faults were divided into two lever, low severity lever, and high severity lever categories. The most informative variables were selected to make the decision, and the robustness of the fault detection and identification from noise is enhanced. By analyzing fault current in different working conditions, a fault diagnosis method was proposed for an Neutral Point Clamped (NPC) inverter in [2]. The authors designed a filter to isolate faults, especially in the switches process. A big data processing method was also adopted for the fault diagnosis area in [3]. The author cut the fault detection into three phases: signal separation, sensor fusion and fault detection. Fault detection was realized based on signal separation and sensor fusing. The author of [4] addressed a new reliability model based on energy dissipation by considering performance degradation behaviors. In view of the shortcomings of traditional fault diagnosis methods, a fault diagnosis method based on fault tree was proposed by NI Shao-xu [5]. Based on the analysis and judgment of system fault phenomena, according to the relationship between task and function, a fault tree model, including each functional unit, was established, the minimum cut set of the system was obtained, the key importance of the unit was calculated, and then the fault diagnosis could be completed [6]. LEI Yaguo [7] proposed a method of deep migration diagnosis for mechanical equipment faults, where their knowledge of migration fault diagnosis that had been accumulated in a laboratory environment was applied to engineering equipment. Firstly, a domain-shared deep residual network was constructed to extract the migration fault features from the monitoring data of different pieces of mechanical equipment; then, a domain adaptation regular term constraint was applied in the training process of a deep residual network to form a deep migration diagnosis model. Author [8] proposed a fault detection method on benchmark problem. Firstly, in order to reach good classification performances, a selection of important features is done. Then a fault database is established. Based on the database, a faulty observation is built. Observation based method is still an effective way for fault diagnosis.

Based on the working principle of solenoid valves, this article establishes a solenoid valve system model, designs an input feedforward system to correct the first-order time lag effect of the sensor, and then use a genetic algorithm combined with the output big data of the actual sensor to identify the model parameters. By using the established model, the fault observer of an actual sensor is designed. The output of the fault observer is proven under different kinds of sensor faults about which enterprises are concerned at present. The system block diagram of fault diagnosis and identification is shown in Figure 1.

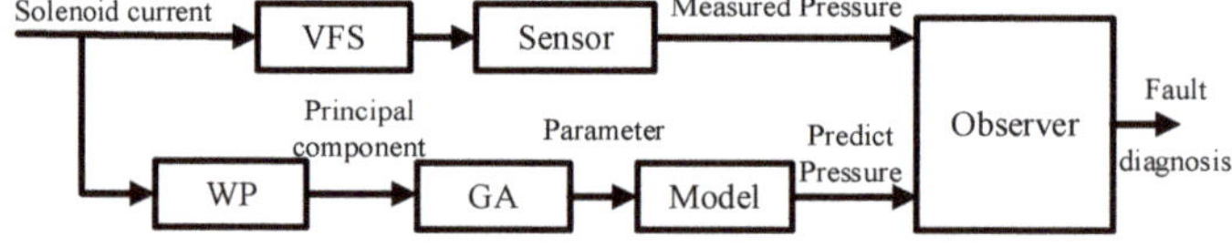

Figure 1. Block diagram of fault diagnosis system.

2. Modeling the Clutch Pressure System

2.1. Establishment of Mathematical Model

The clutch pressure system of Dual Clutch Transmission (DCT) is composed of a VFS (variable force solenoid) valve, a two-position three-way slide valve, and a corresponding oil hydraulic circuit. The structure diagram is shown in Figure 2:

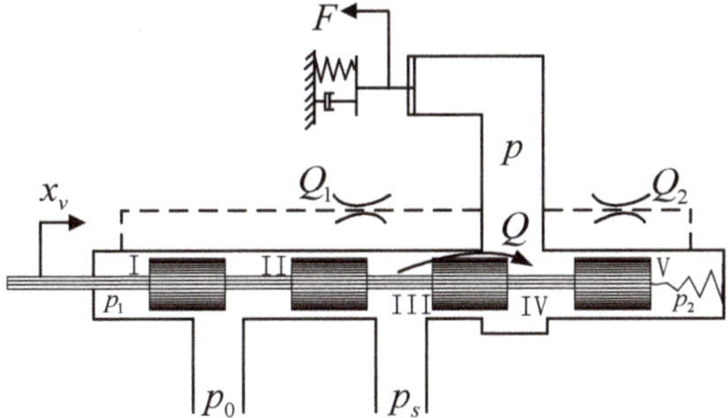

Figure 2. VFS (variable force solenoid) valve structure diagram.

The oil chamber III is directly connected to the oil inlet, and the oil chamber IV is connected to the clutch pressure plate. As the two-position, three-way valve piston moves to the right, the oil return circuit is gradually sealed and the oil inlet circuit is gradually opened. The pressure oil flows from chamber III into chamber IV. In practical work, the leakage flow in a VFS is approximately zero.

From relevant knowledge of fluid mechanics [9] and hydraulic control system [10], the flow rate into the hydraulic cylinder can be obtained so that the flow rate and pressure from the III chamber to the IV chamber meet the throttling Formula (1):

$$Q = C_d w x_v \sqrt{\frac{2}{\rho} \Delta p} \tag{1}$$

where Q is the flow rate, C_d is the flow coefficient of slide valve, w is the port area gradient of the slide valve, x_v is the valve core moving distance, ρ is the hydraulic oil density, and Δp is the chamber pressure difference of both sides.

Assuming that the back-flow pressure is zero, the hydraulic oil is an incompressible fluid, and the throttle orifice is symmetric and matched, the flow in each chamber can be obtained as:

$$Q = C_d w x_v \sqrt{\frac{2}{\rho} (p_s - p)} \tag{2}$$

$$Q_1 = C_d w_1 x_1 \sqrt{\frac{2}{\rho} (p - p_1)} \tag{3}$$

$$Q_2 = C_d w_2 x_2 \sqrt{\frac{2}{\rho} (p - p_2)} \tag{4}$$

Formula (2) can be written as:

$$Q = f(x_v, p) \tag{5}$$

The telescopic distance of the VFS valve is about 1.8 mm. It can be considered that it works in a small range near the operating point; as such:

$$x_v = x_{v0} + \Delta x_v \tag{6}$$

$$p = p_0 + \Delta p \tag{7}$$

Formula (5) becomes:

$$Q = f(x_v, p) = f(x_{v0} + \Delta x_v, p_0 + \Delta p) \tag{8}$$

Expanding Formula (5) into a second-order Taylor series results in:

$$\Delta Q = Q - Q_0 = f(x_v, p) - f(x_{v0}, p_0) = \left[\frac{\partial f}{\partial x_v} \Delta x_v + \frac{\partial f}{\partial p} \Delta p \right]_{\substack{x_v = x_{v0} \\ p = p_0}} \tag{9}$$

The original working point of the VFS valve can be approximated to zero, that is:

$$x_{v0} = p_0 = Q_0 = 0 \tag{10}$$

then:

$$Q = K_q x_v - K_c p \tag{11}$$

where K_q is the flow gain of the slide valve and K_c is the flow pressure coefficient of the spool valve.
Then, Formula (11) becomes:

$$x_v = \frac{Q}{K_q} + \frac{K_c}{K_q} p \tag{12}$$

The pressure in the orifice can be approximately proportional to its orifice area:
From Formulas (3) and (4):

$$C_d w_1 x_1 \sqrt{\frac{2}{\rho}(p - p_1)} \approx C_d w_2 x_2 \sqrt{\frac{2}{\rho}(p - p_2)} \tag{13}$$

From Formula (14):

$$A_1^2(p - p_1) = A_2^2(p - p_2) \tag{14}$$

where $A_i = w_i x_i$
Due to:

$$p \gg p_1 - p_2 = \Delta p \tag{15}$$

Thus, Formula (15) can be approximated as:

$$(A_1^2 - A_2^2)p = \left(\frac{A_1 + A_2}{2} \right)^2 \Delta p \tag{16}$$

Based on Newton's second theorem, the force balance equation of the slide valve can be established as:

$$F_{spool} + p_1 S - p_2 S = m_{mass} \ddot{x}_v + k_\zeta \dot{x}_v + k_{spring} x_v \tag{17}$$

where m_{mass} is the quality of the solenoid armature, k_ζ is partial damping coefficient in the solenoid valve, k_{spring} is elastic coefficient of the return spring in the electromagnet, x_v is the displacement of the servo valve core, F_{spool} is the electromagnetic force on the system, and S is the force area of the piston of the slide valve

Combining the practical working environment of the slide valve and Formulas (12), (16), (17) results in:

$$\frac{\mu_0 S_s}{8 r^2} I^2 = m_{mass} \left(\frac{K_c}{K_q} p \right)'' + k_\zeta \left(\frac{K_c}{K_q} p \right)' + k_{spring} \left(\frac{K_c}{K_q} p \right) - \left(\frac{4(A_1^2 - A_2^2)}{(A_1 + A_2)^2} \right) p S \tag{18}$$

where μ_0 is the vacuum permeability, S_s is the magnetic flux area, r is the winding radius, and I is the energizing current of the solenoid valve.
A Laplace transform of Formula (18) gives:

$$G(s) = \frac{p(s)}{I^2(s)} = \frac{c}{s^2 + as + b} \tag{19}$$

where a, b, and c are parameters that are to be determined.

2.2. Input Feedforward Design

2.2.1. Input Feedforward

Considering that a sensor has a certain delay effect, it is functionally similar to an inertial link [11], so a feedforward system was designed to compensate for that offset:

$$G_{sen}(s) = \frac{\bar{b}}{s + \bar{a}} \tag{20}$$

The zero point of the system was designed according to the time coefficient of the sensor inertia link to correct the system performance.

Formula (19) can be converted into a state space equation:

$$\dot{X} = AX + BU \tag{21}$$

$$Y = CX \tag{22}$$

where $A = \begin{bmatrix} 0 & 1 \\ -b & -a \end{bmatrix}$, $B = \begin{bmatrix} 0 \\ 1 \end{bmatrix}$, $C = \begin{bmatrix} 1 & 0 \end{bmatrix}$.

The observability and controllability of system are proven as follows:

$$\begin{bmatrix} C \\ CA \end{bmatrix} = \begin{bmatrix} 1 & 0 \\ 0 & 1 \end{bmatrix} \tag{23}$$

$$\begin{bmatrix} B & AB \end{bmatrix} = \begin{bmatrix} 0 & 1 \\ 1 & -a \end{bmatrix} \tag{24}$$

The block diagram of the input feedforward system is shown in Figure 3.

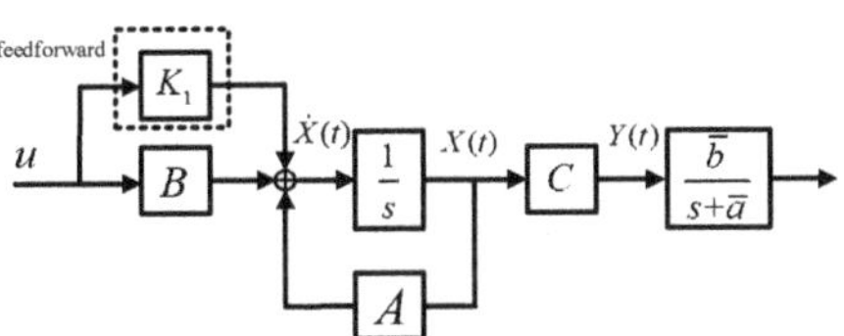

Figure 3. Input feedforward system block diagram.

With input feedforward, the system state space equation becomes:

$$\dot{X} = AX + BU + K_1U = AX + (B + K_1)U \tag{25}$$

$$Y = CX \tag{26}$$

where $K_1 = \begin{bmatrix} k_1 \\ 0 \end{bmatrix}$.

It can be seen from the formula that the input feedforward does not change the C matrix, so the system observability remains unchanged, and the controllability is proven as follows:

$$\begin{bmatrix} B + K_1 & A(B + K_1) \end{bmatrix} = \begin{bmatrix} k_1 & 1 \\ 1 & -bk_1 - a \end{bmatrix} \tag{27}$$

Thus, the system can be controlled.

Therefore, by introducing input feedforward into the original system, the controllability and observability of the system remain unchanged.

At this point, the system transfer function is:

$$G(s) = C[sI - A]^{-1}(B + K) = \frac{k_1 s + (k_1 a + 1)}{s(s + a) + b} \tag{28}$$

If $k_1 = \frac{1}{\bar{a} - a}$, then the zero of the system can be configured and the gain of the transmission function can be adjusted to correct the inertial link of the sensor.

2.2.2. Proof of System Stability

Lyapunov's second law can be used to prove the stability of the system.

When the system input is zero, the state space equation can be expanded to:

$$\dot{x}_1 = x_2 \tag{29}$$

$$\dot{x}_2 = -bx_1 - ax_2 \tag{30}$$

Select:

$$V(x) = \frac{1}{2}(bx_1^2 + x_2^2) \tag{31}$$

Then:

$$\dot{V}(x) = bx_1 x_2 + x_2(-bx_1 - ax_2) = -ax_2^2 \tag{32}$$

According to Lyapunov's second law, if system stability requires $V(x) > 0$, $\dot{V}(x) < 0$, then it requires $b > 0$ and $a > 0$.

Figure 4 shows the steps of the modeling process.

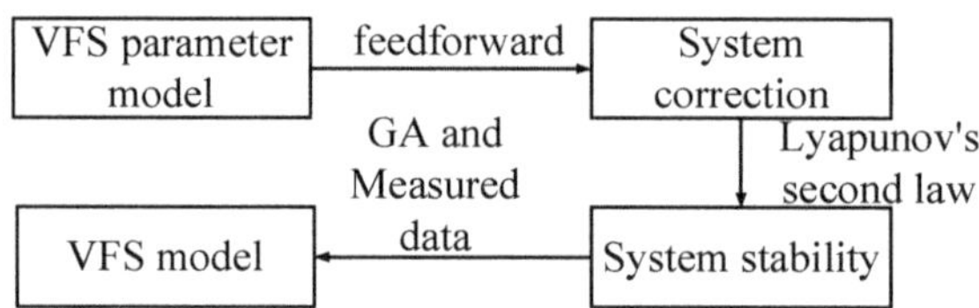

Figure 4. Steps of modeling.

2.3. Model Parameter Identification Based on GA Algorithm

2.3.1. Input Signal Principal Component Extraction Based on Wavelet Packet Transform

Wavelet packet analysis is a kind of effective signal processing theory. The approximate coefficients and detail coefficients of layer 1 are obtained by filtering the original signal through a low pass filter and a high pass filter, respectively. By filtering the approximate coefficients and detail coefficients of layer 1, the approximate coefficients and detail coefficients of layer 2 are gained. The process is then continued until the requirement is met. The signal can be expressed as:

$$a^{m-1} = G^* a^m + H^* d^m \tag{33}$$

where G^* is the conjugate matrix of the low pass filter and H^* is the conjugate matrix of the high pass filter. The original signal can be expressed as a^0.

Here, wavelet packet analysis was performed on the solenoid valve current signals to extract principal component information. The data used for identification were collected when the throttle pedal was at its 30% opening value. The solenoid valve input current signal is shown in Figure 5. The wavelet packet decomposition of the current signal is shown in Figure 6

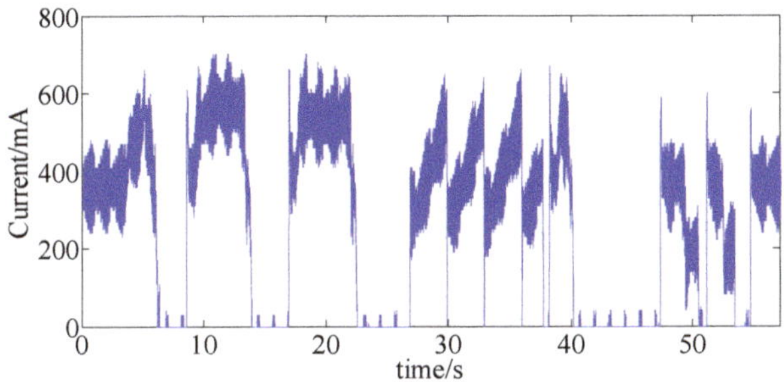

Figure 5. Control current of a solenoid valve.

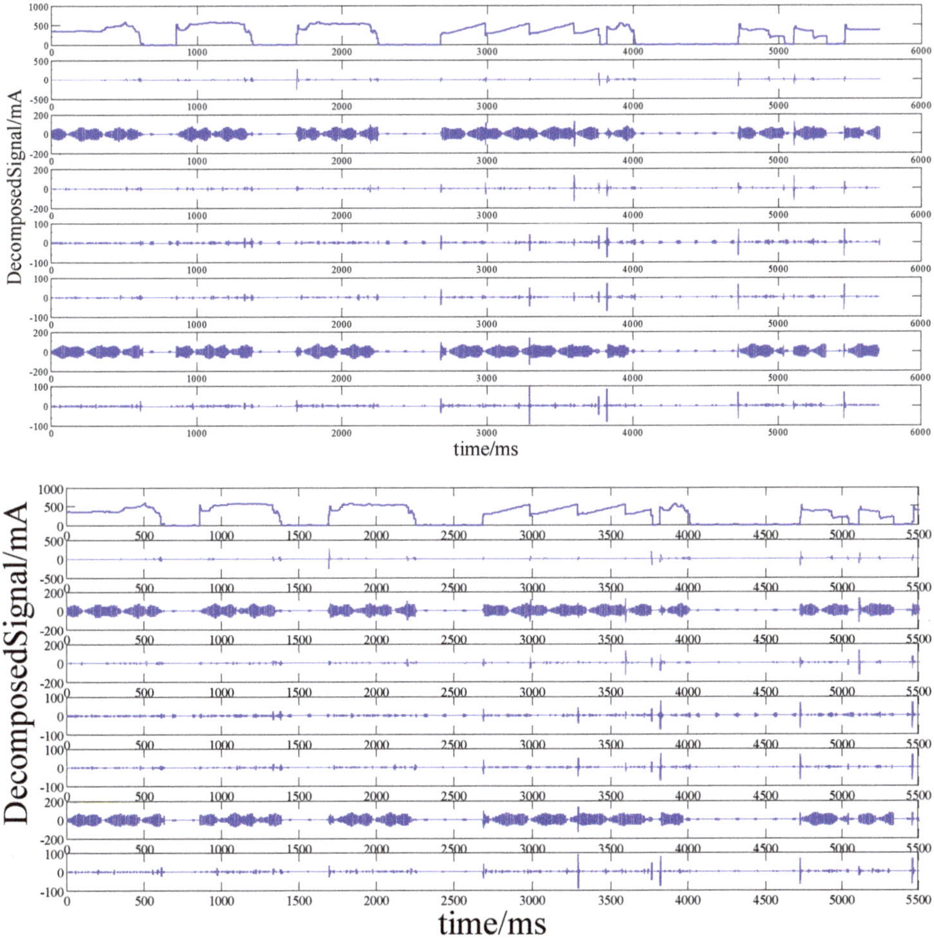

Figure 6. Schematic diagram of wavelet packet signal decomposition.

The root mean square value and two-norm value of all eight-layer wavelet signals that were decomposed by the wavelet packet were obtained, and the ratio of the wavelet packet signal in each layer to the original signal was calculated according to Formulas (34) and (35). The result is shown in Figure 6.

$$y_{1j} = \frac{Norm_j}{\sum\limits_{i}^{8} Norm_i} \tag{34}$$

where $Norm_j$ is the two-norm of the j layer signal.

$$y_{2j} = \frac{Std_j}{\sum\limits_{i}^{8} Std_i} \tag{35}$$

where Std_j is the j layer signal's standard deviation.

Figure 7 shows the signals, except for the first layers are all high-frequency signals, which are different from the frequency of the control signal and closer to the noise. Figure 7 also shows the standard deviation and two-norm values of all the layers. The seventh layer signal's deviation and its two-norm value were 63% and 73% of the original signal, respectively. As shown in Figure 8, the energy of the original signal was also mainly concentrated in the signal of layer 7, so the signal of layer 7 was selected as the input control signal.

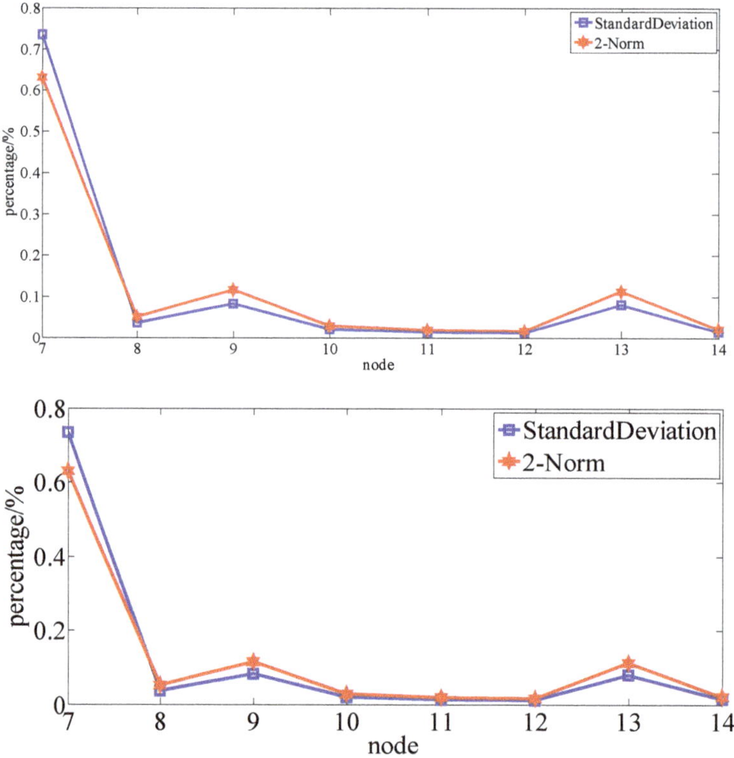

Figure 7. Standard deviation and two-norm of each data node.

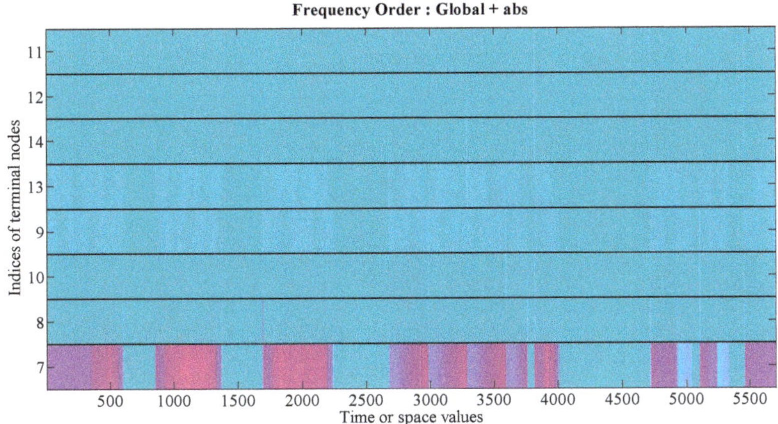

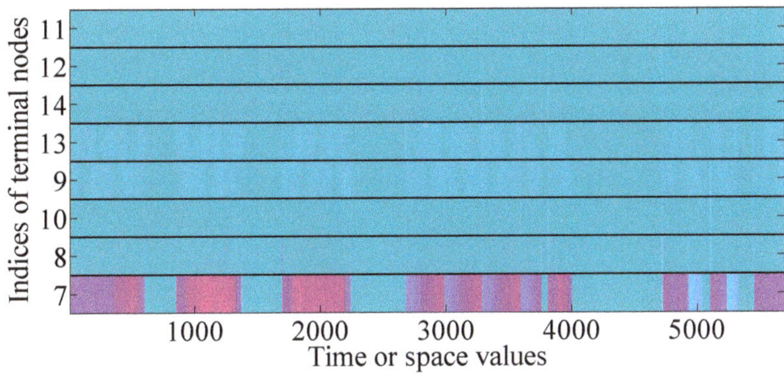

Figure 8. Data analysis of each data node.

2.3.2. Genetic Algorithm

Through feedforward correction, the system could be approximated to a second-order system with three parameters. The parameters of the model could be determined by using the data that were measured by the actual sensor and the genetic algorithm.

A genetic algorithm is a kind of randomized search algorithm that uses natural selection and natural genetic mechanisms for reference. A genetic algorithm simulates the phenomenon of reproduction, crossover, and gene mutation in the process of natural selection and natural heredity. This algorithm keeps a group of candidate solutions in each iteration, selects the better individuals from the solution group according to a certain index, and uses genetic operators (selection, crossover and mutation) to combine these individuals to generate a new generation of candidate solution groups. This process can be repeated until a certain income meet the requirements.

A genetic algorithm is composed of the following four parts:

(1) Coding (generating initial population).
(2) Fitness function.
(3) Genetic operator (selection, crossover, and mutation).
(4) Operation parameters.

The coding of the target is the first step of the algorithm, then the fitness function is used to calculate the fitness value for selecting the better individuals. After the selection, crossover, and mutation processes, a new population is generated. The iteration of the algorithm does not stop until the target meets the requirement.

In this paper, the genetic algorithm and data measured from actual sensor were used to identify the parameters of the solenoid valve model. Binary code was chosen, the original population number was set as 30, the genetic algebra was set as 300, and the fitness function was selected as the root mean square error of the actual sensor output and the model output of each generation. The fitness function is shown in Formula (36):

$$fit = \sqrt{\frac{\left(\widetilde{y}_i - \widehat{y}_i\right)^2}{n}} \tag{36}$$

where $\widetilde{y}_i$ is the model output of the i generation and $\widehat{y}_i$ is the actual sensor output.

As shown in Figure 9, the best and average fitness values of the system for each generation were declined. The best fitness function value after 300 generations of algorithm iteration was about 1.66e + 11. The model parameters were selected at this time as $a = 54$, $b = 993$, and $c = 2$.

According to the result of stability analysis, the model that was optimized by the genetic algorithm was stable.

The data that were used for pressure tracking were collected when the throttle was at 50% of its opening value. Layer 7 of the control current of the solenoid valve was tracked. The output of the model was compared with the actual pressure. The result is shown in Figure 10. It can be seen from the figure that the identified system had a high tracking accuracy for the original signal.

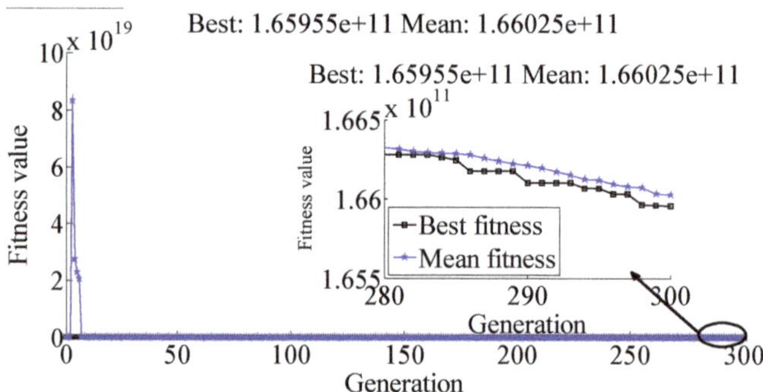

Figure 9. Fitness value of each generation.

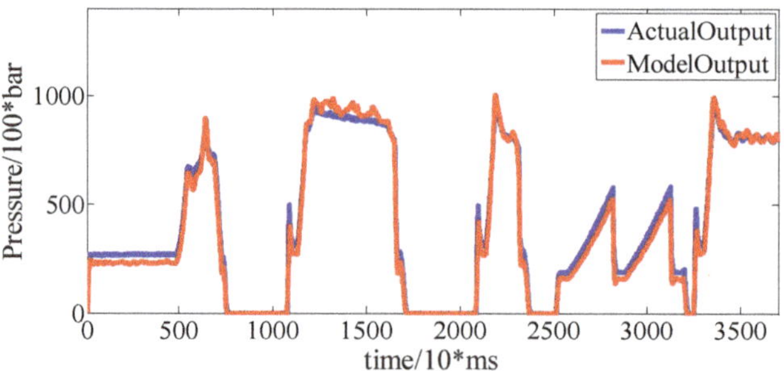

Figure 10. Pressure tracking.

3. Fault Observer

The model was used to design a fault observer of the original solenoid valve system. The structure diagram of the fault observer is shown in Figure 11.

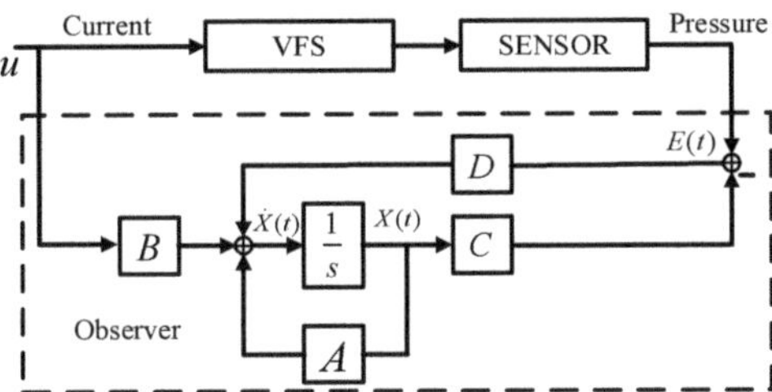

Figure 11. Diagram of the fault observer.

The system equation is:

$$\dot{\mathbf{X}} = A\mathbf{X} + B\mathbf{u} \tag{37}$$

$$\mathbf{Y} = C\mathbf{X} \tag{38}$$

The observer was designed as:

$$\dot{\hat{\mathbf{X}}} = A\hat{\mathbf{X}} + B\mathbf{u} + D\left[\mathbf{Y}(t) - \hat{\mathbf{Y}}(t)\right] \tag{39}$$

$$\hat{\mathbf{Y}} = C\hat{\mathbf{X}} \tag{40}$$

The state error was defined as:

$$e(t) = \mathbf{X}(t) - \hat{\mathbf{X}}(t) \tag{41}$$

The output error was defined as:

$$E(t) = \mathbf{Y}(t) - \hat{\mathbf{Y}}(t) \tag{42}$$

When the sensor fails:

$$\mathbf{Y}(t) = C\mathbf{X}(t) + e_{mj}n(t) \tag{43}$$

$$\begin{aligned}\dot{e}(t) &= \dot{\mathbf{X}}(t) - \dot{\hat{\mathbf{X}}}(t) = A\mathbf{X}(t) + B\mathbf{u}(t) - \left\{A\hat{\mathbf{X}}(t) + B\mathbf{u}(t) + D\left[\mathbf{Y}(t) - \hat{\mathbf{Y}}(t)\right]\right\} \\ &= (A - DC)e(t) - d_j n(t)\end{aligned} \tag{44}$$

$$E(t) = \mathbf{Y}(t) - \hat{\mathbf{Y}}(t) = Ce(t) + e_{mj}n(t) \tag{45}$$

where e_{mj} is the fault vector. In the case that a certain sensor fails, $n(t)$ is the fault manifestation.

Solve Equation (44):

$$\dot{e}(t) - (A - DC)e(t) = -d_j \cdot n(t) \tag{46}$$

$$e(t) = C_e \cdot e^{\int (A-DC)dt} + e^{\int (A-DC)dt} \cdot \int -d_j \cdot n(t) \cdot e^{\int -(A-DC)dt} dt \tag{47}$$

Simplify:

$$\begin{aligned}e(t) &= C_e \cdot e^{(A-DC)t} - e^{(A-DC)t} \cdot \int d_j \cdot n(t) \cdot e^{-(A-DC)t} dt \\ &= C_e \cdot e^{(A-DC)t} - \left[\sum_{i=1}^{j} \frac{d_j \cdot n^{(i-1)}(t)}{[-(A-DC)]^i} + e^{(A-DC)t} \int \frac{1}{[-(A-DC)]^i} e^{-(A-DC)t} d_j \cdot n^{(i)}(t) dt\right]\end{aligned} \tag{48}$$

where $A - DC = -\delta I$ and $\delta > 0$ is the Scalar constant. If the fault type is n-order fault, Formula (48) is converted to:

$$e(t) = -\left[\sum_{i=1}^{j} \frac{d_j \cdot n^{(i-1)}(t)}{[-(A-DC)]^i}\right] \tag{49}$$

where $n^{(i-1)}(t)$ is the $i-1$ order derivative of the fault signal.

Output error:

$$E(t) = -C\left[\sum_{i=1}^{j} \frac{d_j \cdot n^{(i-1)}(t)}{[-(A-DC)]^i}\right] + e_{mj} \cdot n(t) \tag{50}$$

According to Formula (50), the form of output error is determined by the form and the order derivative of the fault signal. Different fault forms have specific output results. By monitoring the state error and output error, system fault diagnosis can be completed. The output error and state error are close to zero when no component fails.

4. Verification

4.1. Constant Output Fault

In this stage, three kinds of sensor faults modes—0 V constant output, 3 V constant output, and 5 V constant output—were selected. The driving conditions of an automobile are classified into four categories: the launching, upshift, fixed gear driving, and downshift conditions. This model assumes that all driving conditions have been covered.

In the launch condition, three kinds of failure are set. The output signal of the clutch pressure sensor is as shown in Figure 11, and the output of fault observer is as shown in Figure 12.

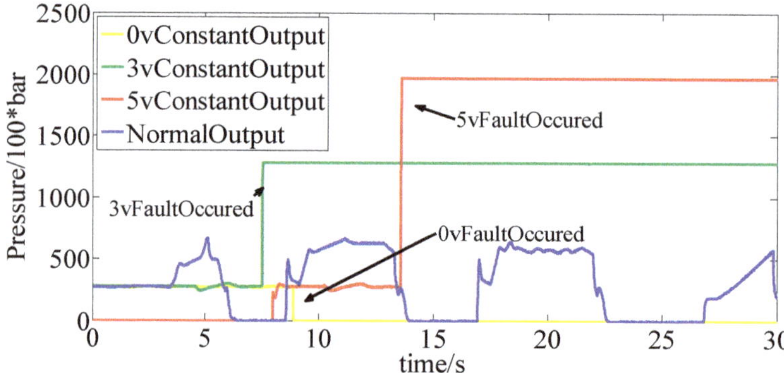

Figure 12. Sensor output under launch condition.

As shown in Figure 12, the outputs of sensor changed to 0, 3 and 5 V at certain time points under launch condition. The fault observer's output also varied, as shown in Figure 13. According to the results, the form of the fault observer's output varied according to different kinds of faults. When there was no fault in the system (the blue line in the figure), the output value fluctuated near 0, indicating that there was no big deviation between the output of the observer and the actual output.

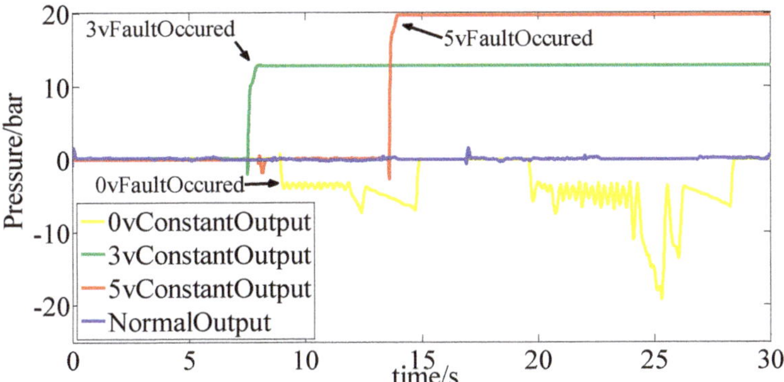

Figure 13. Observer output under launch condition.

When the sensor had a 0 V constant value fault (yellow line in the figure), the fault observer output a signal with a negative amplitude, and the trend of the output was closed to the control current of the solenoid valve. Because the model predicted the clutch pressure value according to the control current of the solenoid valve but the output of the sensor was 0, so the output form of the fault observer followed the difference between the actual sensor and the model system.

When a 3 V constant output fault occurred in the system (green line in the figure), the fault observer output an observation result with a positive amplitude, and the form was consistent with the fault shape of the sensor.

From the system point of view, at this time, the sensor output is a constant non-zero value and feeds back to the controller of the solenoid. The controller of the solenoid generates new control rate according to this received signal. Because the signal is non-zero and constant, the controller's output is surely affected by this value. From the mathematical point of view, the form of fault is constant value fault and the first and above order derivative of fault signal are all zero, so Formula (50) becomes:

$$E(t) = e_{mj} \cdot n(t) \tag{51}$$

Thus, the output of the fault observer and the sensor fault have the same trend.

When a 5 V constant output fault occurred in the system (red line in the figure), the fault observer output an observation result with a positive amplitude, and the form was consistent with the fault shape of the sensor. The cause analysis was consistent with the 3 V constant output fault. As shown in Figure 13, the observer output of different kinds of faults under different driving conditions was confirmed. The threshold for fault diagnosis could be selected for different conditions. The threshold varied with different kinds of faults. In this paper, the threshold for the normal mode and the 3 V constant output fault could be selected as 5 bar, the thresholds for the 3 and 5 V constant output faults could be selected as 17 bar, and the thresholds for the 0 and 3 V constant output faults could be selected as -5 bar. As shown in the Figure 13, the fault was able to be well-diagnosed.

In the upshift condition, three kinds of faults are set. The output signal of the clutch pressure sensor is as shown in Figure 14, and the output of the fault observer is as shown in Figure 15. The threshold could be selected accordingly.

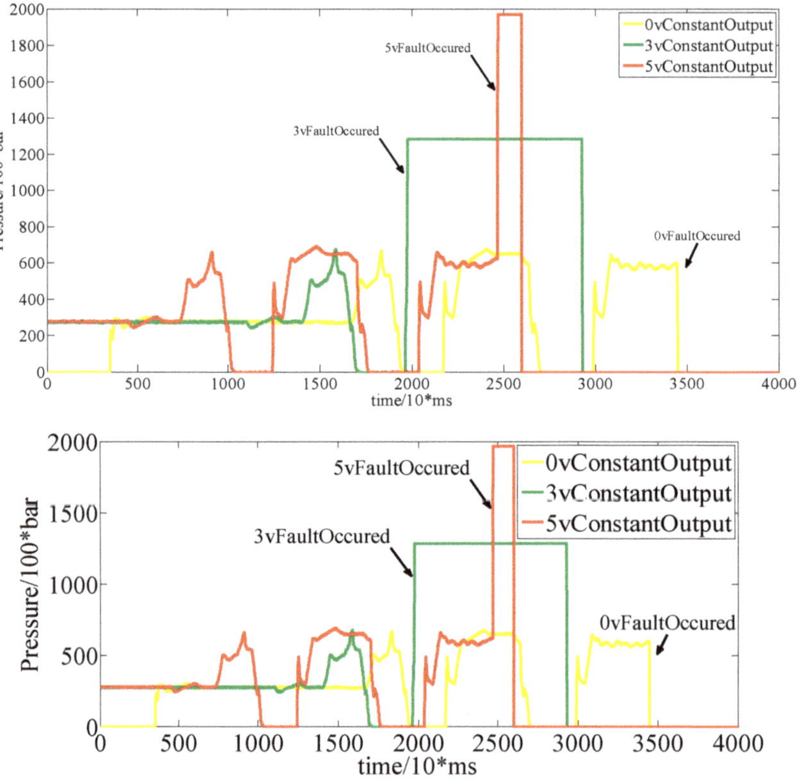

Figure 14. Sensor output under upshift condition.

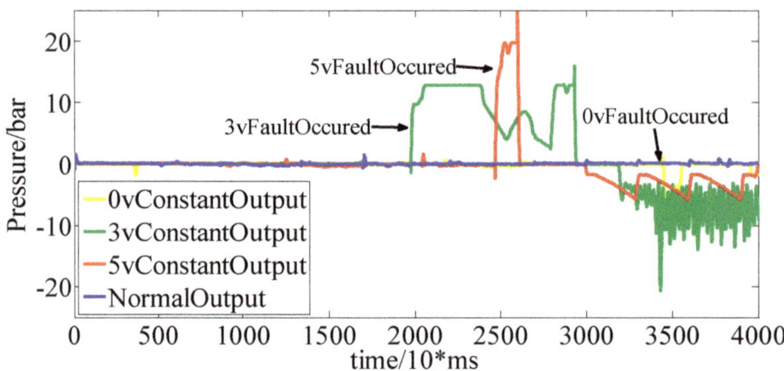

Figure 15. Observer output under upshift condition.

In the fixed gear condition, three kinds of faults are set. The output signals of the clutch pressure sensor are as shown in Figure 16, and the output of the fault observer is as shown in Figure 17. The threshold can be selected accordingly.

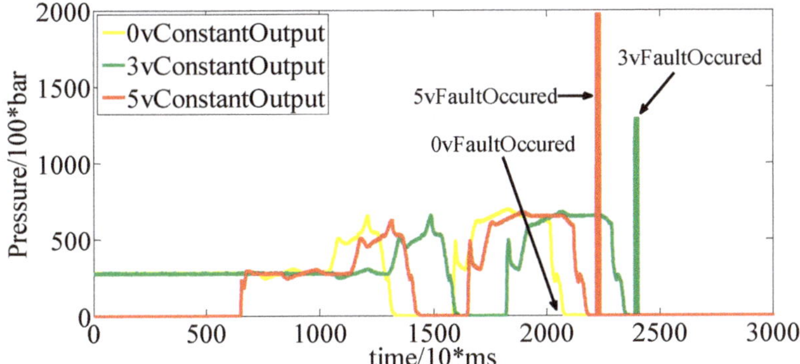

Figure 16. Sensor output under the fixed gear condition.

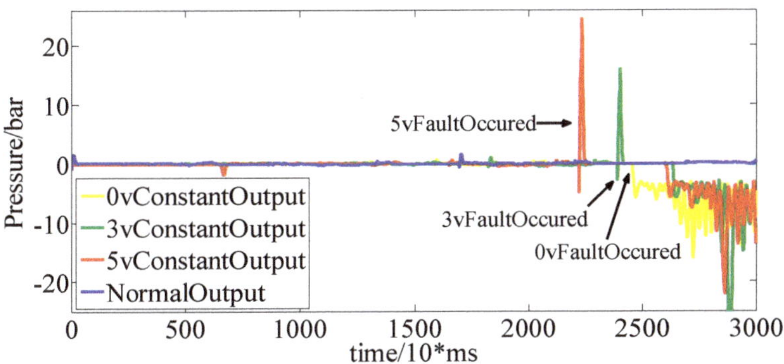

Figure 17. Observer output under the fixed gear condition.

In the downshift condition, three kinds of faults are set. The output signal of the clutch pressure sensor is as shown in Figure 18, and the output of the fault observer is as shown in Figure 19. The threshold can be selected accordingly.

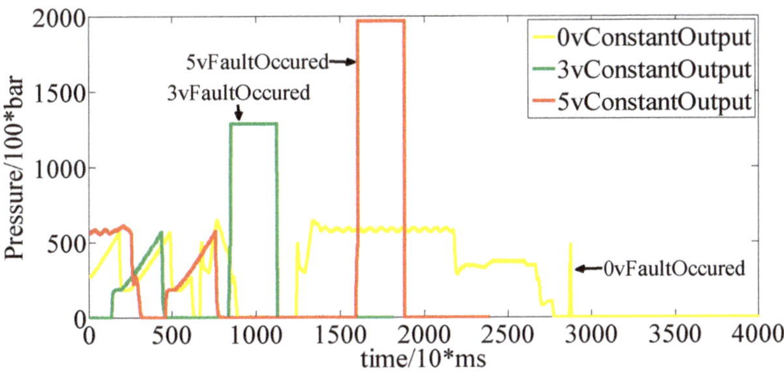

Figure 18. Sensor output under the downshift condition.

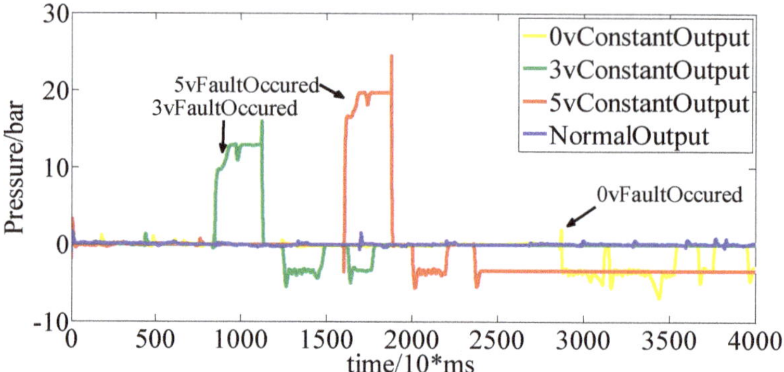

Figure 19. Observer output under the downshift condition.

The observer's output of three kinds of faults under the upshift, fixed gear and downshift conditions are synthetically considered and the result are consistent with that in the launch stage. Because it was impossible to completely shield all the fault-tolerant control strategies and replacement strategies of mature industrial production when collecting the actual sensor signals, the output signals of the sensors were not always consistent under some driving conditions when doing the test, but the diagnosis results were not affected. According to Figures 14–19, the output of the fault observer varied according to different faults, as expected under different driving conditions.

4.2. Pulse Fault

In this stage, the pulse fault was tested. A pulse fault is divided into two kinds; the first one lasts for 0.1 s, and the other is lasts for 1 s, and the enterprise is concerned with both. The amplitude is a certain value. The driving conditions of an automobile are classified into four categories: the launching, upshift, fixed gear driving and downshift conditions. This model assumes that all driving conditions have been covered.

As shown in Figures 20–23, the output of the fault observer varied according to the different types of pulse faults, as expected under different driving conditions. The threshold could be selected accordingly.

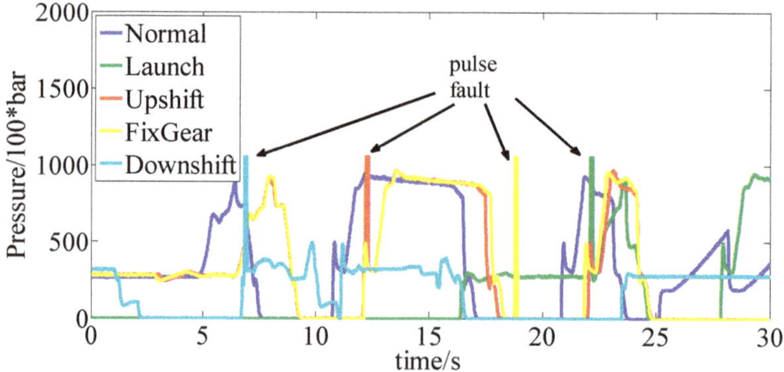

Figure 20. 0.1 s pulse fault sensor.

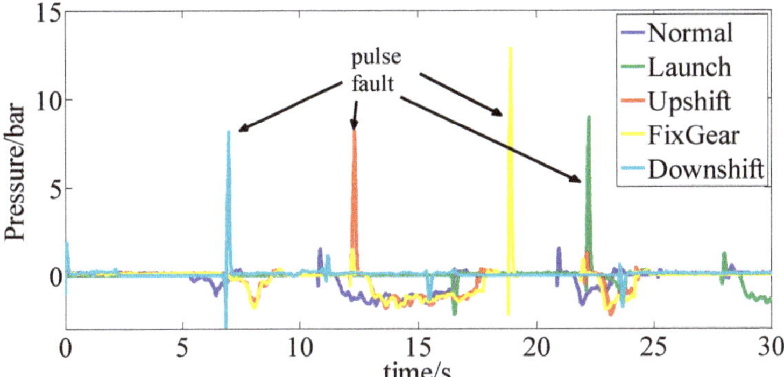

Figure 21. Observer output of the 0.1 s pulse fault sensor.

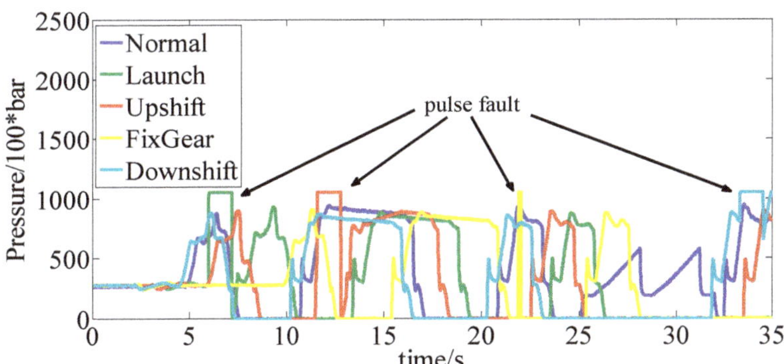

Figure 22. 1 s pulse fault sensor.

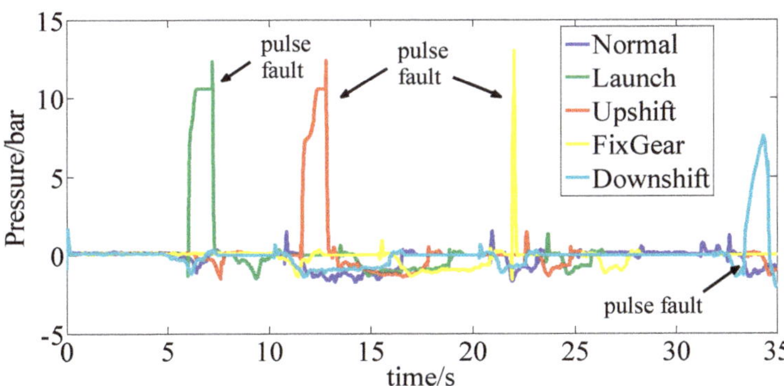

Figure 23. Observer output of 1 s pulse fault sensor.

4.3. Bias Fault

In this stage, the bias fault was tested. Bias fault is a common fault that deviates in sensor output. The driving conditions of an automobile are classified into four categories: the launching, upshift, fixed gear driving, and downshift conditions. This model assumes that all driving conditions have been covered.

As shown in Figures 24 and 25, the output of bias fault was expected and effective under different driving conditions. The threshold could be selected accordingly.

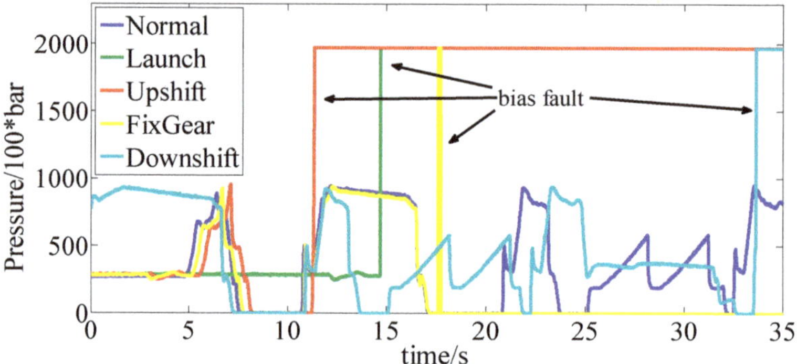

Figure 24. Bias fault sensor.

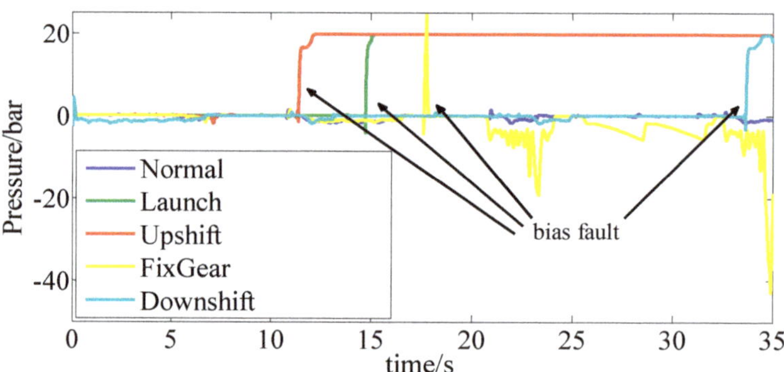

Figure 25. Observer output of bias fault sensor.

5. Conclusions

This paper proposed a fault diagnosis method for the pressure sensor of DCT. Firstly, the model of a VFS valve was established, and then the first-order inertial link of the sensor was corrected with the design of a feed-forward system. The parameters of the model were identified by using the actual sensor data of transmission and a GA algorithm. Then, the fault observer was designed by using the identified model. From the theoretical point of view, different types of faults have different kinds of observation outputs. Here, the form of observation output varied with the form of fault. In the verification stage, the fault observer outputs were verified by the measured fault signals of the pressure sensor. Constant output faults of 0, 3, and 5 V, the pulse fault, and the bias fault, all of which the enterprise is concerned by, are generated. The results showed that the designed fault observer had the expected diagnosis effect and function, can diagnose different types of faults, and can output specific results according to different types of faults. By identifying fault types according to the fault observer output, a fault diagnosis method has established, and this lays a foundation for later fault-tolerant control.

Author Contributions: Conceptualization, G.W. and Z.L.; methodology, Z.L.; validation, G.W.; project administration, G.W.; funding acquisition, G.W. All authors have read and agreed to the published version of the manuscript.

Funding: This research was funded by National Natural Science Foundation, grant number is U1764259.

Conflicts of Interest: The authors declare no conflict of interest.

References

1. Lucke, M.; Stief, A.; Chioua, M.; Ottewill, J.R.; Thornhill, N.F. Fault detection and identification combining process measurements and statistical alarms. *Control Eng. Pract.* **2020**, *94*, 104195. [CrossRef]
2. Choi, U.M.; Lee, J.S.; Blaabjerg, F.; Lee, K.B. Open-Circuit Fault Diagnosis and Fault-Tolerant Control for a Grid-Connected NPC Inverter. *IEEE Trans. Power Electron.* **2016**, *31*, 7234–7247. [CrossRef]
3. Jaradat, M.A.; Langari, R. A hybrid intelligent system for fault detection and sensor fusion. *Appl. Soft Comput.* **2009**, *9*, 415–422. [CrossRef]
4. Cui, X.; Wang, S.; Li, T.; Shi, J. System Reliability Assessment Based on Energy Dissipation: Modeling and Application in Electro-Hydrostatic Actuation System. *Energy Fuels* **2019**, *12*, 3275. [CrossRef]
5. Ni, S.X.; Zhang, Y.F.; Yi, H.; Liang, X.F. Intelligent Fault Diagnosis Method Based on Fault Tree. *J. Shanghai Jiaotong Univ.* **2008**, *42*, 1372–1375.
6. Barik, M.A.; Gargoom, A.; Mahmud, M.A.; Haque, M.E.; Al-Khalidi, H.; Oo, A.M. A Decentralized Fault Detection Technique for Detecting Single Phase to Ground Faults in Power Distribution Systems with Resonant Grounding. *IEEE Trans. Power Deliv.* **2018**, *33*, 2462–2473. [CrossRef]
7. Lei, Y.; Yang, B.; Du, Z.; Lu, N. Deep Transfer Diagnosis Method for Machinery in Big Data Era. *J. Mech. Eng.* **2019**, *55*, 1–8. [CrossRef]
8. Sylvain, V.; Teodor, T.; Abdessamad, K. Fault detection and identification with a new feature selection based on mutual information. *J. Process Control* **2008**, *18*, 479–490.
9. Yu, P. *Hydrodynamics*, 2nd ed.; Science Press: Beijing, China, 2016; pp. 53–74.
10. Li, H. *Hydraulic Control System*, 1st ed.; National Defense Industry Press: Beijing, China, 1990; pp. 53–70.
11. Eriksson, L.; Nielsen, L. *Modeling and Control of Engines and Drivelines*, 1st ed.; Wiley: Hoboken, NJ, USA, 2014.

Article

Mechanical Reliability Assessment by Ensemble Learning

Weizhen You [1], Alexandre Saidi [2], Abdel-malek Zine [3] and Mohamed Ichchou [1,*]

[1]   LTDS, Ecole Centrale de Lyon, 69134 Écully, France; weizhen.you@ec-lyon.fr
[2]   LIRIS, Ecole Centrale de Lyon, 69134 Écully, France; alexandre.saidi@ec-lyon.fr
[3]   ICJ, Ecole Centrale de Lyon, 69134 Écully, France; abdel-malek.zine@ec-lyon.fr
*   Correspondence: mohamed.ichchou@ec-lyon.fr

Received: 31 December 2019; Accepted: 10 February 2020; Published: 14 February 2020

Abstract: Reliability assessment plays a significant role in mechanical design and improvement processes. Uncertainties in structural properties as well as those in the stochatic excitations have made reliability analysis more difficult to apply. In fact, reliability evaluations involve estimations of the so-called conditional failure probability (CFP) that can be seen as a regression problem taking the structural uncertainties as input and the CFPs as output. As powerful ensemble learning methods in a machine learning (ML) domain, random forest (RF), and its variants Gradient boosting (GB), Extra-trees (ETs) always show good performance in handling non-parametric regressions. However, no systematic studies of such methods in mechanical reliability are found in the current published research. Another more complex ensemble method, i.e., Stacking (Stacked Generalization), tries to build the regression model hierarchically, resulting in a meta-learner induced from various base learners. This research aims to build a framework that integrates ensemble learning theories in mechanical reliability estimations and explore their performances on different complexities of structures. In numerical simulations, the proposed methods are tested based on different ensemble models and their performances are compared and analyzed from different perspectives. The simulation results show that, with much less analysis of structural samples, the ensemble learning methods achieve highly comparable estimations with those by direct Monte Carlo simulation (MCS).

Keywords: structural reliability; uncertainties; ensemble learning

---

1. Introduction

Reliability describes the probability that the object realizes its functions under given conditions for a specified time period [1]. As a way to improve the quality of products, reliability assessment is carried out by companies such that they can make product planning and implement preventive maintenance. For a mechanical structure subjected to stochastic excitation, an important task is to evaluate the risk of structure failures, in other words, the failure probability of the structure. Mathematically, the failure probability is determined by a multi-dimension integral over the spaces of all possible variables (factors) involved, i.e., [2]

$$P_f = \int_{g(\mathbf{z}) \leq 0} p_Z(\mathbf{z}) d\mathbf{z} = \int_{\mathbf{z} \in \Omega_Z} I_f(\mathbf{z}) p_Z(\mathbf{z}) d\mathbf{z} = E_{\mathbf{z} \in \Omega_Z}[I_f(\mathbf{z})], \tag{1}$$

where $\mathbf{z}$ is a vector that consists of the variables involved in the excitation. $g(\mathbf{z})$ is a defined performance function used to identify structural failures, that is to say, $g(\mathbf{z}) \leq 0$ when $\mathbf{z}$ falls into failure region. Ω_Z is the uncertainty space of $\mathbf{z}$; $I_f(\mathbf{z})$ is an indicator function that equals 1 when $g(\mathbf{z}) \leq 0$ and 0, otherwise; $E[\cdot]$ denotes the mathematical expectation. In fact, due to various factors (manufacturing, environment, fatigue ...), the structural properties become uncertain. Hence, the assessment of failure probability involves the uncertainties in both structural parameters and the excitation. Considering the relative independence between the uncertainties of the structural properties and those of the excitation, the conditional failure probability can be defined and formulated as [3]

$$P_f^c(\mathbf{x}) = \int_{g(\mathbf{x},\mathbf{z}) \leq 0} p_Z(\mathbf{z})d\mathbf{z} = \int_{\mathbf{z} \in \Omega_Z} I_f(\mathbf{z}|\mathbf{x}) p_Z(\mathbf{z})d\mathbf{z} = E_{\mathbf{z} \in \Omega_Z}[I_f(\mathbf{z}|\mathbf{x})], \tag{2}$$

where $\mathbf{x}$ is a vector that consists of the uncertain properties of the object structure. $I_f(\mathbf{z}|\mathbf{x})$ is a 'conditional' indicator function of $\mathbf{z}$ in terms of $\mathbf{x}$. Generally, in the analysis of structural dynamic responses, the stochastic excitation process is discretized so that a discrete response process is achieved. Then, the characterization of the integrals in Equations (1) and (2) may involve hundreds of random variables in the context of stochastic loading. Therefore, the current reliability problem forms actually a high dimensional problem.

Compared with the dimension of structural properties, the dimension of the excitation is always very high. Direct MCS demands a huge number of samples to ensure a high accuracy of the estimated CFP, but this will be very time-consuming. In this aspect, surrogate models become a good alternative. Surrogate models are developed to deal with highly nonlinear or implicit performance functions. They are introduced to reduce computational burden in reliability analysis. Assuming $\mathbf{x} = (x_1, x_2, \ldots, x_n)$ is a n-dimension input vector, y is the output. For a data set with N samples, $\mathbf{X} = (\mathbf{x}^1; \mathbf{x}^2; \ldots; \mathbf{x}^N)$, the corresponding responses are $\mathbf{Y} = (y^1, y^2, \ldots, y^N)^T$. Assume that the function between the input variables and output response takes the following form $y(\mathbf{x}) = \hat{y}(\mathbf{x}) + \varepsilon$, where $y(\mathbf{x})$ is the unknown response function, $\hat{y}(\mathbf{x})$ is a certain surrogate function seen as an approximation of the real function. ε is the error induced by the surrogate function. A surrogate model technique mainly consists of three parts. Firstly, a number of samples are taken from the input space $\mathbf{X}$; secondly, the output responses $\mathbf{Y}$ of these samples are determined through numerical models such as finite element analysis (FEA), resulting in a training data set $[\mathbf{X}_{N \times n}, \mathbf{Y}_{N \times 1}]$ for the surrogate model; finally, the training data are employed to train a surrogate model. See the illustration in Figure 1.

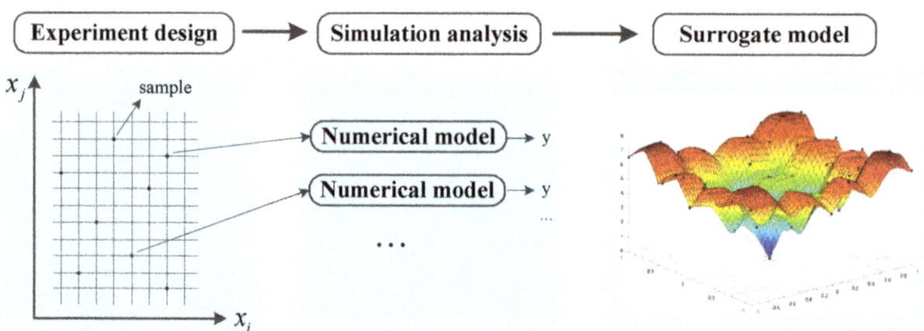

Figure 1. Theoretical framework to estimate failure probability based on an ML model.

Response surface method (RSM) [4] is among the most popular surrogate models. Some researchers combine RSM with other approaches to refine the model parameters so that the model becomes more efficient [5]. Support vector machine (SVM) has also gained much attention recently. Pan and Dias [6] combined an adaptive SVM and MCS to solve nonlinear and high-dimensional problems in reliability analysis. Other surrogate models such as metamodel, ANNs, and Kriging have been proposed by other researchers [7–9]. However, they cannot avoid shortcomings in all situations. The disadvantages in ANNs mainly include the complex architecture optimization, low robustness, and enormous training time [10]. SVM is time-consuming for large-scale applications and sometimes shows a large error in sensitivity calculation of reliability index. RSM has been popular; however, it may be time-consuming to use the polynomial function when the components are complex and the number of random variables is large [11]. Moreover, the approximate performance function is lack of adaptivity and flexibility, and we cannot guarantee that it is sufficiently accurate for the true one.

In the authors' viewpoint, the evaluation of CFP in Equation (3) can be seen as a regression problem that takes a realization of the structural uncertain properties as input and the CFP as output. To improve reliability assessment considering structural uncertainties, more attention should be paid to the non-parametric statistical learning methods, such as RF [12] and GB [13] et al. According to the current research state, we believe that the explorations of ML methods are far from enough. The rest of the paper is organized as follows. In Section 2, the theories pertaining to CFP and its estimation are presented. In Section 3, the framework of failure probability estimation based on ML models is introduced. In Section 4, the principles of ensemble learning methods are presented. In Section 5, numerical simulations on different structures are applied, and discussion of the results are made. Section 6 makes some concluding remarks.

2. Failure Probability Estimation: ML-Based Framework

As mentioned before, the calculation of the overall failure probability considers both uncertainties of the stochastic excitation and those of the structural properties. In this aspect, the target failure probability is estimated according to the formula below:

$$
\begin{aligned}
P_f &= \int_{g(\mathbf{x},\mathbf{z}) \leq 0} p_X(\mathbf{x}) p_Z(\mathbf{z}) d\mathbf{x} d\mathbf{z} \\
&= \int_{\mathbf{x} \in \Omega_X, \mathbf{z} \in \Omega_Z} I_f(\mathbf{x}, \mathbf{z}) p_X(\mathbf{x}) p_Z(\mathbf{z}) d\mathbf{x} d\mathbf{z} \\
&= \int_{\mathbf{x} \in \Omega_X} \left\{ \int_{\mathbf{z} \in \Omega_Z} I_f(\mathbf{z}|\mathbf{x}) p_Z(\mathbf{z}) d\mathbf{z} \right\} p_X(\mathbf{x}) d\mathbf{x} \\
&= \int_{\mathbf{x} \in \Omega_X} P_f^c(\mathbf{x}) p_X(\mathbf{x}) d\mathbf{x} \\
&= E_{\mathbf{x} \in \Omega_X}[P_f^c(\mathbf{x})],
\end{aligned}
\tag{3}
$$

which means that the target failure probability can be seen as the expectation of CFP over the uncertainty space of structural properties. From an ML perspective, training data are firstly prepared to train a model, then this model is employed to make predictions on a new data. The training data consist of several samples of the input vector and the output values while the new data only contain the samples of the input vector. When the CFPs are available for a few samples of the uncertain structural properties, an ML model is trained that takes the samples of **x** as inputs and the corresponding CFPs as outputs. Then, this model is employed to predict the outputs on the new data. Machine learning is a very powerful tool to find potential relationships within the data and make predictions. In this research, only a few hundred samples are needed to train an accurate ML model, then the model can be used to make immediate predictions on the new samples.

Figure 2 shows the framework of the proposed method. Firstly, a set of random samples are generated from the structural uncertainty space; secondly, the samples are split into two subsets, one for training (10%) and the other (90%) for making predictions; thirdly, MCS is employed to estimate target values (CFPs) of the training set; fourthly, an ML model is trained from the training data; fifthly, the ML model is used to make predictions on the prediction set. Finally, the predictions are averaged to obtain the estimation of overall failure probability. It is noticed that the pre-processing of the random samples of the structure is necessary. In this research, the samples are adjusted so that they follow truncated distributions. In Figure 2, the samples to train the ML model are different from the samples for making predictions.

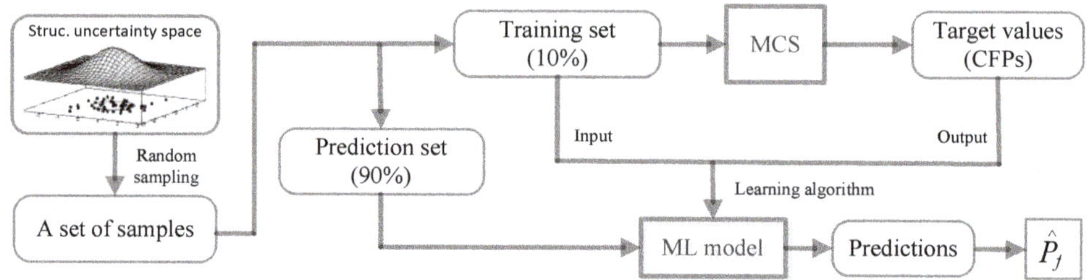

Figure 2. Theoretical framework to estimate failure probability based on the ML model.

3. MCS to Prepare the Training Data

3.1. General MCS to Estimate CFPs

In the perspective of first-excursion probability [2], it is general to compare the responses of interest $y_i(t), i = 1, \ldots, n_r$ against acceptable threshold levels y_i^* within the time duration T of the stochastic excitation. A failure takes place whenever the response y_i exceeds its corresponding threshold. From this point of view, failure implies not meeting the predefined conditions (this does not necessarily imply collapse). In terms of performance function, the failure event F is defined as:

$$F = \{\mathbf{x} \in \Omega_X, \mathbf{z} \in \Omega_Z : g(\mathbf{x}, \mathbf{z}) \leq 0\} \tag{4}$$

The value of $g(\cdot)$ is equal to or smaller than zero whenever a response exceeds its prescribed threshold. Considering the different DOFs of the structure as well as the sampled time points, the performance function $g(\mathbf{x}, \mathbf{z})$ is formulated as [2]

$$g(\mathbf{x}, \mathbf{z}) = 1 - \max_{i=1,\ldots,n_r} \left(\max_{k=1,\ldots,n_T} \left(\frac{|y_i(t_k, \mathbf{x}, \mathbf{z})|}{y_i^*} \right) \right), \tag{5}$$

where $|\cdot|$ denotes absolute value; t_k is the kth time point within the time interval $[0, T]$; y_i is the response of ith degree of freedom (DOF) of the structure; y_i^* is a defined threshold for the ith DOF. By direct MCS, the $P_f^c(x)$ can be approximated as

$$P_f^c(\mathbf{x}) = E[I_f(\mathbf{z}|\mathbf{x})] = \frac{1}{N} \sum_{i=1}^{N} I(g(\mathbf{z}_i|\mathbf{x}) < 0), \tag{6}$$

where $I(\cdot) = 1$ if $g(\cdot) < 0$ and 0, otherwise. The N realizations of the stochastic excitation, $\mathbf{z}_1, \mathbf{z}_2, \ldots, \mathbf{z}_N$, are treated as individual inputs of a finite element code that describes the structural responses.

As already known, the estimations by direct MCS can be treated as a standard reference. However, it is not computationally efficient for estimating very small failure probabilities (for example, <1%) since the number of samples required to achieve a given accuracy is inversely proportional to the scale of P_f. In other words, estimating small probabilities requires information from rare samples that induce structural failures. On average, it requires many samples before one such failure sample occurs. To ensure a demanded accuracy of the estimation, the number of samples needed to be analyzed is $n = (1/P_f - 1)/c^2$ [14], where P_f is the actual failure probability, and c is the coefficient of variation. For example, if $c = 0.1$ and the actual P_f is 10^{-1}, then only $(1/10^{-1} - 1)/0.1^2 = 9 * 10^2$ samples need to be analyzed. However, if the actual P_f value is 10^{-3}, then at least $(1/10^{-3} - 1)/0.1^2 = 9.99 * 10^4$ samples are needed, which consumes much more CPU time. Therefore, the standard MCS is only practically suitable to calculate relatively larger P_f (for example, >1%).

In view of this, the importance sampling (IS) method is a good choice to compute small P_f values. In principle, an IS method tries to adjust the sampling density so that more samples from the failure region F can be obtained. The efficiency of the method relies on the construction of the importance sampling density (ISD), for which the knowledge about the failure region is inevitably required.

3.2. Importance Sampling to Estimate Very Small CFPs

Consider a linear structural system represented by an appropriate model (e.g., a finite element model) comprising a total of n DOFs. The system is subjected to a Gaussian process excitation with zero mean, $z(t)$. Then, the system response can be evaluated by the convolution integral [15],

$$y_i(t, \mathbf{x}) = \int_0^t h_i(t - \tau, \mathbf{x}) z(\tau) d\tau, \tag{7}$$

where $h_i(t - \tau, \mathbf{x})$ is the unit impulse response function for the ith DOF of the structure at time t due to a unit impulse applied at time τ. $\mathbf{x}$ is a vector that consists of the structural properties. Generally, the zero initial condition at $t = 0$ is assumed. As a result of linearity, the response of interest is actually a linear combination of the contributions from each input $z(\tau)$. Considering that the excitation is modeled as a Gaussian process, its discrete form can be approximated as K-L expansions [16], i.e.,

$$z(t) = \mu(t) + \sum_{i=1}^{M} \sqrt{\lambda_i} u_i \phi_i(t), \tag{8}$$

where $\mu(t)$ is the mean function and $u_i, i \in N^+$ are standard normal variables. M is the number of truncated terms to keep. The eigenpair $\{\lambda_i, \phi_i\}$ is the solution of the eigenvalue problem $\int_B C_{HH}(x, x') \phi_i(x') dx' = \lambda_i \phi_i$, which is a Fredholm integral equation of the second kind. The kernel $C_{HH}(x, x')$, being an autocovariance function, is symmetric and positive definite. The set of eigenvalues is, moreover, real, positive, and numerable. By Equations (7) and (8), we obtain

$$y_i(t, \mathbf{x}) = \int_0^t h_i(t - \tau, \mathbf{x}) \sum_{l=1}^{M} u_l f_l^{KL}(\tau) d\tau = \sum_{l=1}^{M} u_l a_l(t, \mathbf{x}) = \mathbf{a}(t, \mathbf{x})^T \mathbf{u}, \tag{9}$$

where $f_l^{KL}(t) = \sqrt{\lambda_l} \phi_l(t)$, and $\mathbf{a}(t, \mathbf{x}) = [a_1(t, \mathbf{x}), \dots, a_M(t, \mathbf{x})]^T$, where

$$a_l(t, \mathbf{x}) = \int_0^t h_i(t - \tau, \mathbf{x}) f_l^{KL}(\tau) d\tau, l = 1, \dots, M \tag{10}$$

From Equation (9), the response process is found to be the scalar product of two vectors: the random vector $\mathbf{u}$ and the deterministic basis function vector $\mathbf{a}(t, \mathbf{x})$, whose elements are convolutions of the basis functions $f_l^{KL}(t)$ of the excitation and the unit impulse response function $h_i(t - \tau, \theta)$ of the system.

Now, consider the excitation that leads to the event $y_i(t_j) \geq y_i^*$ at time $t = t_j$, where y_i^* is a designed threshold for the ith DOF. This corresponds to realizations of $\mathbf{u}$ that satisfy the condition

$$y_i^* - |\mathbf{a}_i(t_j, \mathbf{x})^T \mathbf{u}| \leq 0 \tag{11}$$

In the space of $\mathbf{u}$, these realizations lie in a pair of symmetric subspaces bounded by the hyper-planes $y_i^* = \pm \mathbf{a}_i(t_j, \mathbf{x})^T \mathbf{u}$ that have the same distances $\beta_i(t_j, \mathbf{x}) = y_i^* / ||\mathbf{a}_i(t_j, \mathbf{x})||$ from the origin. In addition, $g_i(\mathbf{u}) = y_i^* - |\mathbf{a}_i(t_j, \mathbf{x})^T \mathbf{u}|$ is the performance function, $g_i(\mathbf{u}) = y_i^* - |\mathbf{a}_i(t_j, \mathbf{x})^T \mathbf{u}| = 0$ is the limit-state surface; $\beta_i^+(t_j, \mathbf{x}) = \beta_i^-(t_j, \mathbf{x}) = \beta_i(t_j, \mathbf{x})$ are the reliability indices for the event $g_i(\mathbf{u}) \leq 0$ at time t_j. The failure probability of the ith DOF at time t_j is denoted as [15]

$$P(g_i(\mathbf{u}) \leq 0) = 2 * \Phi(-\beta_i(t_j, \mathbf{x})), \tag{12}$$

where $\Phi(.)$ denotes the standard normal cumulative probability function. Figure 3 provides an illustration of an elementary failure region $F_{i,j}(\mathbf{x})$ with $n_r = 1$, $n_T = 1$, $n_{KL} = 2$.

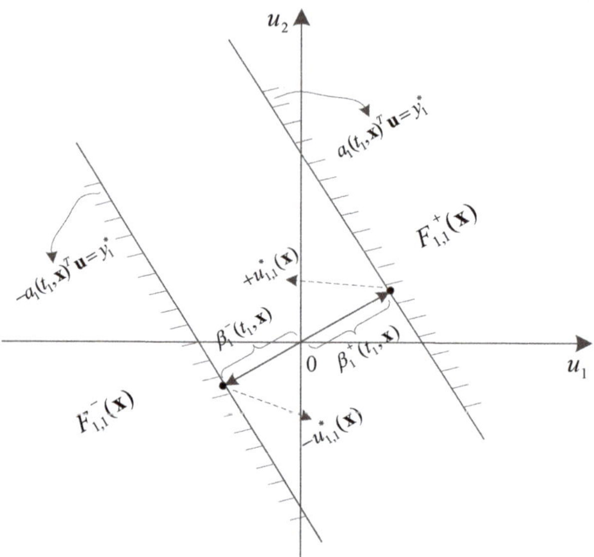

Figure 3. Illustration of a symmetric elementary failure region [3].

To apply the IS technique to evaluate the CFP, the integral in Equation (2) is re-written as

$$\hat{P}_f^c(\mathbf{x}) = \int_{u\in\Omega_u} I_F(u|\mathbf{x})\frac{p_u(u)}{p_{IS,u}(u)}p_{IS,u}(u)du = E_{u\in\Omega_u}[I_f(u^v|\mathbf{x})\frac{p_u(u^{(v)})}{p_{IS,u}(u^{(v)})}] \tag{13}$$

where $p_{IS,u}(u)$ is the Importance Sampling density (ISD) function and $u^{(v)}, v = 1, \ldots, and N$ are samples of uncertain vector U obtained via the ISD $p_{IS,u}(u)$. The most important part for implementing the IS procedures is the design of the ISD function that is able to obtain more samples in the failure region, meanwhile ensuring a low variability of the estimated probability, i.e., failure samples are drawn frequently while the variability of the ratio involving $p_u(u)$ and $p_{IS,u}(u)$ is low. Based on the elementary failure regions, a well defined ISD function is employed in this paper. See more details in Au and Beck [2].

4. Train the ML Model by Ensemble Learning

The ensemble learning methods mainly include Bagging, Boosting, and Stacking. In principle, Bagging adopts Bootstrap sampling to learn independent base learners and takes the majority/average as the final prediction. Boosting updates weight distribution in each round, and learns base models accordingly, then combines them according to their corresponding accuracy. Different from these two approaches, Stacking [17] learns a high-level model on top of the base models (classifier/regressor). It can be regarded as a meta-learning approach in which the base models are called first-level models and a second-level model is learned from the outputs of the first-level models. A short description of the three methods is introduced below:

- Bagging method generally builds several instances of a black-box estimator from bootstrap replicates of the original training set and then aggregates their individual predictions to form a final prediction. This method is employed as a way to reduce the variance of a base estimator (e.g., a decision tree) by introducing randomization into its construction process. Random Forest is representative among bagging methods.

- Boosting is a widely used ensemble approach, which can effectively boost a set of weak classifiers to a strong classifier by iteratively adjusting the weight distribution of samples in the training set and learning base classifiers from them. At each round, the weight of misclassified samples is increased and the base classifiers will focus on these more. This is equivalent to inferring classifiers from training data that are sampled from the original data set based on the weight distribution. Gradient Boosting is a mostly used boosting method.
- Stacking involves training a learning algorithm to combine the predictions of several other learning algorithms. First, all of the other algorithms are trained using the available data; then, a combiner algorithm is trained to make a final prediction using all the predictions of the other algorithms as inputs. Stacking typically yields a performance that is better than any single trained models.

4.1. Random Forest

As a representative bagging method, RF consists of many individual trees called classification and regression tree (CART), each of which is induced from a bootstrap sample. The CARTs are aggregated to make predictions for a future input; see Figure 4. An underlying assumption is that the base learners are independent. As the trees become more correlated (less independent), the model error tends to increase. Randomization helps to reduce the correlation among decision trees so that the model accuracy is improved. Two kinds of randomness exist in a tree learning process: bootstrap sampling and node splitting. A CART has a binary recursive structure. The tree learning process is actually a recursive partitioning process of the data. The root node corresponds to the whole training data.

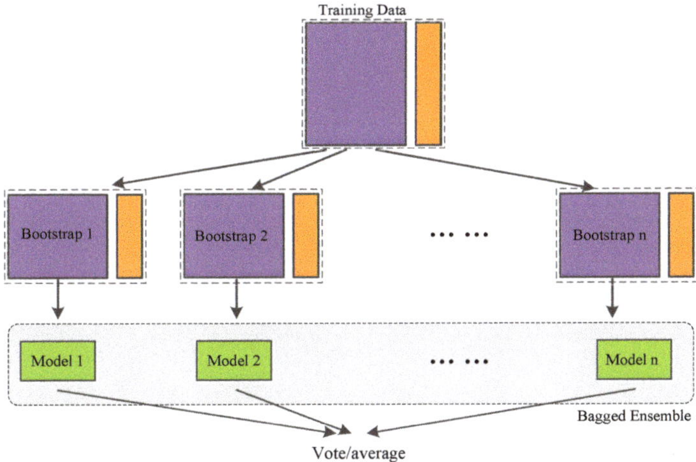

Figure 4. Process of bootstrap aggregating.

When a node is split into two child nodes, the data set is simultaneously divided into two subsets according to the splitting point determined by minimizing the sum of squared errors (SSE), i.e., $\min_{i,j}[\sum_{x \in R_1} (y_i - c_1)^2 + \sum_{x \in R_2} (y_i - c_2)^2]$ where $R_1 = \{x | x_i \leq x_{i,j}\}$ is the region that satisfies the condition $x_i \leq x_{i,j}$, $R_2 = \{x | x_i > x_{i,j}\}$ is the region that satisfies $x_i > x_{i,j}$. In addition, c_m is the average of the output values falling into the region, $c_m = E(y | x_i \in R_m), m = 1, 2$. See the illustrative node splitting process in Figure 5.

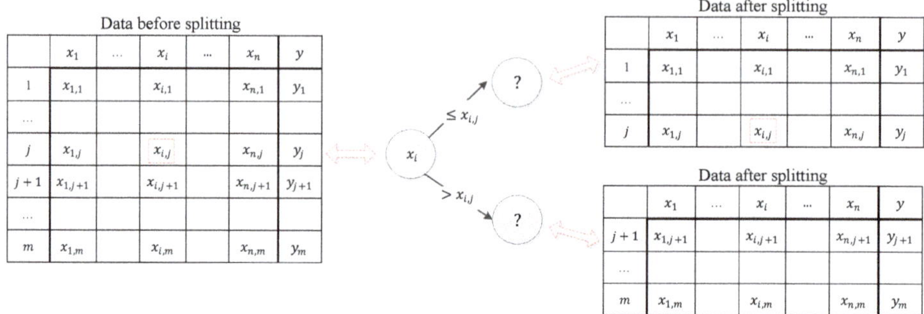

Figure 5. An illustrative node splitting process. The symbol '?' means that the variable used to carry out the next splitting needs to be determined.

Similar in principle to RF, another ensemble learning method called Extra-Trees has further randomness in node splitting. Both of them consist of multiple decision trees. In the tree inferring process, both of them use a random subset of the features as the candidate splitting features. The mainly different operations are listed below: (1) the way to prepare the training set for each tree. RF uses a bootstrap replica to train each tree. In contrast, ETs employ the whole training set to train each tree; (2) the way to find the splitting point. In RF, a decision tree firstly finds the best splitting point for each candidate feature, then chooses the best one; however, in an ET, a random splitting point is chosen for each candidate feature; then, the best one is chosen. See Figure 6.

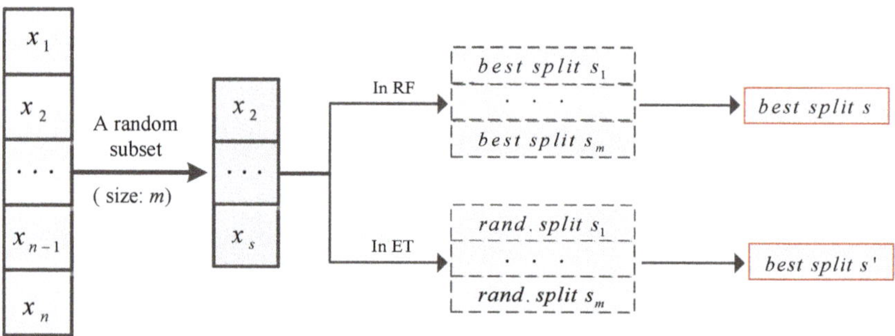

Figure 6. Comparison of RF and ETs in node splitting.

4.2. Gradient Boosting

Gradient Boosting is a machine learning meta-algorithm. It can be used in conjunction with many other types of learning algorithms to improve their performance. The output of the other learning algorithms ('weak learners') is combined into a weighted sum that represents the final output of the boosted classifier. AdaBoost is adaptive in the sense that subsequent weak learners are tweaked in favor of those instances misclassified by previous classifiers. The individual learners can be weak, but as long as the performance of each one is slightly better than random guessing, the final model can be proven to converge to a strong learner. Weak classifiers can be Decision tree/Decision stump, Neural Network, Logistic regression, or even SVM. To determine the coefficient for each classifier, we initialize the weight distribution of the samples in the dataset by a uniform distribution, and update this distribution by considering the misclassified samples in each loop. Meanwhile, we calculate the coefficient for the current classifier.

Finally, we have a certain number of weak classifiers as well as their weights, then the strong classifier is obtained as the weighted sum of these classifiers,

$$f(x) = \sum_{i=1}^{M} \alpha_t h_t(x) \tag{14}$$

where α_t is the coefficient for the 'weak' classifier $h_t(x)$. For GB, we use all training data, instead of bootstrapping, to build each tree. The first tree is built from an initialized constant model (i.e., a constant value or a one-node tree) [18]. From the second tree, to build each tree, we need to calculate the errors of the former trees (as a weighted sum) on all training points until the stopping criteria (number of trees, error threshold, $\cdots$) are satisfied. See Figure 7.

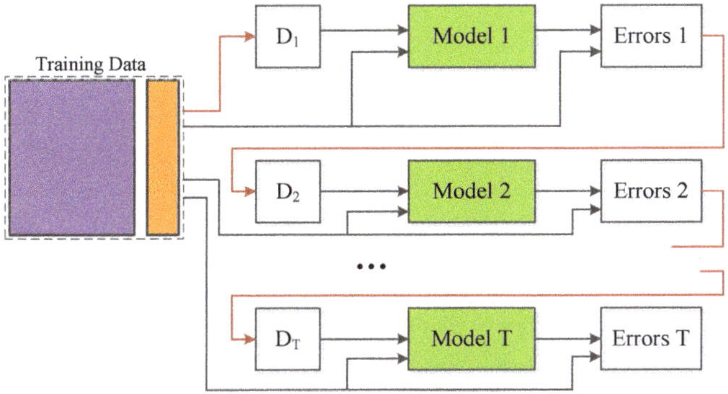

Figure 7. Inducing process of Gradient Boosting.

4.3. Stacking

Stacking is an ensemble learning technique that learns a meta-learner based on the output of multiple base learners. In a typical implementation of Stacking, a number of first-level individual learners are generated from the training data set by employing different learning algorithms. Those individual learners are then combined by a second-level learner that is called meta-learner. See Figure 8.

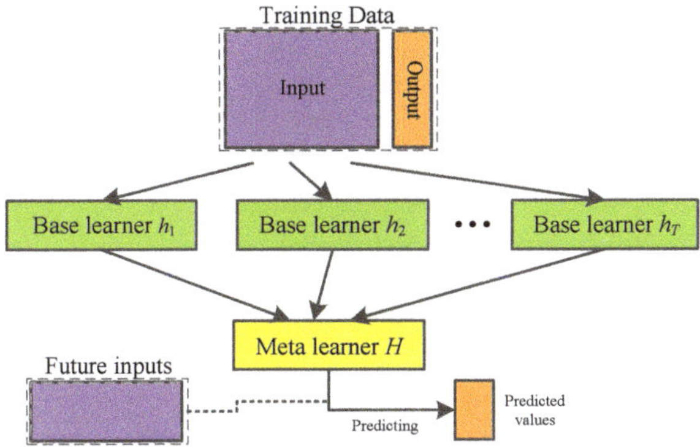

Figure 8. General framework of Stacking.

Table 1 shows the Pseudo-code of Stacking, which demands three steps. Firstly, learn first-level (base) learners based on the original training data. There are several ways to learn base learners. We can apply a Bootstrap sampling technique to learn independent learners or adopt the strategy used in Boosting. Secondly, construct a new data set based on the output of base learner. Assume that each example in D is $(\mathbf{x}_i, y_i)$. We construct a corresponding example $(\mathbf{x}'_i, y_i)$ in the new data set, where $\mathbf{x}'_i = h'(h_1(\mathbf{x}_i), h_2(\mathbf{x}_i), \ldots, h_T(\mathbf{x}_i))$. Thirdly, learn a second-level (meta) learner from the newly constructed data. Any learning method could he applied to learn the meta learner.

Table 1. Pseudo-code for Stacking [17].

Input: Training data $D = \{\mathbf{x}_i, y_i\}_{i=1}^{m}$, $\mathbf{x}_i \in \mathbf{X}$, $y_i \in y$; a Stacking algorithm. **Output:** A meta learner H.

Step1: Induce T base learners, i.e., $h_1, h_2, \ldots, h_T$, from the training set.
 for $t \leftarrow 1$ to T
      Learn a base learner h_t from D.
 end for
 Step2: Construct a new dataset D', where $D' = \{(\mathbf{x}'_i, y_i)\}_{i=1}^{m}$. Here,
      $\mathbf{x}'_i = [h_1(\mathbf{x}_i), h_2(\mathbf{x}_i), \ldots, h_T(\mathbf{x}_i)]$.
 Step3: Build a meta-learner H from D'. Output H.

Once the meta-learner is generated, for a test example $\mathbf{x}$, its predictions are $h'(h_1(\mathbf{x}), h_2(\mathbf{x}), \ldots, h_T(\mathbf{x}))$, where $(h_1, h_2, \ldots, h_T)$ are base learners and h' is the meta-learner. We can see that Stacking is a general framework. We can plug in different learning approaches or even ensemble approaches to generate first or second level learner. Compared with Bagging and Boosting, Stacking "learns" how to combine the base learner instead of voting. In Table 2, notice that the same data D is used to train base learners and make predictions, which may lead to over-fitting. To avoid this problem, 10-fold cross validations (CVs) is incorporated in stacking.

Table 2. Parameters of different models in a 1-DOF case.

Parameters	RF	GB	ETs
nTrees	20	20	20
nFeatures	3	3	3
maxFeatures	2	2	2

5. Numerical Examples

This section presents some numerical examples that illustrate the capabilities of the proposed method for estimating first excursion probabilities. In these examples, the proposed approaches are compared with the direct MCS for estimating structural failure probabilities in order to demonstrate the accuracy and efficiency. Firstly, Section 5.1 provides three basic simulations on 1-DOF, 2-DOF and 3-DOF structures. In Section 5.1, a 10-DOF structure is studied as a high-dimensional uncertainty case.

5.1. Three Test Examples

The structure is a 1-DOF oscillator, and its parameter values are introduced in Elyes et al. [19], with the nominal values of mass $m = 1.0$ kg, damping coefficient $c = 0.03$ Ns/m, and stiffness $k = 696.4$ N/m. The ground acceleration is modeled by a Kanai–Tajimi (K–T) filter, whose natural frequency and damping ratio are $8\,\pi$rad/s and 0.4 correspondingly. As the input of the K–T filter, the white noise process has the power spectral density (PSD) $S_0 = 0.031$ m^2/s^3. The failure criterion is $y^* = 0.16$ m. In this example, the uncertainties exist in the three structure properties and take the form of normal distributions with standard deviations (SD) $\sigma_{m_s} = 0.1$ kg, $\sigma_{c_s} = 0.003$ Ns/m,

$\sigma_{k_s} = 69.64$ N/m. Moreover, in the simulations, the random samples of structural properties are censored to avoid the deviation from the mean being larger than five times the SD. The observation time is defined as [0, 20 s] with time step $\Delta t = 0.05$ s. According to the proposed framework, an ensemble model is trained on a training set that takes the samples of structural properties as inputs and CFPs as outputs. Four kinds of ensemble methods are employed to build the models including RF, GB, ETs, and Stacking.

The parameters of the models RF, GB, and ETs are listed in Table 2. 'nTrees' is the number of random trees in the model; 'nFeatures' is the number of features (variables) in the training set; 'maxFeatures' is the number of sampled features in each node split. Notice that the Stacking model is denoted as 'RF&SVM-GB', which means that RF and SVM are employed as base learners and GB the meta-learner. The RF and GB in Stacking have the parameters listed in Table 2; for the meta-model GB, nTrees = 10, maxFeatures = 1. RMSE (Root mean square error) is used to evaluate the performances of the models in fitting CFPs. Based on the proposed framework, the overall failure probability is estimated. See results in Figure 9.

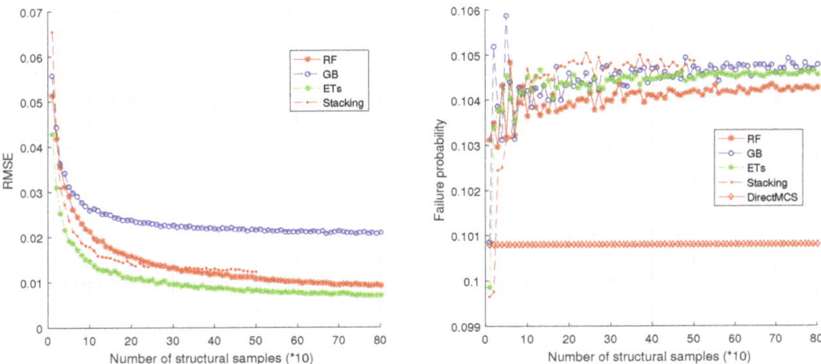

Figure 9. Simulation results on a 1-DOF structure.

It can be seen from Figure 9 that the errors of the ensemble models decrease when more samples are added into the training set. The ensemble models RF, ETs, and Stacking have similar RMSEs even though they are slightly different. The GB model always results in higher RMSEs than other three models. By direct MCS, the reference value of failure probability is determined as 0.1008 (see the right part of Figure 9). This reference value is then compared with the result failure probabilities from the four ensemble models. It is found that the four models converge rapidly and converge at $20 \times 10 = 200$ samples. In addition, RF is slightly better than other three methods. One reason is that RF is good at reducing overfitting by randomization of the training process. Actually, the four methods have very close estimations that is about 0.1045. Compared with the reference value, this value is very close to the reference value.

Figure 10 shows the simulation results on a 2-DOF oscillator. The two DOFs have the nominal values of mass $m_1 = m_2 = 4.6$ kg, damping coefficient $c_1 = c_2 = 62$ Ns/m, and stiffness $k_1 = k_2 = 6500$ N/m. The uncertainties exist in all structural properties and take the form of normal distributions with SDs as 10% of the nominal values. The random samples of the uncertain structural properties are truncated in the same way as in the first simulation. The failure criterion is $y_1^* = y_2^* = 0.024$ m. The same ensemble learning methods (with same parameters) are employed as those in a 1-DOF case. It is found that the stacking method outperforms the other three methods in modeling CFPs. The reference value of failure probability is 0.0684 (see the right part of Figure 10). Similar to the evolutions in Figure 9, the four methods converge rapidly to the value 0.0730. Therefore, the four methods still achieve relatively high accuracy in failure probability estimations.

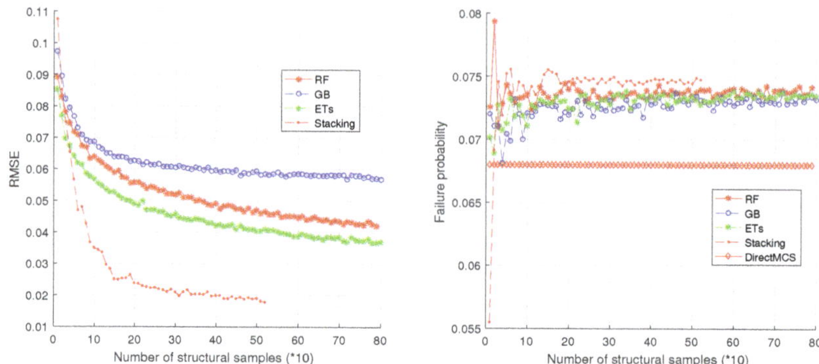

Figure 10. Simulation results on a 2-DOF structure.

For the 3-DOF structure, the structural parameters of each DOF and their uncertainties are the same as those in the 2-DOF case. The same ensemble learning methods (with same parameters) are employed as those in the 2-DOF case. The simulation results (see Figure 11) are similar to those on the 2-DOF structure. These results show that the ensemble learning methods give us highly accurate reliability estimations. In addition, from Figures 9–11, it is found that the four ensemble methods converge at the point where the number of structural samples is about 200. In contrast, by direct MCS (see Figure 12), at least 5000 samples are needed to reach a convergence, thus 96% samples are saved.

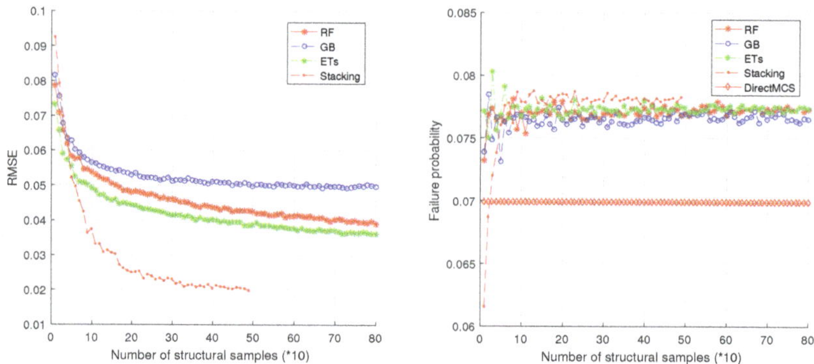

Figure 11. Simulation results on a 3-DOF structure.

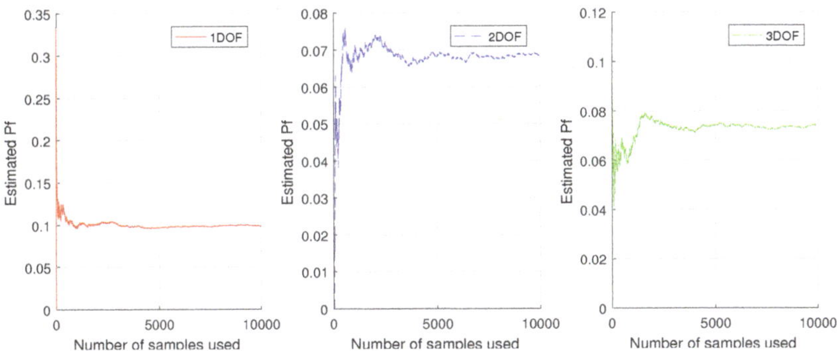

Figure 12. Convergence process of direct MCS.

5.2. A Benchmark Example: 10-DOF Duffing Oscillator

As a benchmark problem introduced in Schueller [20], the ten-DOF Duffing type oscillator has been widely used by researchers in structural reliability domain. In this study, we focus on the linear random structures under stochastic excitation. The statistical properties of the structural parameters and their constraints are listed in Table 3. Moreover, the Gaussian samples of structural properties are censored in the same way as in Section 5.1. The stochastic excitation $p(t)$ is modeled by a modulated filtered Gaussian white noise:

$$p(t) = \Omega_{1g}^2 v_{f1}(t) + 2\zeta_{1g}\Omega_{1g}v_{f2}(t) - \Omega_{2g}^2 v_{f3}(t) - 2\zeta_{2g}\Omega_{2g}v_{f4}(t), \tag{15}$$

where the state space function with respect to the state vector $v_f(t) = [v_{f1}(t), v_{f2}(t), v_{f3}(t), v_{f4}(t)]^T$ of the filter is

$$\dot{v}_f(t) = \begin{bmatrix} 0 & 1 & 0 & 0 \\ -\Omega_{1g}^2 & -2\zeta_{1g}\Omega_{1g} & 0 & 0 \\ 0 & 0 & 0 & 1 \\ \Omega_{1g}^2 & 2\zeta_{1g}\Omega_{1g} & -\Omega_{2g}^2 & -2\zeta_{2g}\Omega_{2g} \end{bmatrix} v_f(t) + [0, w(t), 0, 0]^T \tag{16}$$

Here, $w(t)$ stands for a modulated Gaussian white noise with auto-correlation function $E(w(t)w(t+\tau)) = I\delta(\tau)h^2(t)$ and I denotes the intensity function of the white noise. $h(t)$ has the following form:

$$h(t) = \begin{cases} 0, & t = 0, \\ t/2, & t \in [0, 2s] \\ 1, & t \in [2s, 10s] \\ exp(-0.1(t-10)), & t \in [10s, 20s] \end{cases} \tag{17}$$

$\delta(t)$ is the dirac delta function that equals $+\infty$ at $t = 0$ and 0 at $t \neq 0$. The values $\Omega_{1g} = 15.0$ rad/s, $\zeta_{1g} = 0.8$, $\Omega_{2g} = 0.3$ rad/s, $\zeta_{2g} = 0.995$, and $I = 0.08$ m^2/s^3 are used to model the filter. The input excitation of the filter is a shot noise that consists of a series of independent normally distributed impulses arranged at each time step. The magnitude of the impulse at time $t_k = k\Delta t$ has mean 0 and standard deviation $h(t)\sqrt{I/\Delta t}$.

Table 3. Statistical properties of the structural parameters ($r = SD/\mu$).

Variables	Mean (μ)	SD	Ratio (r)	Range Scope
$m_1,\ldots,m_{10}$	10×10^3 kg	1.0×10^3 kg	0.1	$\mu \pm 5\mu r$
k_1,k_2,k_3	40×10^6 N/m	4.0×10^6 N/m	0.1	$\mu \pm 5\mu r$
k_4,k_5,k_6	36×10^6 N/m	3.6×10^6 N/m	0.1	$\mu \pm 5\mu r$
k_7,k_8,k_9,k_{10}	32×10^6 N/m	3.2×10^6 N/m	0.1	$\mu \pm 5\mu r$
$\zeta_1,\ldots,\zeta_{10}$	620×10^4 N s/m	62×10^4 N s/m	0.1	$\mu \pm 5\mu r$

The failure criterion is defined by the maximum relative displacements between two consecutive DOFs over the time interval [0.0 s, 20.0 s]. The sampling time step is set as $\Delta t = 0.05$ s. The failure probabilities are calculated for different threshold values listed in Table 4. We are interested in the first excursion probability in which the relative maximum displacement of the first DOF is greater than the threshold 0.057 m and 0.073 m; in addition, the probability that the relative maximum displacement between the 9th DOF and 10th DOF exceeds the threshold 0.013 m and 0.017 m is considered.

Table 4. Thresholds of interest to evaluate failure probability [20].

Failure Defined by	Res. Threshold1	Res. Threshold2
First, DOF	0.057 m	0.073 m
Tenth DOF	0.013 m	0.017 m

In this simulation, the estimated failure probability values are very small (below 10^{-3}); therefore, the KL-IS method is employed to calculate the conditional failure probabilities. In applying K-L expansions, the number of K-L terms kept is $n_{KL} = 300$. The parameters of different models are shown in Table 5. The overall failure probability is estimated that is concerned with both structural uncertainties and excitation uncertainties. To make the results more convincing, the estimations by the proposed method are compared with those of other published methods; see Table 6. Notice that '*' means this result comes from the literature. In the simulations, it is found that the ensemble learning based methods all reach a convergence at the point $n\_sample = 500$.

Table 5. Parameters of different models.

Parameters	RF	GB	ETs
nTrees	30	30	30
nFeatures	30	30	30
maxFeatures	6	6	6

It is seen from Table 6 that, for high-dimensional reliability problems with respect to linear stochastic dynamics, a number of methods and their variants exist by which this problem can be solved very efficiently when compared to direct MCS. More importantly, the ensemble learning based methods proposed in this research are highly comparable with the already existing methods. It is noticed that the estimated probabilities are very small (e.g., $P_f < 10^{-4}$), which implies that direct MCS is practically not applicable because a very large number of simulations are required. By the comparisons in Table 6, it is found that the number of structural samples analyzed in the proposed methods is very small (e.g., 500), but the finally estimated failure probabilities are of high accuracy. These results reveal that the ensemble learning based methods are powerful in handling high-dimension small probability problems.

Table 6. Reliability estimation results from different methods.

Method	1st-DOF 0.057m	1st-DOF 0.073m	10th-DOF 0.013m	10th-DOF 0.017m
Standard MCS	1.06E-4	8.07E-7	4.88E-5	2.52E-7
num of samples	2.98E+7	2.98E+7	2.98E+7	2.98E+7
SubsetSim/MCMC [21]	1.20E-4	1.00E-6	6.60E-5	4.70E-7
num of samples	1850	2750	2300	2750
SubsetSim/Hybrid [21]	1.10E-4	1.10E-6	5.90E-5	3.20E-7
num of samples	2128	3163	2645	3680
Complex Modal Ana. [22]	1.00E-4	9.80E-7	6.00E-5	4.60E-7
num of samples	300	300	300	300
Spherical SubsetSim [23]	9.20E-5	8.80E-7	4.60E-5	5.30E-7
num of samples	3070	4200	3250	4900
Line sampling [24]	9.80E-5	9.70E-7	6.00E-5	4.60E-7
num of samples	360	3600	360	360
RF-based	7.6E-5	1.0E-6	4.2E-5	1.1E-7
num of samples	500	500	500	500
GB-based	8.48E-5	9.15E-7	4.24E-5	1.06E-7
num of samples	500	500	500	500
ETs-based	8.73E-5	9.89E-7	4.09E-5	1.10E-7
num of samples	500	500	500	500
Stacking-based	1.0E-4	9.2E-7	4.3E-5	1.1E-7
num of samples	500	500	500	500

6. Conclusions

In this study, an ensemble learning based method is proposed for mechanical reliability assessment. The ensemble learning methods are treated as surrogate models that can be employed to fit the CFPs of the structure. For very small failure probabilities for which direct MCS is practically impossible to realize, an importance sampling technique is employed that constructs the importance sampling density function based on the basic failure events of the mechanical structure. Once the surrogate model is built, the predictions of CFPs can be made on new samples of the structural properties, so that the overall failure probability can be directly estimated. The representative ensemble methods RF, GB, ETs, and Stacking are considered in the proposed method. Numerical simulations are performed on a different number of DOFs structures to examine the performance of the proposed methodology. For low-dimension uncertain structures, i.e., 1-DOF, 2-DOF, and 3-DOF, the four ensemble methods are considered respectively to build the estimation models, and the results reveal that all these models achieve very high accuracies. For high-dimension uncertain structural properties, a benchmark example is introduced from the famous work of G.I. Schueller where different representative reliability methods are introduced and compared. These representative methods are further compared with the proposed method that considers different ensemble models. The simulation results show that, with a very small number of structural samples analyzed, the ensemble leaning based method obtains a significant advantage over direct MCS in terms of efficiency, meanwhile keeping a high accuracy. The comparisons with other published methods make the proposed method highly competitive.

Author Contributions: Conceptualization, W.Y., A.S., M.I., and A.-m.Z.; Methodology, W.Y. and A.S.; Investigation, W.Y.; Writing—Original Draft, W.Y.; Writing—Reviewing and Editing, A.S., A.-m.Z., and M.I.; Visualization, W.Y. All authors have read and agreed to the published version of the manuscript.

Funding: This research received funds from the China Scholarship Council.

Conflicts of Interest: The authors declare no conflict of interest.

References

1. Bertsche, B. *Reliability in Automotive and Mechanical Engineering: Determination of Component and System Reliability*; Springer Science & Business Media: Berlin/Heidelberg, Germany, 2008; pp. 1–2.
2. Au, S.K.; Beck, J.L. First, excursion probabilities for linear systems by very efficient importance sampling. *Probabilistic Eng. Mech.* **2001**, *16*, 193–207. [CrossRef]
3. Valdebenito, M.A.; Jensen, H.A.; Labarca, A.A. Estimation of first excursion probabilities for uncertain stochastic linear systems subject to Gaussian load. *Comput. Struct.* **2014**, *138*, 36–48. [CrossRef]
4. Lee, S.H.; Kwak, B.M. Response surface augmented moment method for efficient reliability analysis. *Struct. Saf.* **2006**, *28*, 261–272. [CrossRef]
5. Hadidi, A.; Azar, B.F.; Rafiee, A. Efficient response surface method for high-dimensional structural reliability analysis. *Struct. Saf.* **2017**, *68*, 15–27. [CrossRef]
6. Pan, Q.; Dias, D. An efficient reliability method combining adaptive support vector machine and monte carlo simulation. *Struct. Saf.* **2017**, *67*, 85–95. [CrossRef]
7. Dubourg, V.; Sudret, B.; Deheeger, F. Metamodel-based importance sampling for structural reliability analysis. *Probabilistic Eng. Mech.* **2013**, *33*, 47–57. [CrossRef]
8. Cardoso, J.B.; de Almeida, J.R.; Dias, J.M.; Coelho, P.G. Structural reliability analysis using monte carlo simulation and neural networks. *Adv. Eng. Softw.* **2008**, *39*, 505–513. [CrossRef]
9. Su, G.; Peng, L.; Hu, L. A gaussian process-based dynamic surrogate model for complex engineering structural reliability analysis. *Struct. Saf.* **2017**, *68*, 97–109. [CrossRef]
10. Nitze, I.; Schulthess, U.; Asche, H. Comparison of machine learning algorithms random forest, artificial neural network and support vector machine to maximum likelihood for supervised crop type classification. In Proceedings of the 4th GEOBIA, Rio de Janeiro, Brazil, 7–9 May 2012; Volume 79, pp. 35–40.
11. Kang, S.C.; Koh, H.M.; Choo, J.F. An efficient response surface method using moving least squares approximation for structural reliability analysis. *Probabilistic Eng. Mech.* **2010**, *25*, 365–371. [CrossRef]
12. Breiman, L. Random forests. *Mach. Learn.* **2001**, *1*, 5–32. [CrossRef]
13. Friedman, J.H. Greedy function approximation: A gradient boosting machine. *Ann. Stat.* **2001**, *29*, 1189–1232. [CrossRef]
14. Jensen, J.J. Fatigue damage estimation in nonlinear systems using a combination of Monte Carlo simulation and the First, Order Reliability Method. *Mar. Struct.* **2015**, *44*, 203–210. [CrossRef]
15. Der Kiureghian, A. The geometry of random vibrations and solutions by FORM and SORM. *Probabilistic Eng. Mech.* **2000**, *15*, 81–90. [CrossRef]
16. Phoon, K.K.; Huang, S.P.; Quek, S.T. Simulation of second-order processes using Karhunen–Loeve expansion. *Comput. Struct.* **2002**, *80*, 1049–1060. [CrossRef]
17. Wolpert, D.H. Stacked generalization. *Neural Netw.* **1992**, *5*, 241–259. [CrossRef]
18. Friedman, J.; Hastie, T.; Tibshirani, R. *The Elements of Statistical Learning*; Springer Series in Statistics: New York, NY, USA, 2001; pp. 355–360.
19. Mrabet, E.; Guedri, M.; Ichchou, M.N.; Ghanmi, S. Stochastic structural and reliability based optimization of tuned mass damper. *Mech. Syst. Signal Process.* **2015**, *60*, 437–451. [CrossRef]
20. Schuëller, G.I.; Pradlwarter, H.J. Benchmark study on reliability estimation in higher dimensions of structural systems—An overview. *Struct. Saf.* **2007**, *29*, 167–182. [CrossRef]
21. Au, S.K.; Ching, J.; Beck, J.L. Application of subset simulation methods to reliability benchmark problems. *Struct. Saf.* **2007**, *29*, 183–193. [CrossRef]
22. Jensen, H.A.; Valdebenito, M.A. Reliability analysis of linear dynamical systems using approximate representations of performance functions. *Struct. Saf.* **2007**, *29*, 222–237. [CrossRef]
23. Katafygiotis, L.S.; Cheung, S.H. Application of spherical subset simulation method and auxiliary domain method on a benchmark reliability study. *Struct. Saf.* **2007**, *29*, 194–207. [CrossRef]
24. Pradlwarter, H.J.; Schueller, G.I.; Koutsourelakis, P.S.; Charmpis, D.C. Application of line sampling simulation method to reliability benchmark problems. *Struct. Saf.* **2007**, *29*, 208–221. [CrossRef]

MDPI
St. Alban-Anlage 66
4052 Basel
Switzerland
Tel. +41 61 683 77 34
Fax +41 61 302 89 18
www.mdpi.com

Vehicles Editorial Office
E-mail: vehicles@mdpi.com
www.mdpi.com/journal/vehicles

www.ingramcontent.com/pod-product-compliance
Lightning Source LLC
LaVergne TN
LVHW070135170726
843515LV00007B/1803